Characterization of Deep Marine Clastic Systems

Geological Society Special Publications
Series Editor A.J. Fleet

GEOLOGICAL SOCIETY SPECIAL PUBLICATION NO. 94

Characterization of Deep Marine Clastic Systems

EDITED BY

A. J. HARTLEY
Department of Geology and Petroleum Geology,
University of Aberdeen, UK

and

D. J. PROSSER
Z & S Geology, Aberdeen, UK

1995
Published by
The Geological Society
London

THE GEOLOGICAL SOCIETY

The Society was founded in 1807 as The Geological Society of London and is the oldest geological society in the world. It received its Royal Charter in 1825 for the purpose of 'investigating the mineral structure of the Earth'. The Society is Britain's national society for geology with a membership of around 7500. It has countrywide coverage and approximately 1000 members reside overseas. The Society is responsible for all aspects of the geological sciences including professional matters. The Society has its own publishing house, which produces the Society's international journals, books and maps, and which acts as the European distributor for publications of the American Association of Petroleum Geologists, SEPM and the Geological Society of America.

Fellowship is open to those holding a recognized honours degree in geology or cognate subject and who have at least two years' relevant postgraduate experience, or who have not less than six years' relevant experience in geology or a cognate subject. A Fellow who has not less than five years' relevant postgraduate experience in the practice of geology may apply for validation and, subject to approval, may be able to use the designatory letters C Geol (Chartered Geologist).

Further information about the Society is available from the Membership Manager, The Geological Society, Burlington House, Piccadilly, London W1V 0JU, UK. The Society is a Registered Charity, No. 210161.

Published by The Geological Society from:
The Geological Society Publishing House
Unit 7, Brassmill Enterprise Centre
Brassmill Lane
Bath BA1 3JN
UK
(*Orders*: Tel. 01225 445046
Fax 01225 442836)

First published 1995

British Library Cataloguing in Publication Data
A catalogue record for this book is available from the British Library.

ISBN 1-897799-35-7

Typeset by LBJ Enterprises Ltd, Aldermaston and Chilcompton

Printed by The Alden Press, Osney Mead, Oxford, UK.

Distributors

USA
AAPG Bookstore
PO Box 979
Tulsa
OK 74101-0979
USA
(*Orders*: Tel. (918) 584-2555
Fax (918) 584-0469)

Australia
Australian Mineral Foundation
63 Conyngham Street
Glenside
South Australia 5065
Australia
(*Orders*: Tel. (08) 379-0444
Fax (08) 379-4634)

India
Affiliated East-West Press PVT Ltd
G-1/16 Ansari Road
New Delhi 110 002
India
(*Orders:* Tel. (11) 327-9113
Fax (11) 326-0538)

Japan
Kanda Book Trading Co.
Tanikawa Building
3-2 Kanda Surugadai
Chiyoda-Ku
Tokyo 101
Japan
(*Orders*: Tel. (03) 3255-3497
Fax (03) 3255-3495)

Contents

Characterization of deep marine clastic systems

ADRIAN HARTLEY[1] & JEREMY PROSSER[2]

[1] *Department of Geology & Petroleum Geology, King's College, University of Aberdeen, Aberdeen AB9 2UE, UK*

[2] *Z & S Geology Ltd, Glover Pavilion, Aberdeen Science & Technology Park, Balgownie Road, Bridge of Don, AB22 8GW, UK*

Abstract: Sandstones deposited in deep marine environments form important hydrocarbon reservoirs in many basins throughout the world. However, despite the plethora of outcrop studies and the development of numerous submarine fan models and classification schemes, very few applied studies at a reservoir scale have been published.

This publication has arisen from the perceived needs of the academic and industrial communities to understand the controls on the architecture and geometry of deep marine clastic reservoirs. A number of areas of concern have been addressed: (1) Are conceptual models applicable to understanding sandstone body development and distribution at a reservoir scale? (2) Do we understand the processes that are active in the formation of deep marine clastic systems and the likely influence of these processes on reservoir quality? (3) How do we correlate and at what scale do correlation mechanisms work within and between deep marine clastic reservoirs? (4) How can we quantify heterogeneity and reservoir quality within these reservoirs?

Production histories for many mature fields in which hydrocarbons are reservoired in deep marine clastic sediments, frequently show the reservoirs to be heterogeneous, with difficulties commonly encountered during production (e.g. Yowlumne Field, California, Berg & Royo 1990; Forties Field, Central North Sea, Kulpecz & van Geuns 1990; Everest Field, Central North Sea, Thompson & Butcher 1991). In particular, problems have arisen during production as a result of poor correlation of individual reservoir units. Whilst it is easy to match log patterns from adjacent wells (**Anderton**, this volume) and to develop models which should permit correlation between wells (e.g. Reading & Richards 1994), in practice resulting correlations are often far from satisfactory. **Anderton** suggests that it is likely that outcrop analogues and facies models will be used to predict reservoir architecture and geometry in proximal and distal areas of deep marine clastic reservoirs. However, due to the inherent variability in sedimentation processes and their interaction with pre-existing topography, modelling of the mid-fan area is best approached using stochastic techniques as it is unlikely that facies development and distribution can be predicted in this setting. **Ouchi et al.** have taken a novel approach to predicting the likely geometry and architecture of deep marine clastic deposits, by simulating sub-aqueous fan development in flume tank experiments. The results, whilst not scaled to natural submarine fans, clearly illustrate the control of slope gradient on fan sand-content. The higher the gradient the more likelihood of developing overlapping sand-rich lobes at slope breaks (i.e. sand can be transported to a slope break prior to deposition), in contrast, lower gradients result in an elongate fan geometry with more isolated sandstone bodies developed across the slope.

Theoretical and experimental modelling of turbidite deposition and its application to understanding potential reservoir geometries are further discussed by **Kneller** and **Hughes et al.** **Kneller** has theorized that the basic models of turbidite deposition should be re-examined based upon simple equations of motion which indicate that deposition can occur beneath flows that are steady or waxing. Five basic types of turbidite sequence could result from different temporal and spatial accelerations within flows and can account for the frequent absence of Bouma or Lowe sequences in deep marine clastic deposits. In addition, **Kneller** has used experimental work to predict the effects of obstacles on sandstone body geometry and illustrates their significance in terms of potential reservoir architecture and geometry. **Hughes et al.** have used experimental work to predict grain fabrics in coarse grained gravity flow deposits, a feature which has implications for palaeocurrent

From Hartley, A. J. & Prosser, D. J. (eds), 1995, *Characterization of Deep Marine Clastic Systems*, Geological Society Special Publication No. 94, pp. 1–3.

determination and hence prediction of heterogeneities within proximal gravity flow deposits.

Experimentally produced soft-sediment deformation features are described by **Nichols** in the context of the potential effects of the three principal liquification processes (fluidization, liquefaction and shear liquification) on the formation and prediction of sandstone body geometry. This study is particularly relevant to a number of Tertiary North Sea reservoirs, as the impact of post-depositional soft- sediment deformation on reservoir architecture has recently been highlighted with improved seismic reflection data and close examination of core data (e.g. Newman *et al.* 1993; Jenssen *et al.* 1993). Previously, many of the complex injection features associated with intrusion of clastic dykes and sills had been ascribed to deposition from debris flows or had simply not been recognized. In particular, problems have arisen from the fact that if core data is absent, the recognition of injection features solely from wireline data is extremely difficult. **Dixon et al.** provide a case study of soft-sediment deformation in early Tertiary submarine fans from the Bruce-Beryl Embayment. They have integrated three-dimensional seismic, wireline and sedimentological data in order to better define sandstone diapirism, clastic intrusion and resultant reservoir geometry. **Brooke et al.** describe a potential causal mechanism for liquification in Tertiary reservoirs by gas injection, drawing on an analogy from Quaternary liquified sands in shallow gas mounds from the Fisher Bank area of the North Sea. **Pauley** has also focused on early Tertiary deep marine clastic systems, highlighting the potential development of unconventional reservoirs in megaturbidites from the Palaeocene of the Central Graben, North Sea.

Correlation, geometry and architecture in deep marine clastic systems have been addressed by a number of authors in this volume. **Cronin** provides a detailed description of an isolated sand-rich channel system from the Miocene Tabernas Basin of SE Spain and discusses the influence of tectonic controls on channel development through comparison with a modern day isolated channel system from the Almeria Canyon off SE Spain. **Browne & Pirrie** highlight the potential for the use of heavy minerals in correlation and provenance studies from a Jurassic submarine fan complex in the Antarctic Peninsula. **Verstralen et al.** illustrate the importance of facies analysis in defining potential reservoir architecture and geometry in Upper Jurassic sediments from the West Shetland Basin.

Two studies focus on potential intra-reservoir heterogeneities. **Watson et al.** document the development of carbonate concretions during early burial of Tertiary deep marine clastic reservoirs (Forth and Balmoral Fields) from the North Sea. Concretion formation in the studied fields is associated with the biodegradation of oil and is likely to significantly compartmentalize the reservoirs. **Prosser et al.** quantify permeability heterogeneity within so-called 'massive' sandstones from the Upper Jurassic reservoir of the Miller Field (South Viking Graben). They show that in terms of permeability variation these sandstones cannot be regarded as homogenous, an observation which has significant implications for reservoir production and simulation studies.

Shale clasts are common features within turbidites; however, little attention has been paid to their origin and potential significance. **Johanssen & Stow** present a novel classification of shale clasts within submarine fan sandstones and illustrate their potential for aiding understanding of transport mechanisms and depositional processes in both core and outcrop.

Conclusions and future work

This volume is by no means (and was never intended to be) a comprehensive documentation of deep marine clastic systems. However, we feel it highlights some of the current avenues and potential ways forward in the study of deep marine clastic systems, particularly with application to hydrocarbon reservoirs. In compiling this volume, we have been led to the conclusion that an improved understanding of deep marine clastic reservoirs would appear to be possible through a combination of outcrop analogue, theoretical and experimental data together with appraisal and production data. Extrapolation of heterogeneities, correlations and production characteristics into the interwell volume could occur through stochastic modelling but should honour all available data. The scope of papers presented in this volume suggests that much future work lies in application of an integrated approach to the quantification of deep marine clastic systems.

This volume is an outgrowth of the Reservoir Characterization of Deep Marine Clastic Systems conference held at the University of Aberdeen in September 1993. We would like to thank the following for reviewing the manuscripts submitted for this volume and helping us to maintain a high quality throughout: J. Alexander, R. Anderton, M. Anketell, S. Brown,

J. Cater, P. Corbett, C. Dodd, R. Dixon, T. Elliott, F. Ethridge, C. Garland, P. Haughton, S. Haszeldine, G. Kelling, G. Kessler, B. Kneller, M. Leeder, S. Leigh, J. Melvin, A. Morton, R. Nicholls, B. Sellwood, P. Shannon, H. Sinclair, R. Smith, S. Sparks, G. Timbrell, N. Trewin, P. Turner, A. Whitam and M. Wilkinson. Financial assistance towards the cost of the publication has been gratefully received from British Gas, Conoco (UK) Ltd, Marathon Oil, Total Oil Marine and the Petroleum Science and Technology Institute.

References

BERG, R. R. & ROYO, G. R. 1990. Channel-fill turbidite reservoir, Yowlumne Field, California. *In*: BARWIS, J. H., MCPHERSON, J. G. & STUDLICK, R.J. (eds) *Sandstone Petroleum Reservoirs*. Springer-Verlag, New York, 467–487.

JENSSEN, A. I., BERGSLIEN, D., RYE-LARSEN, M. & LINDHOLM, R. M. 1993. Origin of complex mound geometry of Palaeocene submarine-fan sandstone reservoirs, Balder Field, Norway. *In:* PARKER, J. R. (ed.) *Petroleum Geology of Northwest Europe: Proceedings of the 4th Conference*. The Geological Society, London, 134–143.

KULPECZ, A. A. & VAN GEUNS, L. C. 1990. Geological modelling of a turbidite reservoir, Forties Field, North Sea. *In:* BARWIS, J. H., MCPHERSON, J. G. & STUDLICK, R. J. (eds) *Sandstone Petroleum Reservoirs*. Springer-Verlag, New York, 489–507.

NEWMAN, M. ST. J., REEDER, M. L., WOODRUFF, A. H. W. & HATTON, I. R. 1993. The geology of the Gryphon Oil Field. *In:* PARKER, J. R. (ed.) *Petroleum Geology of Northwest Europe: Proceedings of the 4th Conference*. The Geological Society, London, 123–134.

READING, H. G. & RICHARDS, M. 1994. Turbidite systems in deep-water basin margins classified by grain size and feeder system. *American Association of Petroleum Geologists Bulletin*, **78**, 792–822.

THOMPSON, P. J. & BUTCHER, P. D. 1991. The geology and geophysics of the Everest Complex. *In:* SPENCER, A. M. (ed.) *Generation, Accumulation, and Production of Europe's Hydrocarbons*. Special Publication of the European Association of Petroleum Geoscientists **1**, 89–98.

Sequences, cycles and other nonsense: are submarine fan models any use in reservoir geology?

R. ANDERTON

BP Exploration, Dyce, Aberdeen AB2 0PD, UK

Abstract: There is no fundamental reason why there should be a relationship between the vertical succession of beds in a submarine fan deposit and their three-dimensional geometry. Parts of some fans behave in such a way that they leave behind sequences that can be used to define bed geometry. However, a significant proportion of the fan sands that are found in hydrocarbon reservoirs were deposited in unstable mid-fan environments that produce chaotic vertical successions. These are best modelled using stochastic techniques.

Submarine fan sandstones are important hydrocarbon reservoirs both in the North Sea and around the world (e.g. Weimer & Link 1991). The reservoir geologist in the petroleum industry is routinely confronted with the problem of understanding the detailed three-dimensional character of such reservoir sandbodies. The basic data at hand to attack this problem consists of all the information that can be gained from well logs and core plus the constraints imposed by the seismic interpretation. The tools available include a wealth of information derived from studies of modern and ancient sediments which is summarized in numerous submarine fan models (e.g. Walker 1978). The question is, do these models really help in solving the problem, and if not, is there any reason to suppose that they ever will?

Submarine fan sediments are particularly amenable to analysis in terms of sequences and cycles. (The term 'sequence' is used here in its original general sense. It was a very useful general term and it is unfortunate that it has been hijacked in recent years by sequence stratigraphy.) Unlike shelf or coastal environments, for example, where there is a complex interplay or several major processes which may produce a bewildering array of facies, submarine fans are dominated by repeated gravity flows which tend to produce accumulations of numerous discrete but similar turbidite beds. They are relatively easy to describe in terms of a few simple parameters such as bed thickness, grain size, percentage of bed amalgamation and sand:shale ratio. Therefore, over the years vast thicknesses of turbidite successions have been logged and analysed. When looking at ancient submarine fan sediments vertical patterns may be seen in these parameters at various scales, in some cases because they really are there (e.g. Mutti 1974; Heller & Dickinson 1985; Lowey 1992) and in others because we have been trained to expect to see them [see Hiscott's (1981) criticisms of cycle recognition].

Facies analysis proceeds, as with the study of any natural phenomena, by breaking down complex phenomena into simpler component parts, classifying them, trying to understand each part and then synthesizing the interpretations to form a model for the whole system (Anderton 1985). When looking at a vertical section or log the end product of the sedimentary evolution of a point is observed, or, strictly speaking, a very small area on the Earth's surface through time. If patterns can be seen in the log, some kind of repetitive element in this evolution may be inferred. But it is possible to do more than that thanks to one of the basic principles of stratigraphic interpretation, Walther's Law (Walther 1894). This states that the facies seen in a certain vertical order must also exist in the same order in a lateral direction. Although this is not always true, for example if major erosion surfaces are present, it opens up the hope that, at least in principle, it may be possible to deduce the lateral distribution of facies from a vertical section or sections. This hope is often converted into the belief that, if we try hard enough, the three-dimensional character of reservoirs can be deduced from vertical sections alone.

Vertical patterns and facies models

The link between the vertical section and the three-dimensional rock body is the sedimentary

From Hartley, A. J. & Prosser, D. J. (eds), 1995, *Characterization of Deep Marine Clastic Systems*, Geological Society Special Publication No. 94, pp. 5–11.

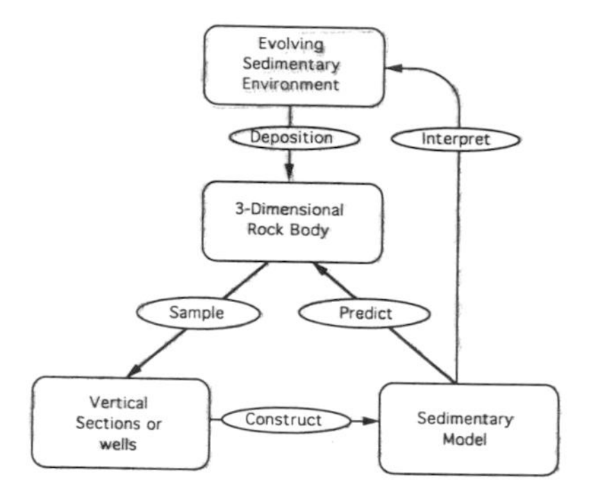

Fig. 1. Flow diagram showing relationships between sedimentary rocks, processes and concepts.

model (Fig. 1). This tells us what a vertical section through a particular type of submarine fan with particular sandbody geometries should look like. So, provided that the right model has been identified, it should be possible to predict what would be found if, for example, a new well was drilled or an existing well allowed to flow. For the reservoir geologist a sedimentary model is just a means to an end, the end being the three-dimensional character of the reservoir; it has no value in itself. For the academic researcher, however, a model may itself be the main product of a piece work or it may be the means to an understanding of some further aspect of the history or evolution of the area (Fig. 1). Most information about ancient submarine fan facies comes from vertical sections logged at outcrop or in cored wells. Good two- and three-dimensional information from large outcrops or high resolution seismic is obviously extremely valuable but is not that common. If it were possible to describe three-dimensional reservoir character directly with the required degree of resolution then the issues raised here would no longer be relevant. Today, submarine fan geometry can be delineated from three-dimensional seismic data in favourable circumstances, e.g. in sand-dominated fans with thick homogeneous sandbodies which show a good acoustic impedance contrast with their surrounding mudstones. Time will show whether it is possible to improve the spacial and lithological resolution of seismic data sufficiently for the internal complexities of more typical fan reservoirs to be mapped routinely. Whether it is seismic or other, yet undiscovered, techniques that eventually enable fan character to be mapped directly, such direct observation is obviously preferable to any kind of conceptual modelling.

In the absence of direct observations, we have to fall back on models that are suggested by our well data. Vertical patterns are one of the main elements in constructing sedimentary models. We can hypothesize about the origins of vertical patterns, although it is seldom possible to validate these thoughts from extensive outcrops. Conceptual models are certainly crucial for producing a generalized picture of the evolving sedimentary environment. It is much more difficult to use them to 'predict' the detailed three-dimensional sandbody geometry.

Patterns may be interpreted from wireline logs or directly from core. Unfortunately, there

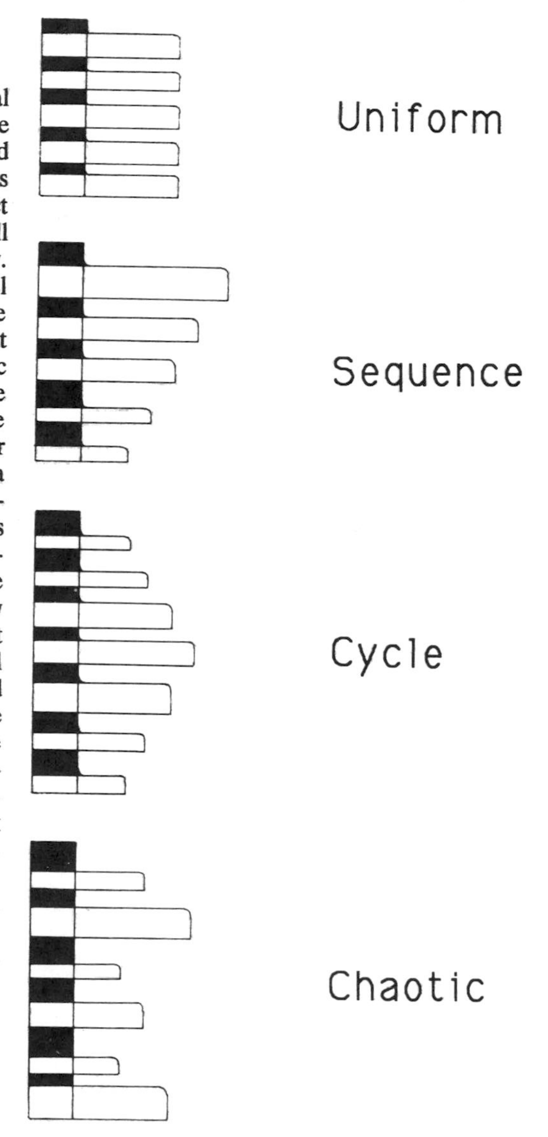

Fig. 2. Types of pattern seen in vertical sections. Sandstone beds shown schematically as white bars, mudstones as black intervals.

is not a unique relationship between log patterns and geology. However, that problem can be put aside by assuming either complete core coverage or a transcendental ability to interpret logs.

If submarine fans are deposited from episodic turbidity currents, there are a limited number of vertical patterns that might be expected to be seen in fan deposits (Fig. 2). If the same process is repeated, a uniform pattern of similar beds is obtained. If things change progressively a sequence is obtained, such as a thickening-upwards sequence. If things change first one way and then the other, a cycle is obtained. A distinctively untypical bed within a sequence or cycle may stand out as an anomalous unit, such as some megabeds. If no pattern can be recognized, the interval is random or chaotic.

In practice, patterns are only rarely clear-cut. There are good examples of turbidite sequences and cycles. These are amenable to elegant description and interpretation and do tend to find their way into the published literature. The much more common, in the author's experience, chaotic sections tend to remain under-represented as it is difficult to make their interpretation interesting enough to satisfy reviewers. The '. . . this 5 km thick turbidite succession shows no order, no structure and makes no sense . . .' sort of paper is unlikely to be published. So we may be guilty, as Hiscott (1981) pointed out, of taking too much notice of the ordered, but unusual, fan successions and not paying enough attention to the more common chaotic examples. To be interested in the former is quite excusable if one is in pursuit of intellectual satisfaction. However, if one has to deal with whatever happens to turn up as a hydrocarbon reservoir, an appreciation of the full range of fan deposits is necessary.

Should patterns be expected?

Vertical patterns are the key to constructing models. But should such patterns be expected in a submarine fan environment? Mutti & Ricci-Lucchi (1972) proposed that channels and lobes are characterized by thinning- and thickening-upwards sequences, respectively. Hiscott (1981) questioned the importance of such sequences and doubted the validity of their proposed depositional processes. The problem is that one can only infer or speculate about depositional processes. Seldom have they been observed directly.

So what can be done? In the absence of a complete understanding of the processes, should ordered or random patterns be expected? Well, some 'thought experiments' can be carried out: consider the proximal part of a fan, the main fan channel or fan valley, where the channel depth is great enough to entirely contain nearly all the turbidity currents that flow down it. Here, sand deposition is largely a function of the magnitude of each current (measured by volume of sediment in the causative slope failure or the velocity, thickness and density of the resulting flow). This is often an area of erosion or bypass. If flows become successively larger, erosion by later flows will remove earlier deposits. Only if flows become successively smaller will sediment be preserved. Then, a sequence of progressively finer sedimentation units will be deposited which may show an initial thickening followed by a thinning-up. Flows which depart from this trend will not destroy the sequence. An unusually large flow may well remove part of the existing sequence, an unusually small flow will interpose a thin bed. But the overall sequence will still be recognizable. In other words, the depositional sequence is relatively stable to random perturbations.

The same is true at the distal end of the fan where the sea floor is very smooth, each bed has a sheet-like geometry and successive flows do not significantly alter the sea floor topography. Here, deposition reflects the changing magnitude of the flows at the fan source and, hence, the overall evolution of the fan. Irrespective of any complex processes going on within the channel-lobe system, if the distal fringe shows a coarsening- and thickening-upwards the fan is prograding. Only the largest flows will deposit sand this far out on the fan, so much of the depositional signal of the fan as a whole is filtered out, removing some of the random 'noise'. So, again, in these areas the depositional sequence is relatively stable.

In the intervening areas of distributary channels and lobes, the mid-fan, the picture is more complex. Imagine an area of flat seabed just downstream from the mouth of a new fan channel where every passing turbidity current is of the same magnitude. As a current is debouched from the channel it expands, decelerates and deposits a roughly teardrop-shaped patch of sand (we are only interested in the fate of the sand here). In the progradational model, as the channel mouth advances, each successive sand unit is displaced downstream resulting, at least in the distal part of the deposit, in a thickening-upwards sequence of beds which collectively form a lobe. Even in this abstractly simple system, the pattern described is unlikely

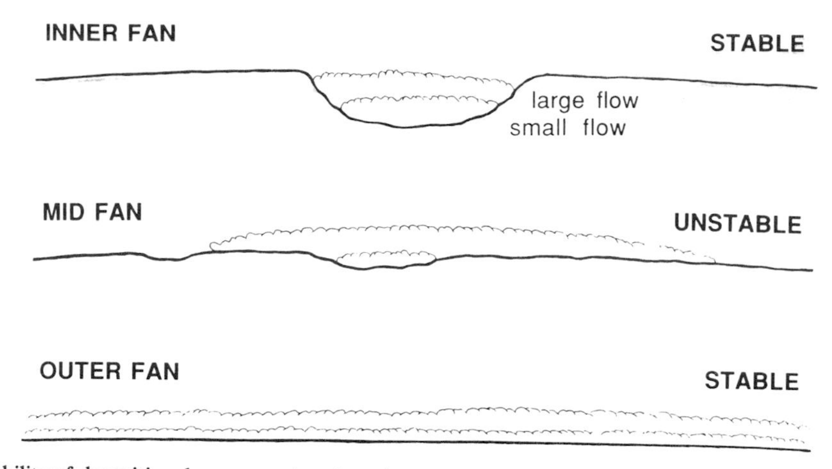

Fig. 3. Stability of depositional processes in submarine fan environments as a function of flow thickness and seabed relief. In the inner fan nearly all flows are contained within the channel. On the outer fan they spread out to cover the whole floor. In the mid-fan the area of deposition is very sensitive to flow thickness.

because the initial bed would not be of uniform thickness and so would change the seabed topography. The second current would see different topography from the first, would have a slightly different flow pattern and so the resulting teardrop would have a different shape. The geometry of each bed would be influenced by every preceding bed. If one adds a more realistic random element to the magnitude of successive flows, a very complex picture starts to develop. A larger than average flow may scour out parts of the preceding deposits. A small flow may dump most of its load near the channel mouth. Thus, the behaviour of each flow will be influenced by the deposits of each of the previous ones, each of which in turn is a consequence of a whole series of random processes going on upstream. This is a very complex type of non-linear behaviour. The result is unpredictable. The system is unstable in the sense that its behaviour is strongly affected by seemingly trivial events or processes. One would be surprised if it produced simple sequences very often.

One can carry out similar 'thought experiments' on the distributary channels where flows may be constrained within, or overtop, channel banks. Here, differences in the magnitude of successive flows can produce a bewildering variety of consequences for channel and overbank deposition.

Hiscott (1981) criticized the use of the prograding lobe model to explain thickening-upwards sequences because it does not appear to apply to the modern Navy Fan (Normark *et al.* 1979). The fact that it is not appropriate in this case does not mean that it does not apply in others. But this is hardly the point at issue here, which is rather that many of the numerous conceivable processes that may occur in mid-fan environments are unstable. In practice there may be any possible combination of progradation, lateral migration, aggradation, avulsion, bypassing and post-depositional movement. Except in certain special cases, it is difficult to see how non-random patterns will emerge. Hiscott (1981) considered that fans are dominated by random processes. This may be overstating the case but I would agree that it may be true of most mid-fan areas. In these, random perturbations are not necessarily damped out by the system but may well be amplified to such an extent that they dominate it. For example, a single large flow may cut a new channel and completely change the subsequent history of the fan. All this is not just to say that the random deposition in the mid-fan is a simple reflection of the random sequence of differently sized slope failures on the delta slope that generate the flows in the first place. The interactions are much more complex than that and inherent in the nature of the depositional system. Other complicating factors such as differential subsidence due to compaction, bank collapse and soft-sediment deformation are ignored here but will add to the general mayhem.

The discussion above suggests that, in terms of depositional processes, a typical submarine fan can be divided into three areas: a stable inner fan; a stable outer fan; and an intervening unstable mid-fan (Fig. 3). The stable areas are likely to produce recognizable depositional sequences. The unstable area will tend to show a chaotic succession. A typical fan should then show a coarsening and thickening-up basal section reflecting outer fan progradation, a chaotic mid-fan section that is very difficult to subdivide and a thinning- and fining-up inner fan channel sequence deposited during final fan abandonment. This, I suggest, is what most submarine fan successions are actually like. This is not to say that real lobes and real channels (as well as numerous other sub-environments) do not exist on the mid-fan surface, but that their deposits do not have any reliable distinguishing features such that it is possible to say that any individual bed seen in a vertical section was deposited in any particular sub-environment and thus has any particular geometry.

The concept of stable and unstable areas is useful in considering the differences between fans. Deep channels and large flat areas are stable, areas of irregular moderate relief are unstable. Muddy fans are areally dominated by stable areas, sandy fans by unstable areas. Recognizable sequences should, therefore, be more common (as a proportion of total thickness) in muddy than sandy fans. To put it another way, muddy fans are more 'ordered', sandy fans more 'chaotic'. Muddy fans are probably over-represented in the literature, because of their editorial appeal. Sandy fans are more important as hydrocarbon reservoirs. There is an obvious danger of inappropriate, ordered models being applied to chaotic reservoirs.

In the last few years outcrop studies of ancient submarine fans have increasingly focused, quite rightly, on lateral facies relationships. There are two quite different goals in such studies. On the one hand one can map out lobes, channels and individual sandstone beds with a view to reaching a deterministic 'understanding' of how submarine fans work and hence being able to predict their geometry from limited (e.g. well) data. The other approach is just to collect large amounts of data on sandbody geometries from which the databases required for stochastic modelling can be built.

The aim of academic workers is, or should be, to reach the former, if rather distant, goal. Oil companies tend to take a more limited and pragmatic approach and would be more than happy to have large databases on fan sandbody geometries. Indeed, this is an area where several companies are actively engaged. Hopefully, much of this data will eventually reach the public gaze. In the meantime, such work as has been published (e.g. Enos 1969), plus the author's experience confirms the view that the geometries of, at least the small-scale elements of mid-fan areas, individual beds and packets of beds are highly unpredictable. If several seemingly similar beds are traced laterally over hundreds of metres they are often seen to terminate in very different ways. They may progressively thin and wedge out, or thin out and then reappear, or thin in a step-like manner or end abruptly showing that they were deposited in shallow scours or channels. One can often explain what has controlled the geometry of any individual bed but it is usually impossible to set up rules to describe the behaviour of a run of beds because of the complex way the controlling factors interact.

The larger scale relationships are easier to map out in the field than individual beds (e.g. Mutti *et al.* 1985; Kleverlaan 1989) and such studies, together with information from modern fans, enable general predictions to be made about the gross geometries and lateral facies relationships of sandbodies from mid-fan sub-environments. What always has to be asked, however, is whether it would be possible to interpret the environment of deposition of such sandbodies from only a limited number of vertical sections?

Models, geometry and correlation

All this brings us back to the question of sandbody geometry. Inner fan channel sandbodies are large enough and have enough of an acoustic impedance contrast with the surrounding muds that they can often be imaged on seismic. Outer fan sands are sufficiently extensive not to pose a great correlation problem. But many clastic submarine fan hydrocarbon reservoirs come from mid-fan channel-lobe areas in sandy fans.

The interpretation of submarine fan facies and its use to predict sandbody geometry should be objective, but it hardly ever is. If you believe that there should be some kind of order in your fan, you will look very hard to find it. The human brain is very good at finding order even when there isn't any. So, it is easy to find patterns where they don't exist and 'shoehorn' a reservoir into an inappropriate model. Similarly, if one does not believe there should be any order, real and useful patterns may be dismissed. For mid-fan successions the danger of the former is generally greater than the latter, in the author's opinion.

In reservoir geology, the construction of the wrong model does not necessarily do any harm if it is only used as the kind of conceptual model used to give a feeling of well-being that results from 'understanding' what is going on. The damage is only done when it is used to drive well correlation. For example, if vertical patterns are used to interpret the existence of alternations of lobe and channel facies, one would go to the relevant wells with the idea that what are considered to be lobe units should show a greater lateral persistence than the channels. A more objective approach would be to take the well logs at face value and try to correlate units without any model in mind. (In practice, this is often done anyway, even when a highly sophisticated model has been constructed.) But here again you are caught in the trap of seeing what you expect to see.

On a coarse scale in marine successions with good biostratigraphy and a host of other modern techniques, correlation is not a problem. (The understanding of submarine fans at this scale is crucial in hydrocarbon exploration, but that is another story.) However, at the reservoir scale, where one is trying to correlate within sandy sections only metres to a few hundred metres thick deposited in fractions of a million years, it comes down again to pattern recognition. Again, the human brain is very good at pattern recognition. It seldom misses a real pattern and can often see patterns in random noise. Thus, objective correlation tends to be over-correlation. A reliable facies model really is of use if it can indicate how many and which units you should be trying to correlate.

In a sandy, chaotic mid-fan system the individual sandbodies may all be smaller than the inter-well distances. Under these circumstances correlations should not be made within the deposit, although it is a difficult temptation to resist. Correlatable surfaces are possible here if some external event is superimposed on the internal fan processes. For example, temporary fan abandonment, a volcanic eruption or a catastrophic river discharge may all produce widespread distinctive markers. These enable the fan deposit to be divided into a series of layers. This is very useful, but gives little information about the architecture of the fan sandbodies themselves.

The use of a model, whether right or wrong, derived from a single well to drive a correlation between wells is a piece of linear logic. It is much more likely that one will look at the similarities and differences between several wells, i.e. the ease of correlation, to help formulate the model. The model can then be used to justify the correlations drawn between the wells. This gives a consistent and comforting picture but it is, of course, a totally circular argument. There is no simple way of breaking out of this circle. Correlation and sedimentary models are intimately tied together, the correlation is justified by the model and the model by the correlation. There may be a hundred and one possible combinations of model and correlation. Picking the right one requires great skill. Perhaps we should not even try. It is more important to consider the whole range of possibilities and examine the consequences, in terms of reservoir behaviour, of each.

Conclusions

Vertical sections through submarine fan deposits can show both ordered patterns and chaotic intervals. The latter are volumetrically much more important than realized from the literature because the study of the former tends to be more satisfying. This discussion has focused on bed thickness patterns, although it applies in a similar way to other parameters such as grain size and structure. There is a danger that ordered facies models may be applied indiscriminately to fan successions, which is especially unfortunate in hydrocarbon reservoirs where chaotic mid-fan systems are common. Facies models are important in guiding correlation and the construction of reservoir models. However, ordered models derived from sequence analysis may be quite misleading when applied to chaotic mid-fan deposits. Mathematical stochastic models may generally be more useful here than models based on deterministic concepts. There is plenty of scope for studying mid-fan behaviour using mathematical models. Although stochastic models do not give us the satisfaction of saying '. . . this sandbody has this size and shape because . . .' they are useful in building reservoir simulation models. For chaotic systems they are much closer approximations to reality than models which assume deterministic relationships between depositional environments and sandbody geometries.

A possible approach when trying to model sandbody geometry in submarine fan deposits on a sub-seismic scale is first to look honestly and objectively at the available cores and logs. Obviously correlatable surfaces produced by external processes should be identified, if present, using all available techniques and used to divide the reservoir into a series of layers. Clearly ordered sequences can be marked out and examined in detail in order to formulate deterministic facies models for those parts. For the rest of the succession a single or a series (if correlatable surfaces are present) of stochastic models can be built using core, well and analogue data. In some deposits the deterministic elements will dominate. At the other extreme some ancient submarine fans will invite an entirely stochastic description. Most will lie somewhere between the two. It is also, of course, a question of scale. A stochastic description may be appropriate to a given fan deposit if there is only widely-spaced well data. Infill drilling may subsequently provide enough detail for a more deterministic treatment.

Models derived from ordered successions are intellectually satisfying and enable specific predictions to be made. For successions that are chaotic or have chaotic intervals, the facies models that can be constructed provide only general predictions that are couched in statistical terms. Submarine fan models are of use in reservoir geology, although models derived from the study of vertical patterns that assume a large degree of internal order may be inappropriate to the understanding of the chaotic sandbody geometry seen in many hydrocarbon reservoirs.

I would like to thank the referees for their helpful criticisms of the original version of the manuscript. The paper is published with the permission of BP. The opinions stated are entirely those of the author.

References

ANDERTON, R. 1985. Clastic facies models and facies analysis. *In:* BRENCHLEY, P. J. & WILLIAMS, B. P. J. (eds) *Sedimentology – Recent Developments and Applied Aspects.* Geological Society, London, Special Publication, **18,** 31–47.

ENOS, P. 1969. Anatomy of a flysch. *Journal of Sedimentary Petrology,* **39,** 680–723.

HELLER, P. L. & DICKINSON, W. R. 1985. Submarine ramp facies model for delta-fed, sand-rich turbidite systems. *American Association of Petroleum Geologists Bulletin,* **69,** 960–976.

HISCOTT, R. N. 1981. Deep-sea fan deposits in the Macigno Formation (Middle–Upper Oligocene) of the Gordana Valley, northern Apennines, Italy – Discussion. *Journal of Sedimentary Petrology,* **51,** 1015–1033.

KLEVERLAAN, K. 1989. Three distinctive feeder-lobe systems within one time slice of the Tortonian Tabernas fan, SE Spain. *Sedimentology,* **36,** 25–45.

LOWEY, G. W. 1992. Variation in bed thickness in a turbidite succession, Dezadeash Formation (Jurassic–Cretaceous), Yukon, Canada: evidence of thinning-upward and thickening-upward cycles. *Sedimentary Geology,* **78,** 217–232.

MUTTI, E. 1974. Examples of ancient deep-sea fan deposits from circum-Mediterranean geosynclines. *In:* DOTT, R. H. JR & SHAVER, R. H. (eds) *Modern and Ancient Geosynclinal Sedimentation.* Society of Economic Palaeontologists and Mineralogists, Tulsa, Special Publication, **19,** 92–105.

—— & RICCI-LUCCHI, F. 1972. Le torbiditi dell' Appennino settentrionale: introduzione all' analisi di facies. *Memorie della Società Geologica Italiana,* **11,** 161–199.

—— REMACHA, M., SGAVETTI, J., ROSELL, R., VALLONI, M. & ZAMORANO, M. 1985. Stratigraphy and facies characteristics of the Eocene Hecho Group turbidite systems, south-central Pyrenees. *In:* MILÀ, M. D. & ROSELL, J. (eds) *6th European Regional Meeting, Lleida, Spain – Excursion Guidebook.* International Association of Sedimentologists, 519–576.

NORMARK, W. R., PIPER, D. J. W. & HESS, G. R. 1979. Distributary channels, sand lobes, and mesotopography of Navy Submarine Fan, California Borderland, with applications to ancient fan sediments. *Sedimentology,* **26,** 749–774.

WALKER, R. G. 1978. Deep water sandstone facies and ancient submarine fans: models for exploration for stratigraphic traps. *American Association of Petroleum Geologists Bulletin,* **62,** 932–966.

WALTHER, J. 1894. *Einleitung in die geologie als historische wissenschaft.* Lithogenesis der Gegenwart. Fisher Verlag, Jena, **Bd 3,** 535–1055.

WEIMER, P & LINK, M. H. 1991. *Seismic Facies and Sedimentary Processes of Submarine Fans and Turbidite Systems. Springer-Verlag, New York.*

Experimental study of subaqueous fan development

SHUNJI OUCHI,[1] FRANK G. ETHRIDGE,[2] EDWARD W. JAMES[2] & S. A. SCHUMM[2]

[1] *College of Science and Engineering, Chuo University, 1-13-27 Kasuga, Bunkyo-ku, Tokyo 112, Japan*

[2] *Department of Earth Resources, Colorado State University, Fort Collins, Colorado 80523, USA*

Abstract: Subaqueous fan-shaped depositional forms were developed during flume experiments, that were conducted to investigate processes of underwater deposition. The flume was filled with fresh water, and clay plus salt water was introduced as a density flow at the head of the underwater slope. A lobe-shaped sandbody (sand lobe) migrated downslope, as a liquified flow, when sand was mixed into the clay plus salt water flow. In a steep-slope experiment (37°) sand lobes were deposited at the toe of the slope and overlapped each other to form a fan-shaped body. Without sand, clay plus salt water flows of small discharge, which showed characteristics of laminar flow, tended to incise a channel on the sandy fan surface. The larger discharge flows, in which some turbulence developed, however, spread into the overlying water body and did not cut a channel. Clay moved further and was deposited on the basin-floor. Sand lobes occasionally migrated on to the basin-floor clays, especially when an incised channel developed on the fan surface. When sand lobes reached the basin-floor clay deposits, movement of the sand accelerated basinward, the sand lobe slid on the clay like the floating plug of a debris flow, and formed crescent-shaped mounds and hummocky topography. The accelerated movement of the distal portion of the sand lobes appeared to cause them to become detached from the main sandbody. With gentler slopes (18°), clay deposition occurred on the slope. Sand moved in the form of an elongate lobe, and the fan itself became more elongate than the one developed on the steeper slope. The elongate nature of the fan deposit on the low slope was apparently related to the higher efficiency of sand movement on the clay that settled on this slope.

Subaqueous fans, on both slopes, collapsed and generated relatively large subaqueous debris flows when the water in the flume was being drained. The failure started on the surface with a fluidized flow, and the entire fan slumped basinward. A reduction of hydrostatic pressure caused by the lowering of the water levels apparently generated a pore-pressure gradient in the deposits. The fan slopes became unstable and pore water probably flowed upward as the pressure gradient developed. This pore-water flow induced the fluidized flow on the surface and then the slump of the entire fan.

The experimental subaqueous fans cannot be compared quantitatively with natural fans because they are not scale models. The general similarity of process, however, might provide insights into depositional and erosional processes in natural systems.

The main purpose of this study is to develop a better understanding of deep sea fan development by providing information on underwater sedimentary processes that are generated by density currents. The movement of sediment in the ocean by turbidity currents and the development of deep sea fans have attracted much attention because of the possibilities that turbidity or density currents cut submarine canyons (Daly 1936; Kuenen 1937, 1938) and carry coarse sediments to the deep sea bottom (Natland & Kuenen 1951; Shepard 1951). Petroleum geologists are particularly interested in the formation of deep sea fans, because coarse-grained deposits of ancient deep sea fans often form oil reservoirs (Walker 1978; Shanmugam & Moiola 1988).

Studies of modern submarine fans and ancient submarine fan deposits have resulted in a large number of models designed to explain fan formation and deposits (Bouma *et al.* 1985; Posamentier *et al.* 1991). Most of the early fan models are based on studies of the rock record (e.g. Mutti & Ricci Lucchi 1972; Walker 1978), but development of bathymetric and seismic techniques, such as improved echo sounders and

From Hartley, A. J. & Prosser, D. J. (eds), 1995, *Characterization of Deep Marine Clastic Systems*, Geological Society Special Publication No. 94, pp. 13–29.

deep-penetration seismic reflection techniques, has made it possible to describe and analyse the morphology and geology of modern deep sea fans (Bouma *et al.* 1985; Twichell *et al.* 1992). These studies resulted in a new series of submarine fan models. There are, however, considerable differences among these models (Nilsen 1980; Walker 1980; Shanmugam & Moiola 1988; Posamentier *et al.* 1991; Normark *et al.* 1993). The problem stems, in large part, from the difference in the perspective of each study. For example, studies of modern deep sea fans mainly concentrate on the morphology and surface geology, whereas studies of ancient deep sea fans involve the detailed investigation of facies and sequences of sedimentary rocks. In addition, a wide variety of both modern and ancient deep sea fan types exist (Bouma *et al.* 1985).

Subaqueous sediment gravity flows (Middleton & Hampton 1973, 1976) are considered as the main agent of deep sea fan formation, and many experimental studies (e.g. Kuenen 1938, 1951; Kuenen & Migliorini 1950; Middleton, 1966a, b, 1967, 1993; Hampton, 1972, 1975; Luthi 1981; Parker *et al.* 1987; Garcia & Parker 1989), and theoretical studies (e.g. Komar 1969, 1972, 1977; Allen 1971; Simpson 1982; Parker *et al.* 1986) have been conducted to determine the nature and character of these flows. These studies, however, concentrate on the behaviour and characteristics of the flow itself, and they usually reveal little about the development of underwater sedimentary landforms. Moreover, the actual flows in the deep sea, which should be the prototypes of the scale-model experiments, have not been well observed or measured, except in a few instances of limited observation (Gould 1951; Heezen & Ewing 1952; Weirich 1984). In summary, the controversial deep sea fan models are almost entirely based on the sedimentary sequences of ancient deep sea fans or on the topography and surface geology of modern deep sea fans, and processes explained in the models are speculations based primarily on analogies with subaerial land forms.

The experiments conducted in this study are not scale models that reproduce the flows in the deep sea, but they are designed to produce miniature land-forms that develop by deposition from sediment gravity flows. This type of analogue experiment is considered to be a useful means of filling the gap between studies of sediment gravity flows and the development of deep sea fans, because similar experiments have been successful in explaining fluvial land-forms (Schumm *et al.* 1987). Application of the results to natural systems should, however, be made with caution. The term sand lobe is used here to describe the shape of bodies of sediment plus salt water that migrated downslope as density currents. The term is also used in the descriptive sense of Normark *et al.* (1993) for the resulting deposit, that is generally lobate in plan view and externally mounded in cross-section.

Equipment and experimental procedure

Equipment

Experiments were conducted in a 5.87 m long, 1.8 m wide and 0.59 m deep flume with a fresh water inlet box (0.65 m × 1.8 m × 1.32 m) immediately upstream of the flume (Fig. 1). Water from the inlet box flows to the main flume through a 10 cm diameter PVC pipe, which helped to keep the water body free of sediment. Two drains, located on the downstream wall of the flume, were also used to maintain water clarity and to control the water level in the flume. One of these drains was located near the floor of the flume and the discharge rate was controlled by a valve located outside the flume. The other drain consisted of a Z-shaped pipe which drained water from the surface. The base of the Z-shaped pipe was rotated to adjust and to maintain a constant water level. A mixture of sand, and a small amount of silt and clay was used to form the simulated continental shelf and slope. In the steep slope experiment (Run 1) the slope was stabilized by the addition of 1–2% cement.

Experimental procedure

Two experiments were performed in which density currents consisting of clay plus salt water with the addition of sand were introduced on very different slopes. The clay plus salt water mixture contained 118 g of kaolinite (about 10 ml by dry volume) and 182 g of salt (9% kaolinite, 14% salt, and 77% water by weight), per litre of water. Specific gravity of this fluid ranged from 1.16 to 1.18 g cm^{-3} depending on temperature, and viscosity was approximately 15 centipoise. The introduction of clay plus salt water generated a density current on the submerged slope, and sand lobes formed and migrated downslope when medium to fine grained sand was added to the flow. Relatively small density-current flow rates (< 50 ml min^{-1}

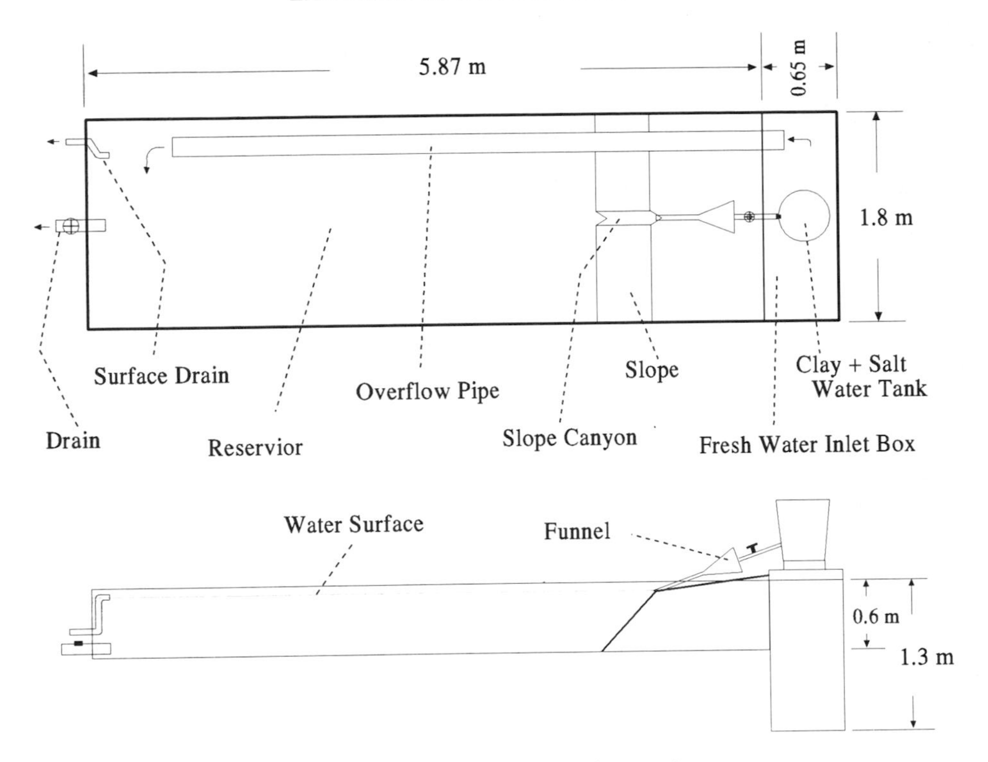

Fig. 1. Plan and longitudinal-section sketch of the flume used in the experiments.

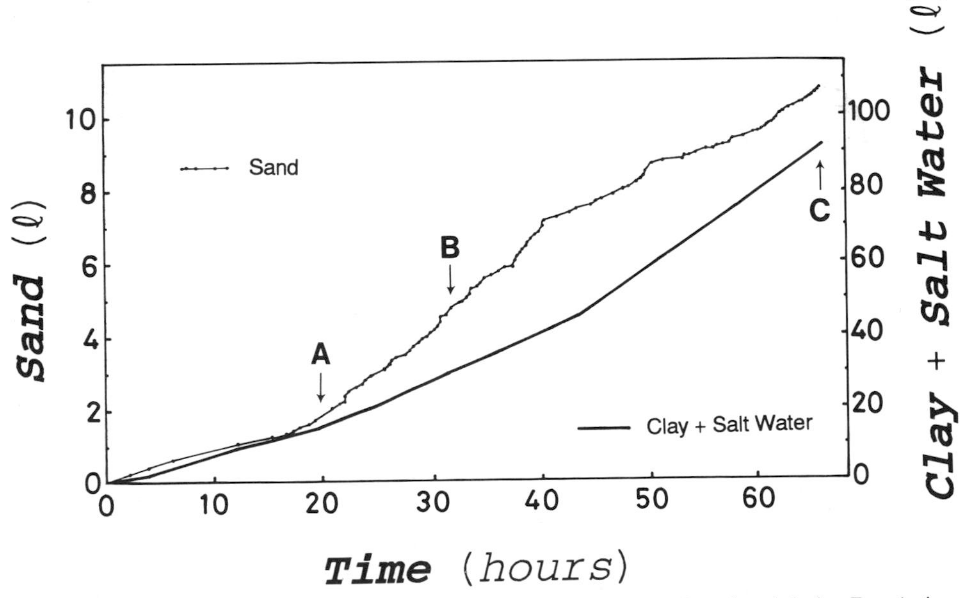

Fig. 2. (**a**) Graph showing cumulative volume of sand and clay plus salt water introduced during Run 1. Arrows A, B and C indicate the times when the subaqueous morphology was measured using a point gauge. Slope and fan morphology at these times are shown in Fig. 4a–c, respectively. (**b**) Graph showing cumulative volume of sand and clay plus salt water introduced during Run 2. Arrows A and B indicate the times when the subaqueous morphology was measured using a point gauge. Slope and fan morphology at these times are shown in Fig. 10a & b, respectively.

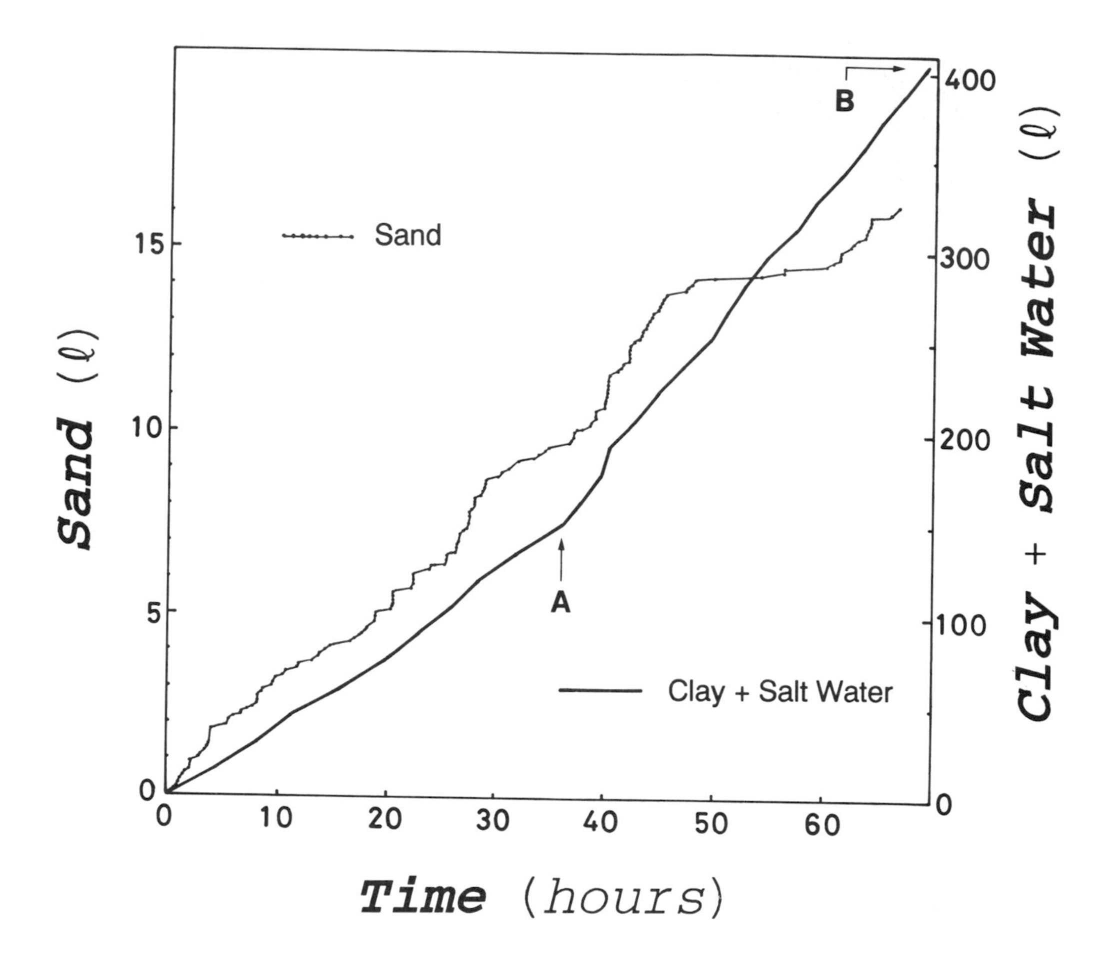

for the steeper slope of Run 1, and less than 100 ml min^{-1} for the gentler slope of Run 2) were used to minimise mixing between the density current and the standing body of fresh water to maintain water clarity, thus allowing observations of the slope and flume bottom.

For Run 1, relatively steep slopes (37°) were constructed with a mixture of sand and a small amount of silt and clay with 1–2% cement powder to increase cohesiveness and to reduce the chances for slope failure. A 2.5 cm deep triangular channel was cut into the slope to facilitate movement of the density currents to the basin floor. For Run 2, a gentler slope of 18° was constructed with sand and a small amount of silt and clay, and no initial channel was cut on the slope.

The clay plus salt water was introduced through a plastic tube (4 mm inner diameter) to a funnel that was connected to a plexiglas tube (91 cm long, 9 mm inner diameter), which extended under the water surface near the top of slope (Fig. 1). Sand was systematically introduced and mixed with the clay plus salt water in the funnel. The cumulative amounts of fluid and sand supply with time in these two runs are shown in Fig. 2. The gentler slope of Run 2 allowed larger discharges to flow down the slope without making the water murky; and therefore, more than twice as much clay and about one and a half times as much sand was delivered to the slope in Run 2 than in Run 1 during a similar time period. The morphology of underwater deposits was measured periodically using a point gauge. At the completion of each run, the flume was drained to dry the deposits. The dried deposits were dissected in a series of longitudinal and transverse sections to examine the three-dimensional sedimentologic and stratigraphic characteristics.

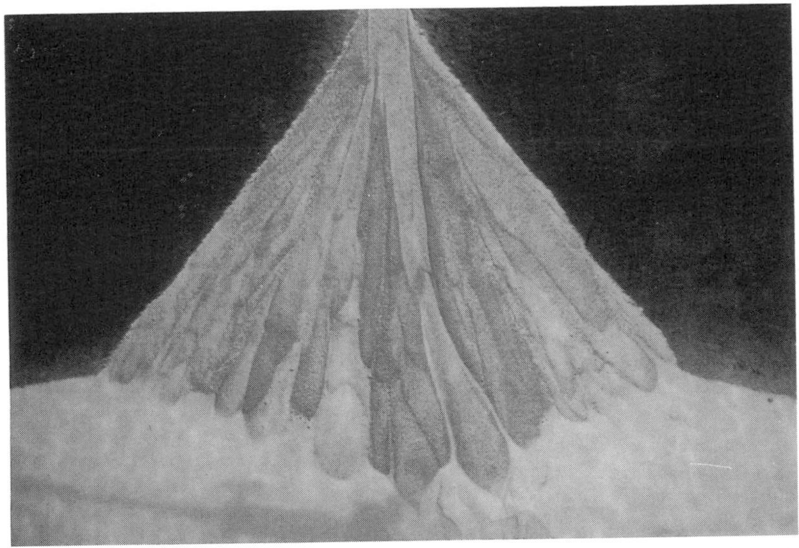

Fig. 3. Subaqueous fan-shaped deposit formed in Run 1. Many sand lobes overlapped to form the fan shape. Maximum width of fan is 60 cm.

Observations and results of experiments with clay plus salt water flows

Run 1

During Run 1 many sand lobes overlapped to form a cone- or fan-shaped deposit at the toe and along the lower portion of the steep, 37° slope (Fig. 3). After 20 h of running and the introduction of 17 l of clay plus salt water, and 1.7 l (2.6 kg) of sand, a fan-shaped sand deposit had formed which was about 36 cm long and 40 cm wide (Fig. 4a). By 31.5 h after the addition of 25 l of clay plus salt water, and 2.9 l (4.4 kg) of sand this fan had grown to about 60 cm long by 65 cm wide (Fig. 4b). After 66 h and the addition of 92 l of clay plus salt water, and 10.8 l (16.5 kg) of sand the fan had grown to about 76 cm by 100 cm (Fig. 4c). Sand lobes on the fan were sometimes bi- or tri-lobed depending on the amount of sand supply and surface topography. The movement of sand lobes indicated that the flow of sand was similar to the liquified flow described by Lowe (1976), in which grains are partially supported by pore fluid (clay plus salt water). After the sand lobe stopped, clay (including the clay in the escaped pore fluid) moved further in suspension and was deposited, as a thin layer on the basin floor, in front of the sandy fan (Fig. 5). The clay deposit spread widely from the toe of slope to 56 cm at 20 h, 88 cm at 31.5 h, and 128 cm at 66 h.

Incised channels developed when the sand supply was cut off. Without the introduction of sand, the clay plus salt water flowed on the fan surface in a sheet for a short time, and then channel erosion started suddenly in the middle to lower portion of the fan with nickpoint formation. A series of nickpoints often migrated upstream in the same channel, and this usually resulted in the formation of terraces in the incised channel. When the channel became sinuous, the flow accelerated on the outer part of a bend, and a new nickpoint developed there. The course of flow became slightly more sinuous and terraces were formed on the inner side of the channel (Fig. 6). The velocity of the clay plus salt water flow varied widely between 2.5 and 10 cm s^{-1} due to the difficulty of keeping the dense fluid discharge constant. A Reynolds number between about 5 and 80 was estimated using a flow depth of from 0.3 to 1.0 cm. These numbers indicate that the flows are well within the range of laminar flow, and this laminar flow had the ability to erode the fan surface. The erosion was sometimes triggered by surges, which developed in the laminar flow and moved downstream serially, except in flows with a Reynolds number < 10 (Simpson 1982). On the other hand, larger discharges (> 50 ml min^{-1}; in which some turbulence developed and mixing into the standing water body occurred), did not trigger the development of incised channels. Clouds of clay spread into the overlying water body and the flow did not erode.

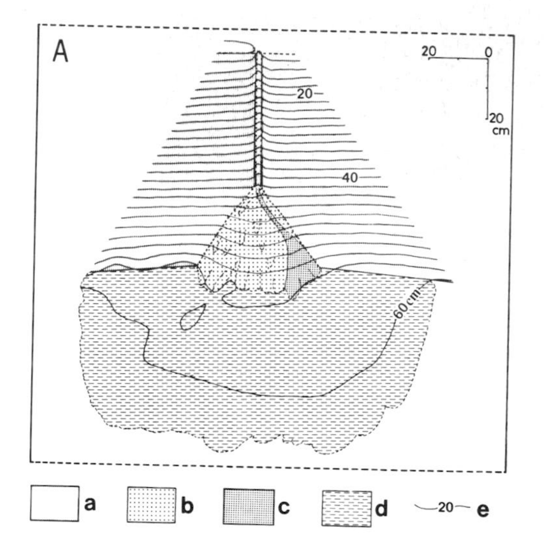

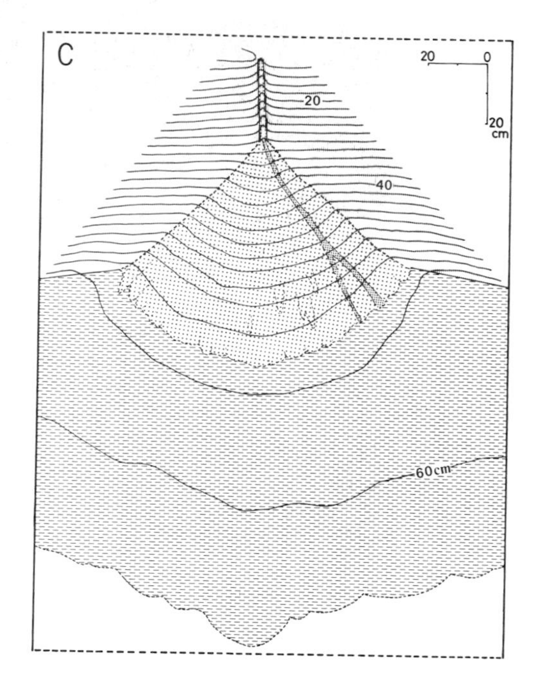

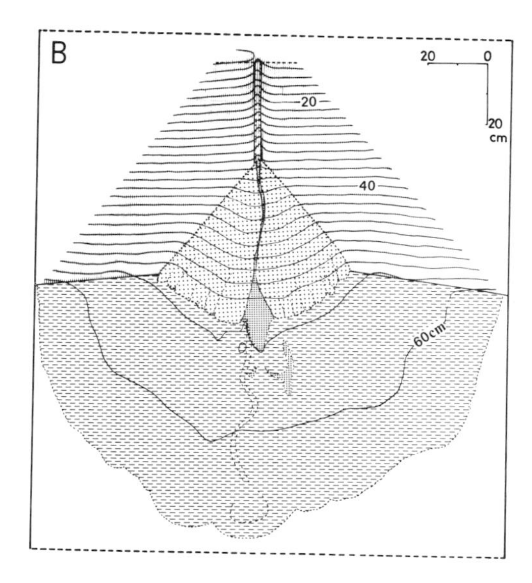

Fig. 4. (**a**) Morphology of subaqueous slope and experimental fan after *c.* 20 h of running with, clay, salt water and sand. Run 1. a, Original slope that was constructed with sand (plus a small amount of silt and clay) and 1–2% cement powder; b, surface of old sand lobe deposition; c, channel and sand lobe active at the time of measurement; d, clay deposition on the basin floor; e, contour lines showing the depth below a fixed water surface. (**b**) Morphology of subaqueous slope and experimental fan after *c.* 31.5 h of running with clay, salt water and sand, Run 1. Note sand lobe that has been deposited on the basin floor (lower centre of figure). (**c**) Morphology of subaqueous slope and experimental fan after *c.*66 h of running with clay, salt water and sand, Run 1.

Sand lobes formed of liquified sand flows sometimes migrated completely down to the basin floor (Fig. 4b), where they encountered clay deposits, especially when an incised channel was well developed and/or the sand supply was large. The sand lobe then slid (the movement accelerated) on the unconsolidated clay and formed a pressure ridge in a crescent-shaped mound, and a hummocky surface formed beyond this mound (Figs 4b & 7). This sand movement appeared to be similar to that of the subaqueous debris flow described by Hampton (1972), in which granular solids are more or less 'floated' by the matrix strength during the movement (Middleton & Hampton 1973, 1976). The sand lobe that migrated on to the clay seemed to behave like a 'floating (or rigid) plug' (Hampton 1972, 780). The flow following the

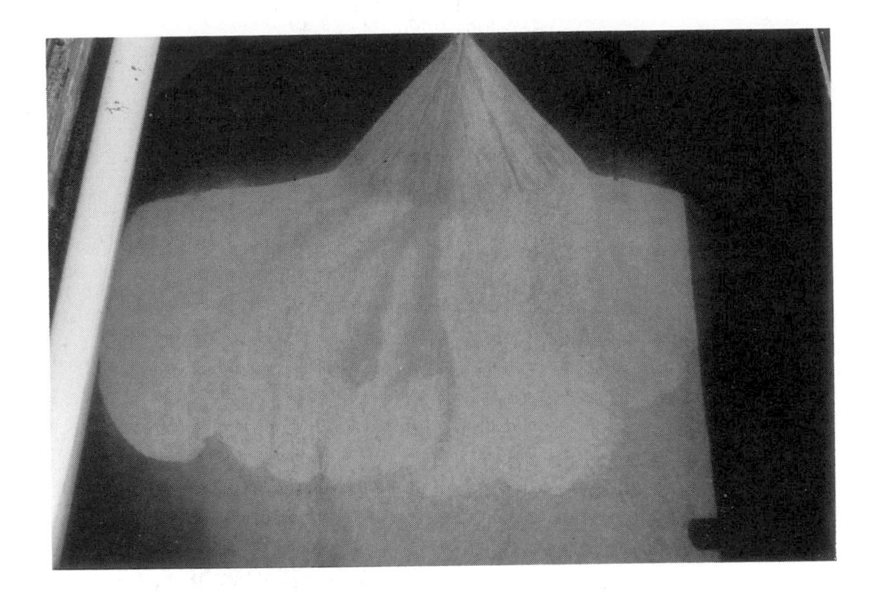

Fig. 5. General view of the experimental subaqueous fan produced in Run 1. Sand was deposited in a fan shape at the toe of slope, and clay was deposited on the basin floor in a thin layer. Maximum width of upper sandy fan is 70 cm in this photograph.

Fig. 6. Aerial view showing small terrace (arrow) in sinuous, incised channel produced during Run 1. Note nickpoint and surges on the flow surface. 10×10 cm grid provides scale.

sand lobe sometimes cut through or was diverted around the crescent mound and formed clay lobes further downbasin leaving a sinuous course behind it. One of these clay lobes contained some fine sand, marked by a dark colour (Figs 4b & 7).

The development of large sand lobes on the fan often induced slides (*c.* 10–15 cm wide) near the toe of fan. Sand slid on the clay deposits, as a floating plug of a subaqueous debris flow (Hampton 1972), which formed a large crescent-shaped mound and hummocky topography on the basin floor (Fig. 8). When the sandbody slid away, a hollow depression formed on the toe of the fan behind it. Upstream of this depression, the steep slope (slide scar) failed, and a nickpoint migrated up-fan, which created a relatively deep incised channel. Sand that was supplied by this erosion formed a small sand lobe that migrated into the hollow.

While water was being drained from the flume, the middle part of the fan surface started to slide and the entire fan slumped down to the basin floor. Slump scars appeared on the slope, and hummocky topography developed widely on the basin floor. This major slump occurred when the water level was *c.* 10–20 cm above the fan head. Pore-water pressure in the sand deposits apparently could not adjust to the hydrostatic pressure decrease, and a pressure gradient developed. The presence of some clay

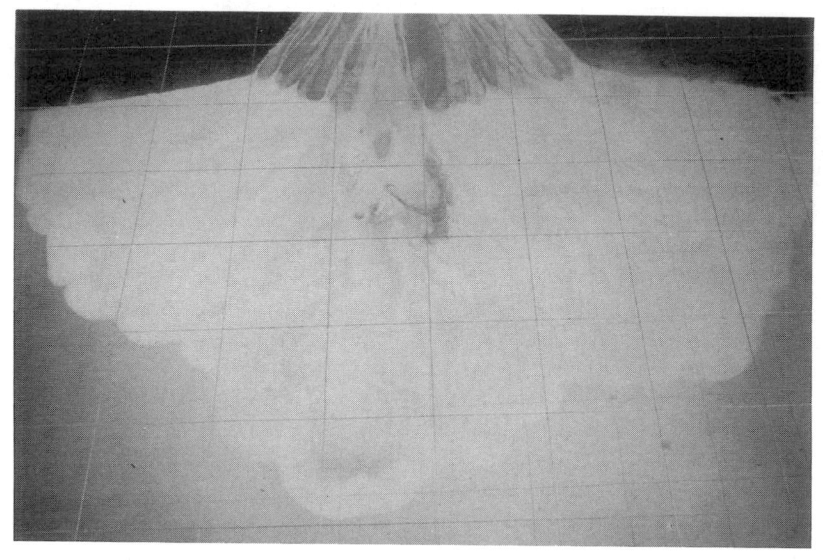

Fig. 7. Aerial view of subaqueous fan produced during Run 1 showing sand lobe that migrated on to the clay layer on the basin floor and which formed a crescent-shaped mound. The flow following the sand lobe was diverted around the mound and formed a clay lobe containing finer sand further down basin. This photo was taken during the time when the fan was mapped (Fig. 4b). 10 × 10 cm grid provides scale.

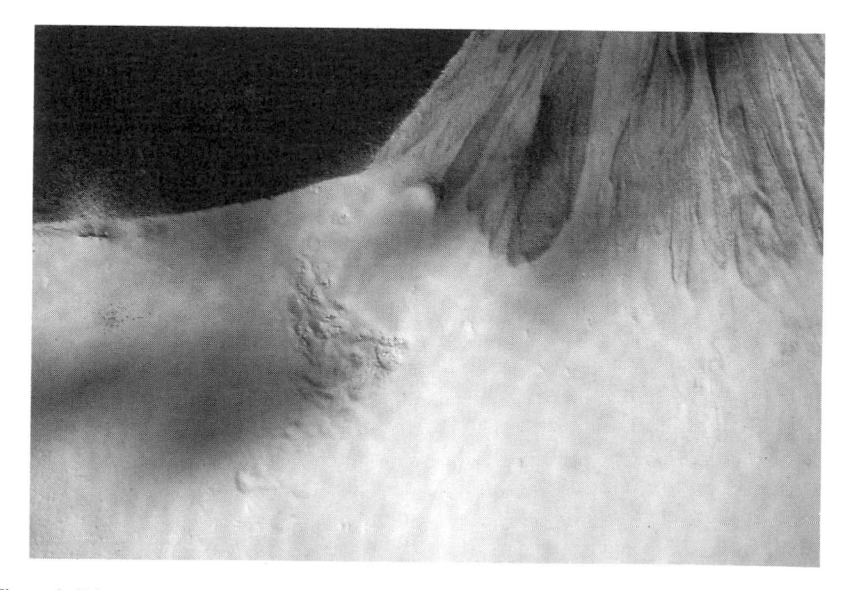

Fig. 8. View of slide that occurred near the toe of the fan during Run 1. Note hummocky surface and the large nickpoint that started to migrate upfan. Width of scarp is *c.* 5 cm.

Fig. 9. Overview of experimental subaqueous fan produced during Run 2. Note that the shape of individual sand lobes and the entire fan are more elongated than that produced during Run 1 (Fig. 3). Width of flume is 1.8 m. The elongate character of the fan produced in Run 2 can also be seen by comparing Figs 4c & 10b.

in the sandy deposits probably played an important role in developing the pressure gradient. The excess pore-water pressure made the fan slope unstable, and the upward pore-water flow probably induced the fluidized flow (Lowe 1976), which lead to the slide of the entire fan.

Run 2

Sand was introduced along with the clay plus salt water to form a liquified flow on the gentle subaqueous slope of 18°. Although processes of deposition were essentially similar to Run 1, the downslope migration of sand occurred as a more narrow elongate lobe, and the underwater fan produced was more elongate than that formed in Run 1 (Figs 9 & 10a). Paths taken by the clay plus salt water flow were more sinuous, and multiple active flow paths were common during the development of this fan. Clay was deposited widely both on the slope and on the basin floor, making it difficult to distinguish the extent of sand deposits. When a sand lobe moved on to the clay on the slope, it slid down in the form of a floating plug of a subaqueous debris flow as in Run 1. Movement of the sand, however, accelerated more than in Run 1 because of clay deposits on the slope. This additional acceleration made individual sand lobes very elongate (Fig. 11). It sometimes appeared that a small sand river was flowing through unconsolidated clay deposits, although clay 'banks' on both sides slowly moved with the sand. Crescent-shaped mounds did not develop, but hummocky topography appeared. After 36 h of running, sandy deposits extended *c.* 120 cm from the fan head, and clay deposits extended 260 cm basinward (Fig. 10a).

When the water level was lowered, sliding of sand deposits occurred in a manner similar to that in Run 1 (this time the water level was a little below the fan head). A series of slump scars appeared in the sandy part of the fan, and a hummocky surface developed widely on the basin floor. At this time clay deposits extended 310 cm basinward from the fan head. The run was then resumed on the slumped surface. Features observed in the resumed run were similar to those in the first phase. The same kind of fan developed on the irregular surface, and clay deposits extended 325 cm basinward from the fan head at the end of the run (Fig. 10b). The final water level lowering was at the very slow rate of 1–3 cm h^{-1}, and major slumps did not occur, although small surface slides did.

Dissected sections of the dried deposits showed that sand layers, which formed the sandy fan body and post-dated the major slide, extended basinward for *c.* 120 cm from the fan head. Some sand deposits from the major slide were observed as far as 250 cm from the fan head. Discontinuous sand deposits with a maximum thickness of 2.5 cm were formed along the same horizon (Figs 12 & 13). On the lower part of the original slope at distances between 70 and 120 cm from the fan head, fan deposits thinned to 1 cm or less, and the amount of sand decreased (Fig. 14). Further upslope sand layers (post major slide) formed alternate layers with clay. These sand layers showed an elongated lenticular shape in cross section (Fig. 15). The sand layers thickened at the expense of the clay layers towards the fan head. Near the fan head the sand deposit was *c.* 10 cm thick. The lower sand bed (*c.* 4 cm thick at the fan head) in this deposit was contorted and variable in thickness indicating the effects of the major slide.

(a)

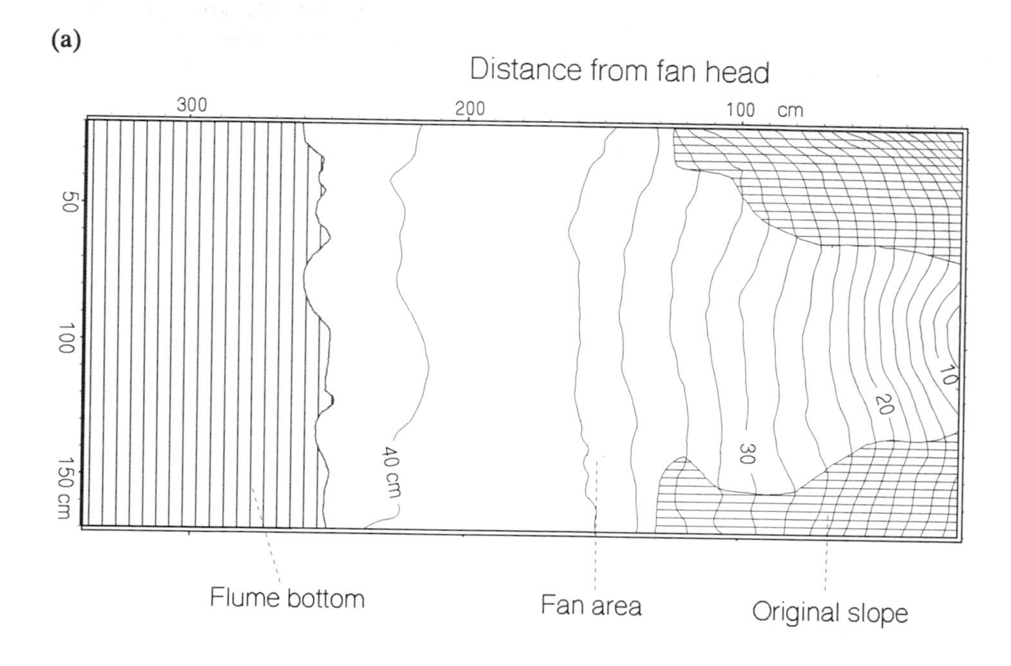

(b)

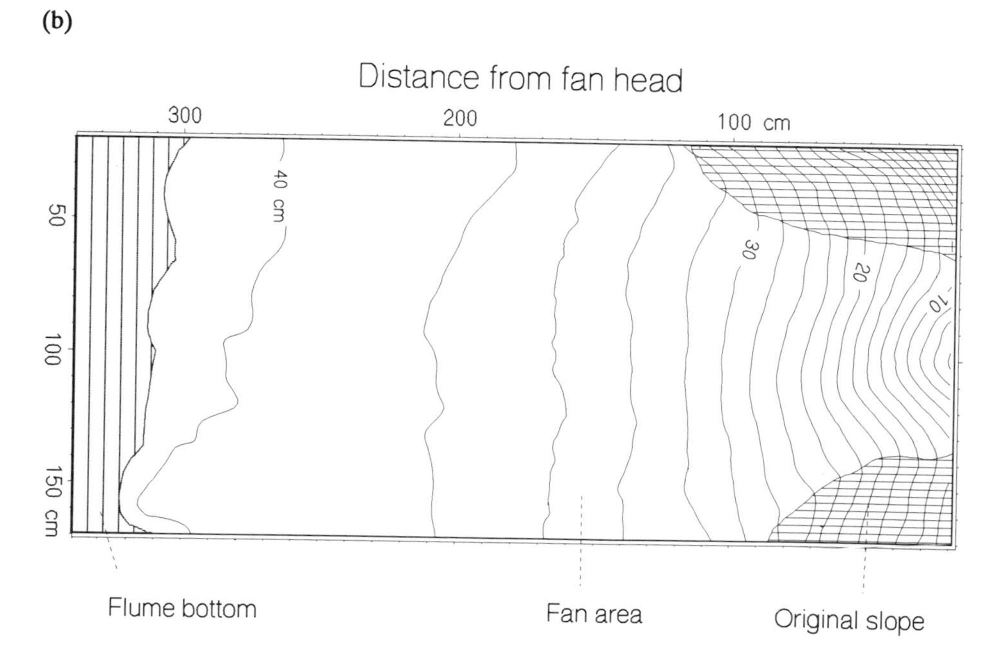

Fig. 10. **(a)** Morphology of subaqueous slope and fan after *c.* 36 h of running with clay, salt water and sand, Run 2. Horizontal lines in lower area labelled flume bottom are not contour lines, but simply illustrate the flat flume floor. Horizontal scale shows distances from the right wall of the flume. The figure does not show the whole width of the flume. The fan collapsed while the water was being drained, just after the data for this figure were taken. **(b)** Morphology of subaqueous slope and fan after *c.* 69 h of running with clay, salt water and sand, Run 2.

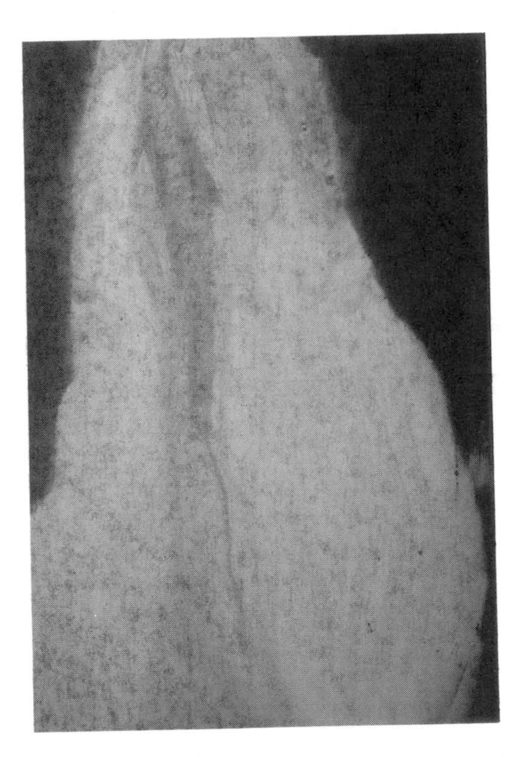

Fig. 11. Sand movement occurred as very elongated bank-like features on the gentler fan of Run 2. Maximum width of sand lobes is 15 cm.

Comparison with deep sea fans

Types of fans

Sand was deposited in a conical fan shape in Run 1 (Fig. 3), but the deposition showed a more elongate form in Run 2 (Fig. 9). Although the size of the experimental underwater fans are much smaller and the slopes are much steeper than those of real deep sea fans, the difference in experimental fan shapes appears superficially similar to the difference between active-margin fans and passive-margin fans (Shanmugam & Moiola 1988). The former is represented by small sand-rich fans, and the latter by large mud-rich fans reflecting the sand percentage in sediments supplied from coastal drainages. Stow *et al.* (1985) also suggest that sediment type and supply determine the type of deep sea fan (i.e. mud-dominated elongate fans with high input rates, sand-dominated radial fans with medium input rates, and slope-apron with low input rates). In the experiments, the contrast of fan shape is apparently due to the differences in the gradient of subaqueous slopes and the different proportions of sand and clay that were delivered to the slope. During Run 1 the ratio of sand to clay was higher than during Run 2.

Sand deposition occurred in the form of small lobes and these are the fundamental components of experimental underwater fans, although the lobes in Run 2 were much more elongate than those in Run 1. In Run 1 sand lobes were first deposited at the sharp slope break and overlapped to form a conical fan shape. Clay moved further and was deposited on the flume bottom. The gentler slope of Run 2 allowed a larger discharge of clay plus salt water to flow down without making the water body murky, and the clay to settle on the slope. The clay on the subaqueous slope made the slope surface 'soupy' (Pickering *et al.* 1992), and this caused sand lobes, which were already moving as liquified flows, to accelerate as floating plugs of

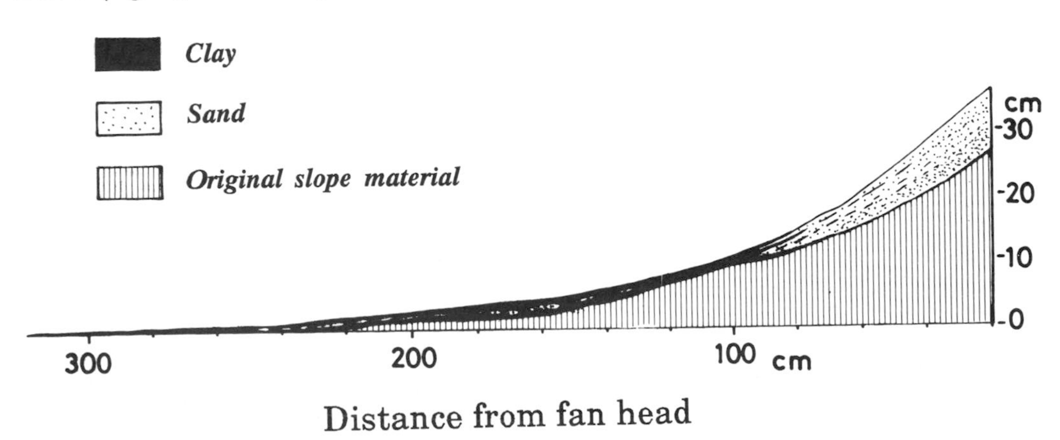

Fig. 12. Sketch showing the longitudinal section of fan deposits of Run 2. The section was taken along a line that was 100 cm from the right wall of the flume.

Fig. 13. Irregular sand lenses in the basin-floor clay layer, which were transported and deformed by the major slide in Run 2. Note coin for scale.

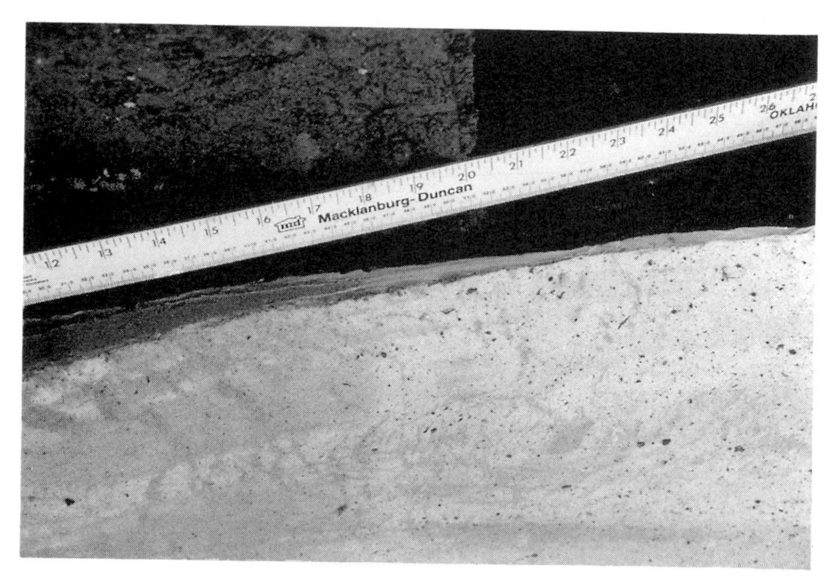

Fig. 14. Longitudinal section showing the distal part of the sandbody that post-dated the major slide in Run 2. Note thin interfingering of sand and clay layers.

debris flows. This may be the major factor that made the underwater fan of Run 2 more elongate. If the slope is too steep for the clay in density flows to settle, most clay will be deposited on the basin floor, and sand will be deposited at the toe of slope forming a cone-shaped deposit. When the slope is gentler, the clay on the slope makes the sand move more effectively. The terms 'highly efficient fan' and 'poorly efficient fan', as applied to large mud-rich and

Fig. 15. Cross-section showing lenticular sand lobe deposits of resumed run in Run 2. Note coin for scale.

small sand-rich deep sea fans, respectively, by Mutti (1979), seems to conform to the experimental observations, although the usage of these terms has been criticized (Shanmugam & Moiola 1985).

Detached lobes

Sand deposits that are detached from the feeder channel, as a result of sediment bypassing are referred to as detached lobes (Mutti 1977, 1985; Shanmugam & Moiola 1985). The sand lobes that reached the clay deposits on the basin floor in Run 1 slid away from the main sand deposits as floating plugs of subaqueous debris flows, which formed a crescent-shaped mound and a hummocky surface on the basin floor clay deposits (Figs 7 & 8). These sand lobes seemed to form deposits similar to the detached lobes described above. Unfortunately, the fan collapsed during draining of the flume and detached lobes could not be seen in the deposits.

Sand deposition of this type did not appear to occur in Run 2. However, the major slide, which occurred during water drainage and which moved a large volume of sand to the basin, caused many irregular detached sand layers in the clay (Fig. 13). These sand layers as a whole were separated from the upper fan sand deposits. The sand deposits associated with the major slide may possibly be regarded as a detached lobe; however, these irregular and disconnected sand patches could not be clearly related to a larger sand lobe.

Slope failure

Large slides, which occurred in Runs 1 and 2 while the water was being drained, generated large subaqueous debris flows. The sand was transported further basinward and deposited in somewhat irregular layers. Water-level lowering apparently generated a pore-pressure gradient in the sand deposits, that induced liquefaction and resultant slope failure. Judging from the fluidized flow first observed on the fan surface, upward pore-water flow in the sediments is most likely to be the trigger. This kind of subaqueous slope failure is known to occur in the shallow marine environment (Prior & Coleman 1984; Robb 1990). For example, Dunlap *et al.* (1979) reported that the pore pressures elevated by storm wave action induced liquefaction in sediments on the bottom of a shallow sea. Wells *et al.* (1980) showed that excess pore pressures in sediments generated by receding tides can be one of the contributory factors to nearshore, very low angle (0.03–0.08°) slope failure.

Excess pore pressure is also generated by rapid submarine sedimentation, forming under-consolidated layers in the deep sea (Morgenstern 1967). This excess pore pressure in under-consolidated deposits is considered to be one of the main factors that causes instability on gentle submarine slopes (Hampton *et al.* 1978; Booth 1979).

Effect of sea-level change

Sea-level change has been considered as one of the most important factors controlling the development of deep sea fans (e.g. Shanmugam & Moiola 1982; Shanmugam *et al.* 1985; Posamentier *et al.* 1991), and the idea that eustatic lowering of sea level is the primary factor in the development of submarine fans has been widely accepted (Shanmugam & Moiola 1988). Daly (1936) hypothesized that the lowering of sea level increased the amount of suspended sediment supplied to the continental slopes and deep marine environments. He speculated that water agitation by waves breaking on the continental shelves puts much sediment into suspension. Today, the explanation is that, when sea level was lowered and most continental shelves became emergent, rivers eroded the shelves and supplied large amounts of coarse material to the heads of submarine canyons to accelerate deep-sea fan development. Posamentier *et al.* (1991) emphasized that during intervals of relative sea-level fall, submarine fans developed farther into the basin than during the subsequent relative sea-level stillstand. Sea-level falls and lowstands have also been characterized as times of increased frequence of sediment failures (Posamentier & Vail 1988; Trincardi & Field 1992).

Although the experiments in this study were not performed to evaluate the effects of water-level changes, nevertheless, the large slides that occurred during water-level lowering, implies that the increase in pore-pressure gradients should be considered as one explanation of deep sea fan expansion during sea-level lowering in areas where active deposition has ceased. Global eustatic lowering of sea level can reduce the external hydrostatic load and induce excess pore pressure in the deposits, and this may result in the upward pore-water flows that may induce failures on gentle subaqueous depositional slopes. These slope failures would generate a large number of subaqueous debris flows, which would move a large volume of sand further basinward on to clay-rich deposits that can accelerate the movement, and therefore, stimulate the expansion of deep sea fans during periods of relative sea-level fall.

The rate of sea level lowering is, of course, much smaller than that in the experiments, and pore pressure may dissipate and adjust to the reduced hydrostatic load without inducing slope failures. However, involvement of a large amount of material, especially clay, may slow the adjustment of pore pressure significantly. Morgenstern (1967) pointed out that the dissipation of pore pressure is less in the larger volume of material. Booth (1979), moreover, estimated the existence of a thick under-consolidated layer (100 m or more) in the submarine slope sediments of the Gulf of Mexico, and assumed that excess pore pressure existed for > 25 000 a. If the dissipation of pore pressure is slow enough, the large amount of eustatic sea-level fall would probably be able to generate enough of a pore-pressure gradient in deep sea deposits to induce slope failures even on very gentle slopes. What rate of sea-level lowering is fast enough, and how much pore pressure should be elevated to induce failures on deep sea depositional slopes is largely unknown. However, it can be said at this point that the slopes formed by underwater deposition may become unstable during water-level lowering. If deep sea depositional slopes became unstable by the development of a pore-pressure gradient during low sea-level stands, slides could be more easily induced by other factors, such as earthquakes, slope steepening, high accumulation of organic matter and low shear strength (Trincardi & Field 1992). This might result in the accelerated growth of deep sea fans during sea-level fall.

Summary and conclusions

In this study we were successful in creating miniature subaqueous fans in a flume and observing the way in which these fans developed. Although these fans and their development processes cannot be compared to natural deep sea fans in detail, some observations provide interesting insights on the problems of deep sea fan development as follows:

(1) The mixture of dense fluid (clay plus salt water) and sand formed liquified flows that moved down an underwater slope, and they created subaqueous depositional forms superficially similar to deep sea fans. Clay plus salt water flows of small discharge without sand, which showed characteristics of a laminar flow, had the ability to incise a channel on the sandy fan surface. Larger flows, in which some turbulence developed, however, spread into the overlying water body and did not cut a channel.

(2) During Run 1 sand lobes were deposited at the toe of the slope and overlapped each

other to form a fan-shaped sandbody. During Run 2, with a gentler slope, clay settled on the slope, and sand moved as an elongate lobe, and the fan itself became much more elongate. The elongate nature of the fan deposit on the lower slope was apparently related to the higher efficiency of sand movement on the clay. The differences in sand movement and the fan shape on the steep and gentle subaqueous slopes in the experiments appear superficially similar to differences noted between small sand-rich active-margin and large mud-rich passive-margin fans.

(3) Sand lobes occasionally migrated on to the basin floor clays from the steeper slopes of Run 1. The movement of sand lobes then accelerated, and the sand lobe slid away like the floating plug of a debris flow (Hampton 1972). This movement of sand may possibly result in formation of detached sand lobe deposits.

(4) Experimental underwater fans, on both steep and gentle slopes, collapsed when the water in the flume was being drained, and this generated relatively large subaqueous debris flows. The reduction of hydrostatic pressure caused by the lowering of the water level apparently generated a pore-pressure gradient. The pore water probably flowed upward as the pressure gradient developed, and this flow induced a fluidized flow on the surface and then slumping of the entire fan. These subaqueous slope failures indicate that, if the dissipation of pore pressure was slow enough (and this is not unrealistic), eustatic sea-level lowering might induce submarine slope instability due to the development of a pore-pressure gradient in natural deposits. This provides another possible explanation of accelerated deep sea fan development during periods of relative sea-level lowstands.

We thank Willem Scott for his assistance in performing the experiments. Financial support for this research was provided by Amoco Production Company, Shell Development Company, and Union Pacific Resources. We specifically wish to acknowledge Norm Haskell, Lee Krystinik and Keith Shanley for their support and encouragement. Constructive reviews by Jan Alexander and Roger Anderton have improved the manuscript significantly. The senior author was on sabbatical leave from Chuo University, which supported his contribution to the research.

References

Allen, J. R. L. 1971. Mixing at turbidity current heads, and its geological implications. *Journal of Sedimentary Petrology*, **41**, 97–113.

Booth, J. S. 1979. Recent history of mass-wasting on the upper continental slope, northern Gulf of Mexico, as interpreted from the consolidation states of the sediment. *In:* Doyle, L. J. & Pilkey, O.H. (eds) *Geology of Continental Slopes.* Society of Economic Paleontologists and Mineralogists, Special Publication, **27**, 153–164.

Bouma, A. H., Normark, W. R. & Barnes, N. E. (eds) 1985. *Submarine fans and related turbidite systems.* Springer-Verlag, New York, 351.

Daly, R. A. 1936. *Origin of submarine 'canyons'. American Journal of Science, 5th series,* **31**, 401–420.

Dunlap, W., Bryant, W. R., Williams, G. N. & Suhayda, J. N. 1979. Storm wave effects on deltaic sediments – results of SEASWAB I and II. *Proceedings of the Conference on Port and Ocean Engineering under Arctic Conditions at the Norwegian Institute of Technology,* **2**, 899–920.

Garcia, M. & Parker, G. 1989. Experiments on hydraulic jumps in turbidity currents near a canyon-fan transition. *Science,* **245**, 393–396.

Gould, H. R. 1951, Some quantitative aspects of lake mead turbidity currents. *In:* Hough, J. L. (ed.) *Turbidity Currents and the Transportation of Coarse Sediments to Deep Water.* Society of Economic Palaeontologists and Mineralogists, Special Publication, **2**, 34–52.

Hampton, M. A. 1972. The role of subaqueous debris flow in generating turbidity currents. *Journal of Sedimentary Petrology,* **42**, 775–793.

—— 1975. Competence of fine-grained debris flows. *Journal of Sedimentary Petrology,* **45**, 834–844.

——, Bouma, A. H., Carlson, P. R., Molnia, B. F., Clukey, E. C. & Sangrey, D. A. 1978. *Quantitative study of slope instability in the Gulf of Alaska.* Proceedings of 10th Annual Offshore Technology Conference, Houston, Texas, OTC #3314, 2037–2314.

Heezen, B. C. & Ewing, M. 1952. Turbidity currents and submarine slumps, and the 1929 Grand Banks earthquake. *American Journal of Science,* **250**, 849–873.

Komar, P. D. 1969. The channelized flow of turbidity currents with application to Monterey deep-sea fan channel. *Journal of Geophysical Research,* **74**, 4544–4558.

—— 1972. Relative significance of head and body spill from a channelized turbidity current. *Geological Society of America Bulletin,* **83**, 1151–1156.

—— 1977. Computer simulation of turbidity current flow and the study of deep sea channels and fan sedimentation. *The Sea,* **6**, 603–621.

Kuenen, Ph. H. 1937. Experiments in connection with Daly's hypothesis on the formation of submarine canyons. *Leidshe Geol. Mededeelingen, Dl,* **7**, 327–351.

—— 1938. Density currents in connection with the problem of submarine canyons. *Geological Magazine,* **75**, 241–249.

—— 1951. Properties of turbidity currents of high density. *In:* Hough, J. L. (ed.) *Turbidity Currents and the Transport of Coarse Sediments to Deep*

Water. Society of Economic Paleontologists and Mineralogists, Special Publication, **2,** 14–33.

—— & Migliorini, C. I. 1950. Turbidity currents as a cause of graded bedding. *Journal of Geology,* **58,** 91–127.

Lowe, D. R. 1976. Subaqueous liquefied and fluidized sediment flows and their deposits. *Sedimentology,* **23,** 285–308.

Luthi, S. 1981. Experiments on non-channelized turbidity currents and their deposits. *Marine Geology,* **40,** M59–M68.

Middleton, G. V. 1966a. Experiments on density and turbidity currents: I. Motion of the head. *Canadian Journal of Earth Sciences,* **3,** 523–546.

—— 1966b. Experiments on density and turbidity currents: II. Uniform flow of density currents. *Canadian Journal of Earth Sciences,* **3,** 623–637.

—— 1967. Experiments on density and turbidity currents: III. Deposition of sediment. *Canadian Journal of Earth Sciences,* **4,** 475–505.

—— 1993. Sediment deposition from turbidity currents. *Annual Reviews of Earth and Planetary Sciences,* **21,** 89–119.

—— & Hampton, M. A. 1973. Sediment gravity flows: Mechanics of flow and deposition. *In:* Middleton, G. V. & Bouma, A. H. (eds) *Turbidities and Deep-Water Sedimentation.* Pacific Section Society of Economic Paleontologists and Mineralogists, Los Angeles, 1–38.

—— —— 1976. Subaqueous sediment transport and deposition by sediment gravity flows. *In:* Stanley, D. J. & Swift, D. J. P. (eds) *Marine Sediment Transport and Environmental Management.* John Wiley & Sons, New York, 197–218.

Morgenstern, N. R. 1967. Submarine slumping and the initiation of turbidity currents: *In:* Richards, A. F. (ed.) *Marine Geotechnique.* University of Illinois Press, Urbana, 189–220.

Mutti, E. 1977. Distinctive thin-bedded turbidite facies and related depositional environments in the Eocene Hecho Group (South-central Pyrenees, Spain). *Sedimentology,* **24,** 107–131.

—— 1979, Turbidites et cones sous-marins profonds. *In:* Homewood, P. (ed.) *Sedimentation Detrique (Fluviatile, Littorale et Marine).* Institute of Geology, Universite de Fribourg, 353–419.

—— 1985. Hecho turbidite system, Spain. *In:* Bouma, A. H., Normark, W. R. & Barnes, N. E. (eds) *Submarine Fans and Related Turbidite Systems.* Springer-Verlag, New York, 205–208.

—— & Ricci Lucchi, F. 1972. Turbidites of the northern Apennines: Introduction to facies analysis. (English translation by Nilsen, T. H. 1978). *International Geological Review,* **20,** 125–166.

Natland, M. L. & Kuenen, Ph.D. 1951. Sedimentary history of the Ventura basin, California, and the action of turbidity currents. *In:* Hough, J. L. (ed.) *Turbidity Currents and the Transportation of Coarse Sediments to Deep Water.* Society of Economic Paleontologists and Mineralogists, Special Publication, **2,** 76–107.

Nilsen, T. H. 1980. Modern and ancient submarine fans. Discussion of papers by Walker, R. G. & Normark, W. R. *American Association of Petroleum Geologists Bulletin,* **64,** 1094–1101.

Normark, W. R., Posamentier, H. & Mutti, E. 1993. Turbidite systems: state of the art and future directions. *Review of Geophysics,* **31,** 91–116.

Parker, G., Fukushima, Y. & Pantin, H. M. 1986. Self-accelerating turbidity currents. *Journal of Fluid Mechanics,* **171,** 145–181.

—— Garcia, M., Fukushima, Y. & Yu, W. 1987. Experiments on turbidity currents over an erodible bed. *Journal of Hydraulic Research,* **25,** 123–147.

Pickering, K. V., Underwood, M. B. & Taira, A. 1992. Open-ocean to trench turbidity-current flow in the Nankai Trough: flow collapse and reflection. *Geology,* **20,** 1099–1102.

Posamentier, H. W. & Vail, P. R. 1988. Eustatic control on clastic deposition, II – sequences and systems tract models. *In:* Wilgus, C., Hastings, B. S. & Kendell, C. G. St.C. (eds) *Sea Level Changes – An Integrated Approach.* Society of Economic Paleontologists and Mineralogists Special Publication **42,** 125–154.

—— Erskine, R. D. & Mitchum, R. M. Jr. 1991. Models for submarine-fan deposition within a sequence-stratigraphic framework. *In:* Weimer, P. & Martin, H. L. (eds) *Seismic Facies and Sedimentary Processes of Submarine Fans and Turbidite Systems.* Springer-Verlag, New York, 127–136.

Prior, D. B., & Coleman, J. M. 1984. Submarine slope instability. *In:* Brunsden, D. & Prior, D. B. (eds) *Slope Instability.* John Wiley & Sons, New York, 419–455.

Robb, J. M. 1990. Groundwater processes in the submarine environment. *In:* Higgins, C. G., & Coates, D. R. (eds) *Groundwater Geomorphology; The Role of Subsurface Water in Earth-Surface Processes and Landforms.* Geological Society of America, Special Paper, **252,** 267–281.

Schumm, S. A., Mosley, M. P. & Weaver, W. E. 1987. *Experimental Fluvial Geomorphology.* John Wiley & Sons, New York, 413.

Shanmugam, G. & Moiola, R. J. 1982. Eustatic control of turbidites and winnowed turbidites. *Geology,* **10,** 231–235.

—— —— 1985. Submarine fan models: problems and solutions. *In:* Bouma, A. H., Normark, W. R. & Barnes, N. E. (eds) *Submarine Fans and Related Turbidite Systems.* Springer-Verlag, New York, 29–34.

—— —— 1988. Submarine fans: characteristics, models, and reservoir potential. *Earth-Science Reviews,* **24,** 383–428.

—— —— 1991. Types of submarine fan lobes: models and implications. *American Association of Petroleum Geologists Bulletin,* **75,** 156–179.

—— —— & Damuth, J. E. 1985. Eustatic control of submarine fan development. *In:* Bouma, A. H., Normark, W. R. & Barnes, N. E. (eds) *Submarine Fans and Related Turbidite Systems.* Springer-Verlag, New York, 23–28.

Shepard, F. P. 1951. Transportation of sand into deep water: *In:* Hough, J. L. (ed.) *Turbidity Currents and the Transportation of Coarse Sediments to Deep Water.* Society of Economic Paleontologists and Mineralogists, Special Publication, **2,** 53–65.

SIMPSON, J. E. 1982. Gravity currents in the laboratory, atmosphere, and ocean. *Annual Review of Fluid Mechanics,* **14,** 213–234.

STOW, D. A. V., HOWEL, D. G. & NELSON, C. H. 1985. Sedimentary, tectonic, and sea-level controls. *In:* BOUMA, A. H. NORMARK, W. R. & BARNES, N. E. (eds) *Submarine Fans and Related Turbidite Systems.* Springer-Verlag, New York, 15–22.

TRINCARDI, F. & FIELD, M. E. 1992. Collapse and flow of lowstand shelf margin deposits: an example from the eastern Tyrrhenian Sea, Italy. *Marine Geology,* **105,** 77–94.

TWICHELL, D. C., SCHWAB, W. C., NELSON, C. H., KENYON, N. H. & LEE, H. J. 1992. Characteristics of a sandy depositional lobe on the outer Mississippi fan from SeaMARC IA sidescan sonar images. *Geology,* **20,** 689–692.

WALKER, R. G. 1978. Deep-water sandstone facies and ancient submarine fans: Models for exploration for stratigraphic traps. *American Association of Petroleum Geologists Bulletin,* **62,** 932–966.

—— 1980. Modern and ancient submarine fans: reply. *American Association of Petroleum Geologists Bulletin,* **64,** 1101–1108.

WEIRICH, F. H. 1984. Turbidity currents: monitoring their occurrence and movement with a three-dimensional sensor network. *Science,* **224,** 384–387.

WELLS, J. T., PRIOR, D. B. & COLEMAN, J. M. 1980. Flowslides in muds on extremely low angle tidal flats, northeastern South America. *Geology,* **8,** **272–275.**

Beyond the turbidite paradigm: physical models for deposition of turbidites and their implications for reservoir prediction

BEN KNELLER

Department of Earth Sciences, The University of Leeds, Leeds SL2 9JT, UK

Abstract: The simple physical models which are popularly used to describe deposition from turbidity currents are based on the notion of unidirectional waning flow, resulting in the familiar Bouma sequence (Ta–e) or its high-density counterpart, the Lowe sequence (S1–3). Most geologists working on turbidite successions know only too well how wide is the range of facies which do not fit into the standard facies models. The application of simple equations of motion shows that deposition can occur beneath flows that are steady or even waxing. The various combinations of different spatial and temporal accelerations produce markedly different vertical and lateral variations in the resulting turbidite. On these grounds alone, we should expect not one but at least five basic types of sequence in turbidite beds, for which likely candidates can be found in many deep water clastic systems. This has important implications for deposit geometry, gradients in reservoir properties, and geological interpretation and correlation of well cores. Since many basins are confined, *a priori* reasoning tells us to expect interactions between turbidity currents and topography, producing anomalous intrabed vertical sequences, multiple current directions and locally enhanced deposition. Flume experiments with radially spreading scaled sediment gravity flows demonstrate how the geometry of sandstone deposits may be controlled by topography. Interactions between the flow and topographic features are dependent upon the shape and orientation of the obstacle, and upon its size relative to the height of the flow. Some of the results are counter-intuitive, and may have significant impact on location of reservoir sands around basin-floor and marginal topography.

It is invariably the case that there are insufficient geological, geophysical and petrophysical data to create an unique reservoir description or reservoir engineering model. Consequently, geological models of a variety of kinds and scales, whose effectiveness is dependent upon the validity of the assumptions incorporated into them – assumptions which are often themselves models – must be relied upon. Central amongst these assumptions concerning deep water depositional systems are models for transport and deposition of sediment by turbidity currents; these influence the way we think about (and make predictions of) a whole range of sedimentary phenomena, from the character of individual beds to the geometry of turbidite systems and their components. They also affect predictions concerning lateral changes in reservoir-significant rock properties such as grain size, and the way correlation of lithofacies and individual beds is approached. Clearly, the reliability of larger-scale models comes into question if these central assumptions are untrue or only partly true. Consequently, correlations may not only be badly wrong, but may also produce reservoir predictions which are grossly in error both in terms of overall geometry and petrophysical gradients.

Most geologists working with deep water systems in core or outcrop will be familiar with the depositional models of Bouma (1962) and Lowe (1982) which, following Kuenen & Migliorini's (1950) explanation of graded bedding, relate the sequences of internal structures within individual turbidite beds (Fig. 1a & b) to deposition from unidirectional waning turbidity flows. However, most workers are also aware of the high proportion of deep water facies which can be interpreted in terms of these models only with difficulty or not at all – massive sandstones (Kneller & Branney 1995), disordered turbidite facies (Branney *et al.* 1990), beds with 'anomalous' or repeated vertical sequences or intra-bed palaeocurrent discrepancies (Pickering & Hiscott 1985; Kneller *et al.* 1991; Edwards *et al.* 1994), abrupt grain-size breaks, large-scale bedforms etc. (for general discussion see Pickering *et al.* 1989; Mutti & Normark 1991; Mutti 1992). The range of facies which do not accord with these simple models exceeds those that do; sequences of uncomplicated normally-graded beds are actually rather uncommon, particularly

From Hartley, A. J. & Prosser, D. J. (eds), 1995, *Characterization of Deep Marine Clastic Systems*, Geological Society Special Publication No. 94, pp. 31–49.

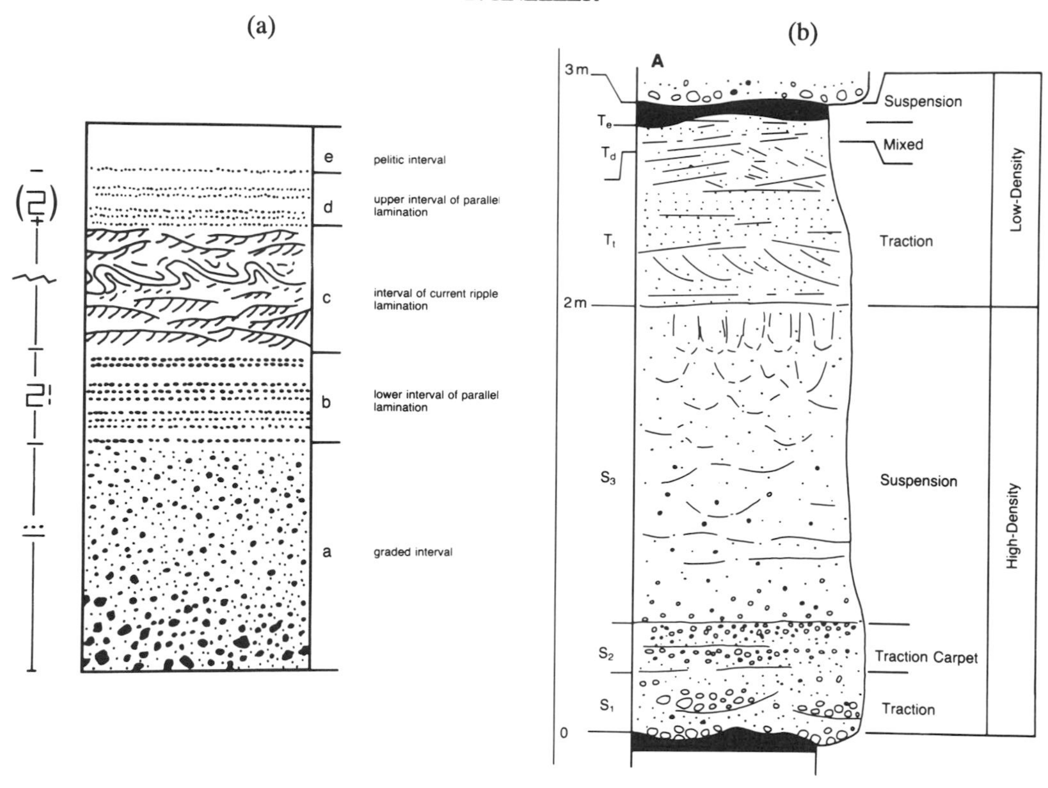

Fig. 1. (a) Idealized sequence of structures in a low-density sandy turbidity current (from Bouma 1962). **(b)** Idealized sequence of structures of structures in a high-density sandy turbidity current (from Lowe 1982).

so in those turbidite sandstone lithofacies likely to form hydrocarbon reservoirs.

Unidirectional waning flow?

The notion of unidirectional waning flow constitutes a paradigm which has dominated thinking on deep water sedimentation for over 30 years. There is, however, no reason to believe that all turbidity currents in the deep marine realm behave as simple waning flows (Stanley *et al.* 1978; Normark & Piper 1991). Furthermore, both deposition from and flow directions within turbidite currents and other related gravity currents are profoundly affected by topography (e.g. Pantin & Leeder 1987; Edwards *et al.* 1994; Muck & Underwood 1990; Kneller *et al.* 1991; Druitt 1992). Many turbidite systems are topographically confined (e.g. Mutti & Ricci-Lucchi 1978; Thornburg *et al.* 1990; Weaver *et al.* 1992), and onlap basin margin or intrabasinal topography (Cazzola *et al.* 1985; Stanley *et al.* 1978) or generate their own depositional topography (Normark *et al.* 1993); thus, evidence of reflections or deflections, and local sand accumulations controlled by the topography, may be expected.

Steadiness and uniformity of flows

Steady flow is defined as a succession of fluid particles through a point fixed in space having identical velocity vectors, the fluid flow remaining unchanged with time (Allen 1985). Waning flow refers to one type of flow unsteadiness, and describes a temporal change in current velocity at a point whereby successive fluid particles have smaller velocities (i.e. where the flow passing a given point gets slower, Fig. 2a). Waxing flow describes a temporal change in current velocity at a point, whereby successive fluid particles have larger velocities. The velocity change in unsteady flows can be described by the term $\partial u/\partial t$, where t is time and u is the mean downstream velocity at that point (i.e. ignoring short-term fluctuations due to fluid turbulence).

A fluid flow characterized by a non-zero spatial acceleration is called non-uniform (Allen 1985). Spatial changes in flow velocity are

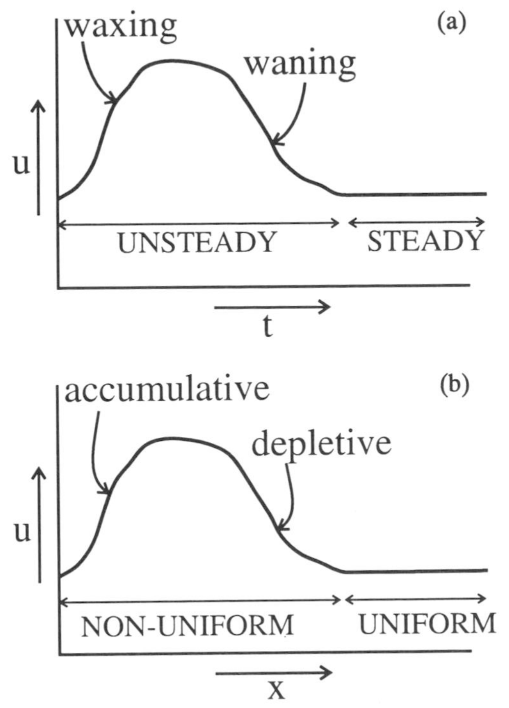

Fig. 2. Definition sketches for: (**a**) steadiness/unsteadiness (Kneller & Branney 1995) and (**b**) uniformity/non-uniformity.

described by the terms uniform and non-uniform. Flow which becomes more rapid downstream is described as accumulative and that which becomes slower is called depletive (Kneller & Branney 1995); a definition sketch is given in Fig. 2b. Figure 3 illustrates various situations in which accumulative or depletive flow may occur in response to guiding topography, i.e. changes in flow constriction or gradient. The velocity change in non-uniform flows can be described by the term $u.\partial u/\partial x$, where x is the streamwise distance.

The net acceleration experienced by a particle moving within the flow (the substantive acceleration, du/dt) is dependent on the sum of these terms:

$$du/dt = \partial u/\partial t + u.\partial u/\partial x$$

The substantive acceleration can be represented as a point on a graph whose axes are the spatial and temporal accelerations (Fig. 4); wherever the substantive acceleration (which is the sum of these two terms) is negative, individual particles within flows experiencing these accelerations will decelerate; this condition obtains over half the field. Take the case of a flow in which the mean downstream velocity is sufficient to carry all the sediment as suspended

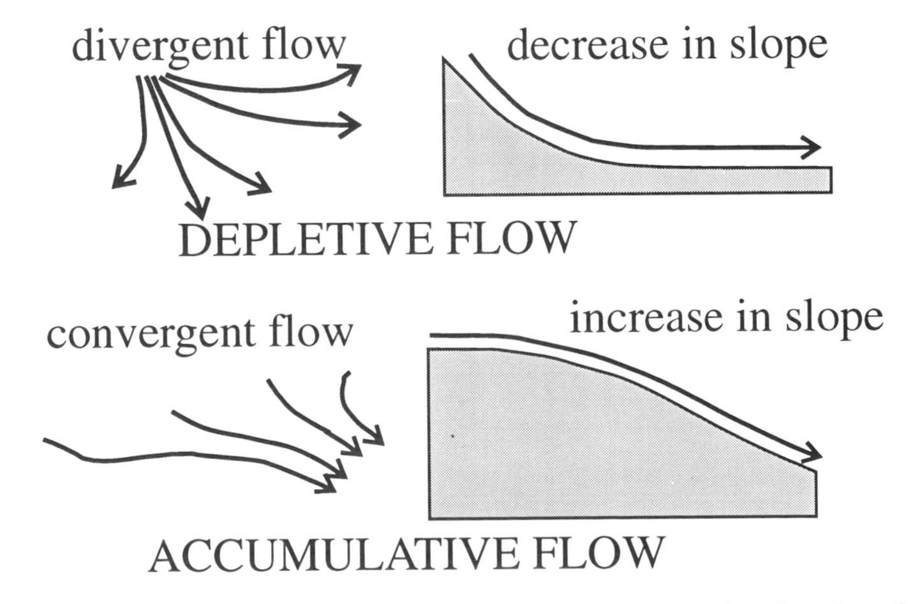

Fig. 3. Situations producing non-uniform flow. Depletive flow will occur at canyon and channel mouths, at changes in gradient at the foot of slopes or associated with constructional topography, where flows exceed bank-full in channelized systems, and against the slopes of obstructing topography. Accumulative flow will occur at increases in slope and where flow is constricted, and is particularly likely around obstacles and where flows pass through interconnected sub-basins such as perched and silled basins, or basins dissected by tectonic or diapiric topography.

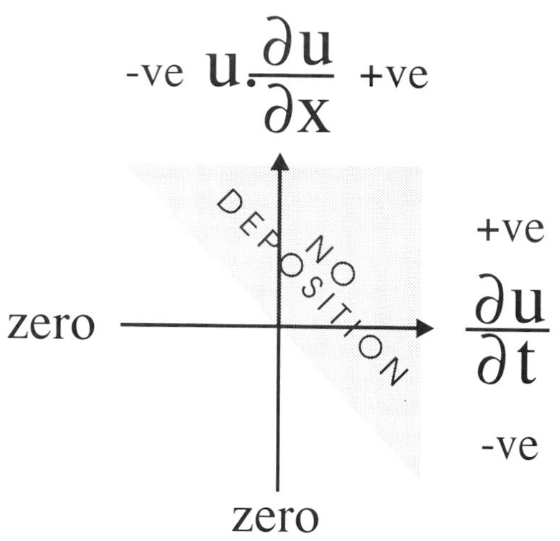

$$\frac{du}{dt} = \frac{\partial u}{\partial t} + u.\frac{\partial u}{\partial x}$$

Fig. 4. Graph to illustrate the conditions under which deposition (du/dt negative) or non-deposition/erosion (du/dt positive) will occur.

load and the substantive acceleration is negative. If this condition is maintained, at some point in time and space the mean downstream velocity will fall below the suspension threshold for the largest grains (or the capacity of the flow will be exceeded by the suspended sediment), and material will begin to transfer to the bed. For material coarser than silt grade, sediment may continue to be moved by traction until the velocity falls below the threshold for bed movement. Whether or not a particle is deposited from the flow depends upon a negative substantive acceleration, which may occur whether the flow is waxing or waning, uniform or non-uniform.

Figure 5 gives a graphical representation of the substantive acceleration. One can represent the velocity at all points and all times within a hypothetical one-dimensional flow by a surface in velocity (u) – distance (x) – time (t) space. The transport history of any particle within the flow is represented by a line on this surface; where the line has a negative slope (i.e. runs 'downhill' in the figure) the substantive acceleration is negative and the particle decelerates as it moves downstream. Suspension and traction thresholds for particular grain sizes can be represented as horizontal planes within this diagram; the volume below the suspension threshold defines the position and duration of deposition, and the volume between these two planes defines the position and duration where traction-plus-fallout might occur in a simple monodisperse (i.e. single grain size) system.

Naturally, this conceptual framework needs to be qualified with various caveats since the

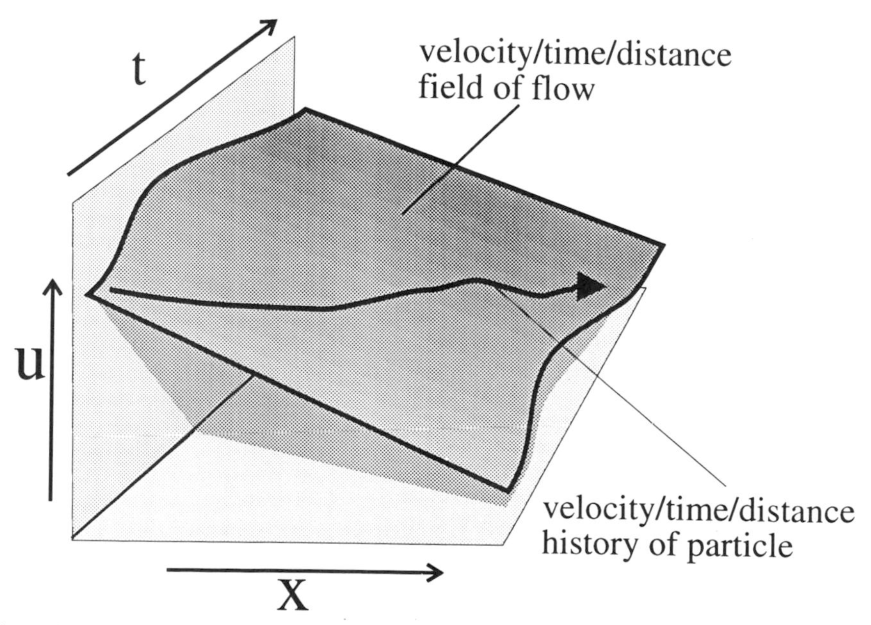

Fig. 5. Three-dimensional surface in velocity/distance/time space representing the velocity at all times and at all points along the length of an hypothetical one-dimensional flow. A particle path is illustrated by the arrow; note that this has a negative slope everywhere, even on the part of the curve representing waxing flow.

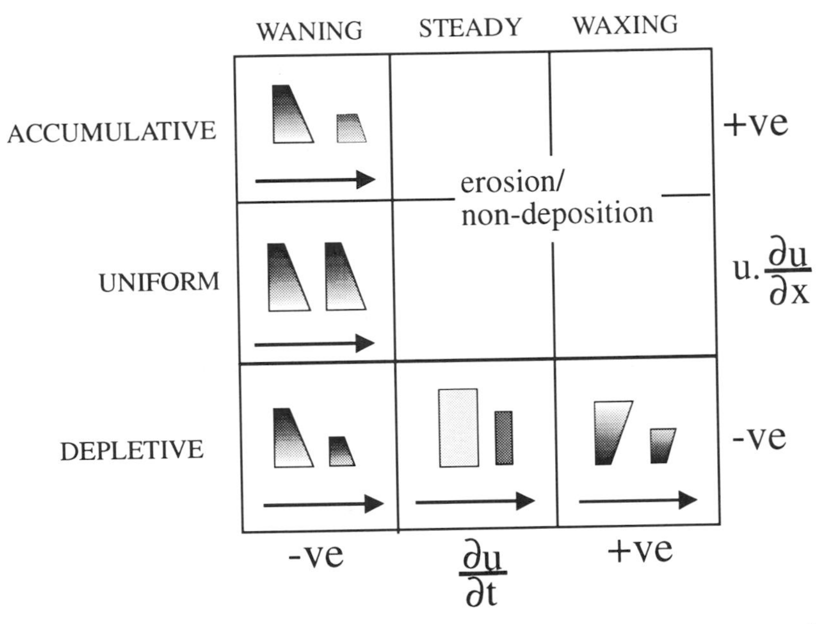

Fig. 6. Acceleration matrix, with illustrative bed sequences for each field showing downstream and vertical changes in relative grain size of the deposits of each field; arrows point downstream.

relationship between mean downstream velocity and deposition is not straightforward. Although the vertical component of turbulence (which is what keeps grains in suspension) is generally a simple function of mean velocity, complexities may arise where the flow regime is changing. Also, the response time of the depositional regime to flow deceleration will be affected by various scalar parameters such as the flow thickness, flow density and settling velocities. Probably the most influential (and least easily investigated) factors are grain-size distribution, fractional concentration of suspended grains, and (in the case of high rates of suspension fallout) processes within the depositional boundary layer (Lowe 1988; Kneller & Branney 1995).

The acceleration matrix

Although the substantive acceleration is the primary control on whether or not deposition takes place, the vertical and lateral variations in the grain size (and types of tractional bedform if any) will depend upon the relative magnitudes and signs of the unsteadiness term (acceleration in time, $\partial u/\partial t$) and the non-uniformity term (acceleration in space, $u.\partial u/\partial x$); this can be visualized as varying shapes of negatively sloping surfaces in Fig. 5. As a simplification, a matrix diagram of spatial acceleration v. temporal acceleration can be produced; this is divided into nine fields, according to whether each of the acceleration terms is positive (waxing flow, accumulative flow), zero (steady flow, uniform flow) or negative (waning flow, depletive flow) (Fig. 6). For four of these fields (steady uniform flow, waxing uniform flow, steady accumulative flow and waxing accumulative flow) there is likely to be erosion or non-deposition, since none of the particles in the flow experience a net decrease in mean streamwise velocity (negative substantive acceleration). For flows with identical starting grain-size distributions, each of the other five fields produces different vertical and lateral variations in grain size within the resultant bed.

In Fig. 6 each of these sequences has been represented schematically by two cartoon graphic grain-size logs, in relative upstream or proximal (left) and downstream or distal (right) positions. These diagrams are highly idealized, and relate only to fully turbulent, non-cohesive flows. The presence or absence of particular sedimentary structures is not implied since the grain size is relative. A full discussion of the modelling will be presented elsewhere, but the qualitative differences between deposits representing each of the five fields is outlined below, with natural examples of the types of beds which may represent these different regimes. In detail,

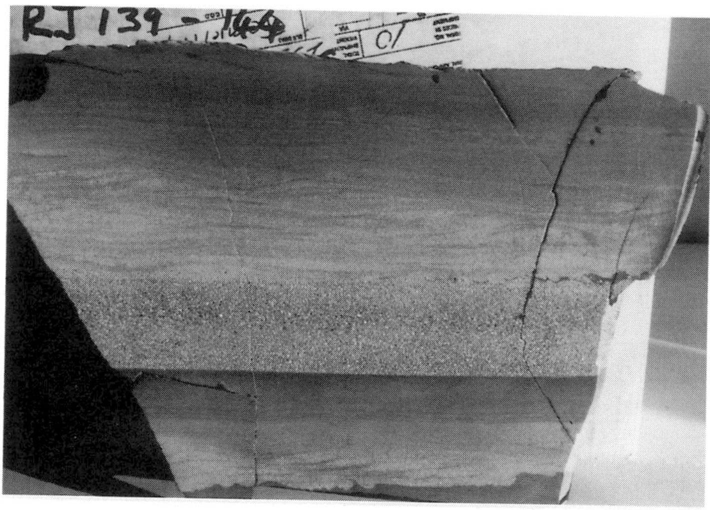

Fig. 7. Normally graded beds produced by deposition from waning flow. The upper bed (Bouma-type Tacde) is 140 mm thick. Southern Highland Group, Vendian-Cambrian, NE Scotland.

the configuration of a particular bed will depend on the relative magnitudes of the two acceleration terms. In addition, the way in which the spatial acceleration term varies with distance (e.g. across a channel mouth or slope break) may mean that the deposit of a particular flow will occupy more than one field.

Depletive waning flow

The field of depletive waning flow is represented by the familiar normally graded sequence (Fig. 7) that becomes thinner and finer-grained downcurrent, with the progressive loss of lower parts of the sequence – the conventional proximal–distal relationship of Bouma-type turbidites (e.g. Walker 1967). Numerous (probably most) turbidites have been interpreted according to this type of model. Although it is rather rare to find sequences where these relationships can be reliably traced in individual beds, there seems little reason to doubt that they are common where periodic surge-type currents flow into unconfined basins (or basins which are large in comparison to the flows) such as large abyssal plains. Surge-type currents are those generated by virtually instantaneous events such as seismogenic slumping (e.g. Heezen & Ewing 1952; Weaver *et al.* 1992; Morgenstern 1967; Garcia & Hull 1994) or, in experiments, the release of a lock gate, and consist of a more or less clearly differentiated head, body and tail.

Uniform waning flow

The field of uniform waning flow is represented by similar graded vertical sequences, but with no downstream fining or thinning; this is the type of sequence generated by 'slab' models (e.g. Allen 1985), which are thus inappropriate for the modelling of deposition of graded beds which thin and fine distally. The deposition of such sequences from surge-type flows demands rather exceptional conditions which might include uniform gradient and constant width between bounding topography.

Depletive steady flow

Depletive steady flows produce beds with no significant vertical variation in grain size but which fine downcurrent; representative deposits might include massive sands (Fig. 8) or thick sequences of climbing ripples (Fig. 9). Deposition in this regime has been discussed by Kneller & Branney (1995). It requires flows which are sustained at relatively constant discharge for long periods, such as may be generated during volcanic eruptions and the consequent remobilization of unconsolidated material (e.g. Lipman & Mullineaux 1981; Kokelaar 1992), by glacial meltwater discharge or by direct fluvial input (Normark & Piper 1991; Wright *et al.* 1990), either constantly (large rivers) or at flood stage only (smaller rivers, e.g. Reynolds 1987). These turbidity currents are qualitatively different from surge-type flows.

Depletive waxing flow

Depletive waxing flow produces coarsening-upwards beds which become finer downstream; time lines intersect the bed downstream, and the upstream depositional limit will migrate downstream with time. Candidates for such deposits in nature include not only the rather unusual example shown in Fig. 10 (which can scarcely be interpreted in any other way), but also the much more widespread inverse graded layers in sand and gravel beds (R2/S2 of Lowe 1982) commonly ascribed to traction carpets (but see also Hiscott 1995). It may also be represented by well-sorted sands with tractional bedforms, initially deposited by rapid suspension sedimentation but followed by a phase of traction under faster currents, with re-suspension of fines, and a sharp upper grain-size break, possibly overlain by a mud drape.

The increase in current velocity associated with the arrival of the head of a surge-type turbidity current is a rapid and singular event, more likely to be associated with the erosion which commonly precedes deposition of sand turbidites than with deposition. Depletive waxing flow is more likely to be produced during the waxing stage of longer-lived flows (e.g. those generated by flood events) or, where repeated (as in repeated inverse grading), by regular fluctuations associated with the passage of turbulent eddies or internal waves, both of which can be demonstrated experimentally (Kneller *et al.* in press).

Accumulative waning flow

The field of accumulative waning flow is represented by normally graded beds with complex proximal–distal relations depending upon the balance of the two acceleration terms, and upon

Fig. 8. Massive sandstone with escape burrow, produced by deposition from quasi-steady flow. Annot Sandstone, Oligocene, SE France.

Fig. 9. Sequence of climbing ripples with constant rate of climb, implying steady flow and constant vertical sediment flux. Southern Highland Group, Vendian-Cambrian, NE Scotland.

Fig. 10. Upward-coarsening sequence at base of thick sand bed, with transition from ripple field through low-angle cross-stratification to massive sand; deposition under waxing flow conditions. Marnoso-Arenacea, Miocene, northern Italy.

Fig. 11. (a) Thick, normally-graded Bouma sequence; (b) detail of middle part showing repeated alternations of planar-laminated and ripple cross-laminated sands show multiple palaeocurrent directions. Peira Cava sandstones, Oligocene, SE France.

the streamwise velocity profile; the most likely configuration is for both base and top of the graded sequence to be cut out downstream. Time lines intersect the bed downstream.

In summary, there are at lease five basic types of sequence that one might expect in turbidite beds, defined by a combination of both vertical and streamwise grain-size variations. There are ample candidates for each of these in the geological record, and they each have implications for the type of flow that produced them and the way in which the flows interacted with topography. Each of these types shows a distinctive pattern of streamwise variation in grain size; this has implications not only for the way beds may be correlated, but also for the way their petrophysical properties vary.

Unidirectional flow?

Multiple palaeocurrent directions within single beds have been reported from many turbidite systems (e.g. Ellis 1982; Pickering & Hiscott 1985; Marjanac 1990; Kneller *et al.* 1991 and refs therein; Pickering *et al.* 1992), and are probably present in many more. These have been interpreted as 'reflection' of the turbidity current from topography at the basic margin; comparable interactions with topography are known from pyroclastic flows (Druitt 1992). Often the changes in current direction occur in association with abrupt reversals in grading (Pickering & Hiscott 1985), and such reversals or repetitions have been used to imply flow reflection. However, the change in current direction may also occur within apparently continuously graded beds that in other respects conform with simple depositional models (Marjanac 1990; Fig. 11).

Pantin & Leeder (1987) and Edwards *et al.* (1994) studied the way in which 'reflections' were generated by orthogonal incidence of saline density currents on ramps; the reverse flow takes the form either of a bore (a moving hydraulic jump) or a series of internal solitary waves. Particle image velocimetry (Edwards 1991) and laser doppler anemometry show that the fluid velocity within the reverse flow is of the same order as that within the forward flow, demonstrating that the reverse flow (the 'reflection') is potentially capable of transporting sediment deposited by the forward flow (Kneller *et al.* in press). Kneller *et al.* (1991) examined the case of oblique incidence, and concluded that the reverse flow propagated normal to the obstructing topography.

The data presented in Fig. 13 show the case of interaction with a lateral ramp. representing a bounding topographic feature such as a fault scarp; the experimental configuration is shown in Fig. 12. A wedge with a slope of 30° was placed into the tank, with the foot of the slope positioned 375 mm to one side of the channelway. The tank was filled with tap water at a temperature of 12°C to a depth of 100 mm. The lock was filled to the same depth with brine (SG = 1030 kg m^{-3}) at the same temperature. This volume of 4.72 l of brine was released from the lock, and the flow was recorded using an SVHS video system mounted above the tank. Velocities of the head and the solitary waves were measured from the video recording.

As in the case of oblique reflection, the internal waves propagate along the interface between the flow and the ambient fluid, normal to the slope. The amplitude of the waves varies with the velocity of incidence of the head of the obverse flow (Fig. 13f); since the velocity of solitary waves is a function of their amplitude, the wave crests rapidly become curvilinear as the waves propagate away from the slope (Fig. 13b–d) although the direction of propagation is normal to the slope. Edwards *et al.* (1994) have shown that for faster, denser flows, the reflection would take the form of a bore or undercutting gravity current rather than solitary waves.

In the case of a sediment gravity flow within a laterally confined turbidite system, current indicators at the base of the bed, produced by

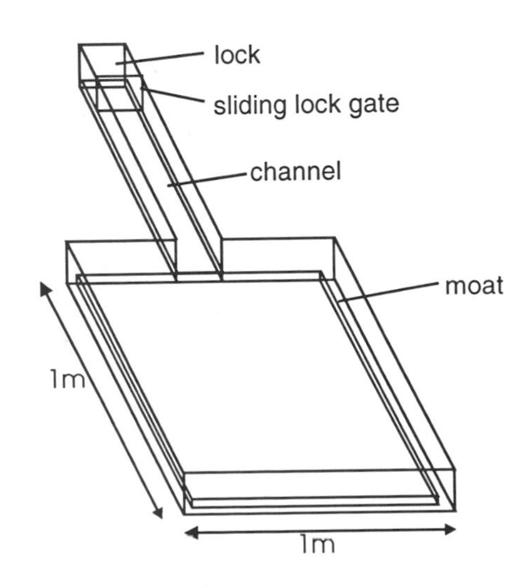

Fig. 12. Schematic diagram of 1 × 1 m Perspex tank used for experiments whose results are figured below.

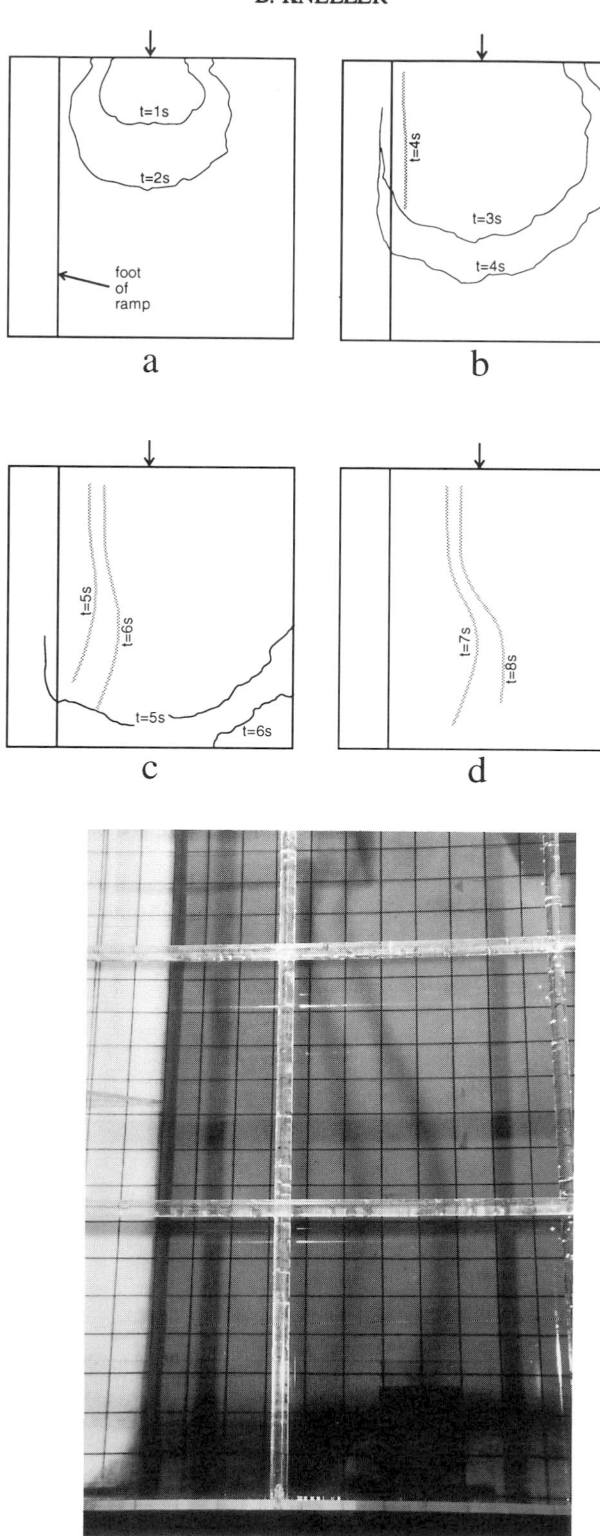

(e)

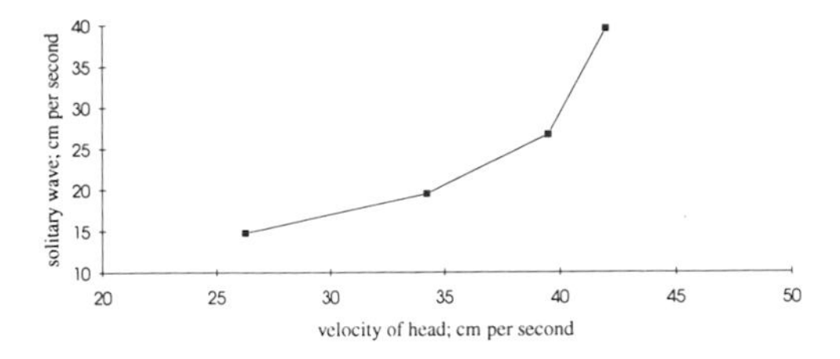

(f)

Fig. 13. Generation of internal solitary waves at a lateral ramp. (**a**) Positions of front of gravity current head at t = 1 and 2 s after current emerged from channel; arrow marks entry point of flow. (**b**) As (a) at t = 3 and 4 s with position of first solitary wave at t = 4 s. (**c**) As (a) at t = 5 and 6 s. (**d**) Positions of first solitary wave at t = 7 and 8 s. (**e**) Oblique overhead photograph of solitary waves at t = *c*. 7.5 s; the foot of the ramp lies along the dark line to the left (separating dyed flow from clear ambient water); the waves are the curved lines of deeper dye colouration to the right centre field. (**f**) Relationship between velocity of gravity current head striking the ramp, and velocity of the first wave generated at that point.

erosion beneath the head of the flow, would indicate radial flow, whereas those at the top of the bed, produced by the effect of internal waves or bores, would indicate flow transverse to the basin (Fig. 14).

Flow and deposition around obstacles

Deposition of sediment is influenced by obstacles in a number of ways over and above the effects of reflection (Alexander & Morris 1994). Firstly, the size and shape of obstructions physically constrain the flow along certain paths, and this limits potential depositional sites; the height of the obstruction relative to the height of the flow determines the degree to which the flow is blocked (i.e. how much of the flow can pass over the obstruction; Simpson 1987; Rottman & Simpson 1989; Muck & Underwood 1990); the shape of the obstruction will determine the deflection of the current. Secondly, the blocking and deflection of flows affects the instantaneous velocity field around obstructions (e.g. Fig. 15), which in turn governs $u.\partial u/\partial x$; where mean velocities decrease rapidly along particle paths, $u.\partial u/\partial x$ will have large negative values, increasing the likelihood of deposition (although this is modified by variations in turbulence intensity which accompany changes in the flow).

Figures 17–25 show the results of flume experiments using scaled, surge-type sediment

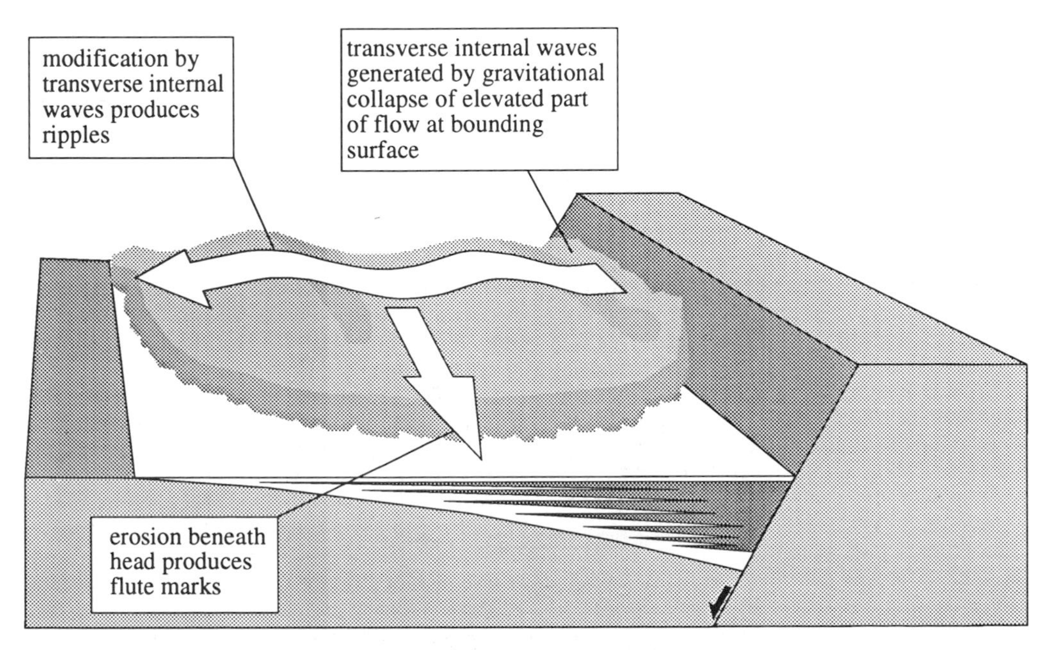

Fig. 14. Geological model for generation of internal waves in a half-graben, producing internally divergent palaeocurrent patterns.

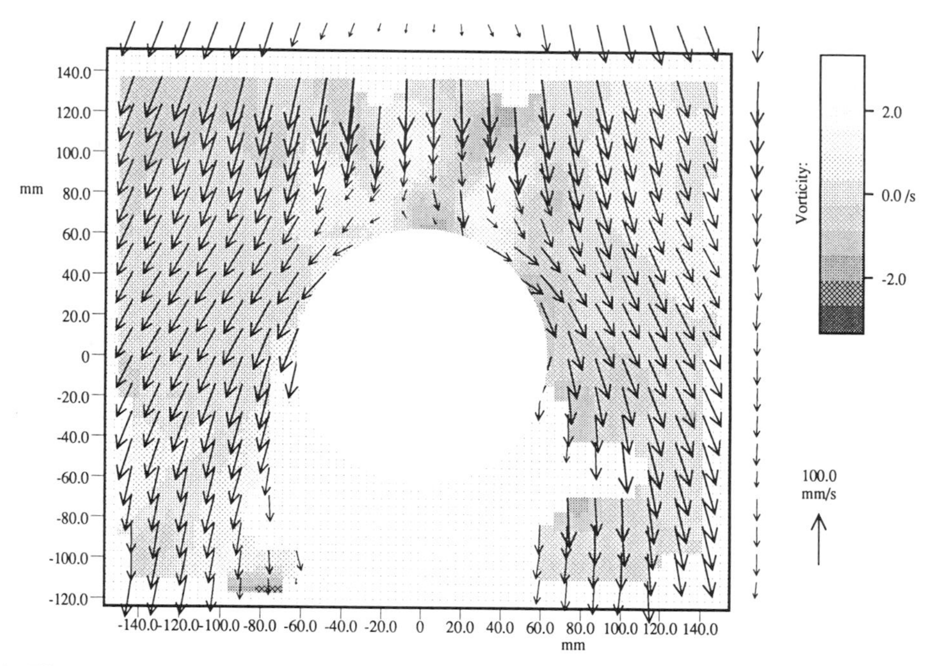

Fig. 15. Illustration of the variation in the velocity field of a gravity current flowing past an obstacle. The experimental configuration was as for the experiment illustrated in Figs 12 & 13, with flow past a 60 mm high hemisphere placed in the centre of the tank. The field was calculated from a video record of the paths of neutrally buoyant particles in the flow using DigImage® image analysis software. Decreases in the magnitude of the velocity vector along particle paths result in negative d*u*/d*x*.

gravity flows, generated using the experimental set-up shown in Fig. 12. The material used as a sediment analogue was polymethyl methacrylate (a low molecular weight form of the acrylic material used in Perspex), with a specific gravity of 1180 kg m^{-3}, and a median grain size of 600 μm. Settling velocities determined by laser phase doppler anemometry ranged from 4 to 28 mm s^{-1} in pure water, with a value of 15 mm s^{-1} for the median (600 μm) grain size; this is in reasonable agreement with a value of 12.3 mm s^{-1} given by Dietrich's (1982) empirically determined equations. Settling velocity was manipulated by using a NaCl solution of specific gravity 1090 kg m^3 as an interstitial medium for the density current (settling velocity for 600 μm grains given by Dietrich's equation = 6.8 mm s^{-1}), with tap water as the ambient fluid. Each flow contained 1.5 kg of sediment in a lock volume made up to 7.08 l with brine. The 1 × 1 m flume was fitted with a raised false floor in order to create a moat to prevent reflections of residual brine from the tank walls. Forward velocities of the head were *c*. 0.2 m s^{-1} along the centreline of the flow; head height (mean of two runs) was measured as 49 mm at 0.11 m from the channel mouth, decreasing to 37 mm at 0.71 m. The flows had densiometric Froude numbers, Fr $\approx$ 0.3, and Reynolds number, Re $\approx 10^4$. These experimental results scale to thin, 'low-efficiency' flows of very coarse sand grade, moving at a few metres per second. The distribution of sediment was assessed by mapping the depth to the sediment surface using an ultrasonic thickness gauge as a bed profiler (Fig. 16; technique described by Best & Ashworth 1994). Thicknesses ranged up to a maximum of *c*. 4.5 mm.

Figure 17 shows the thickness of the deposit of an unobstructed flow, illustrating a lobe-shaped deposit similar to that produced in other studies of sedimentation from unconfined flows (e.g. Luthi 1980). This deposit is used as a basis for comparison with sediment distributions around obstacles; the thickness of this deposit at a given point is referred to as T^u_{xy}. The configurations illustrated in Figs 18–23 were of sinusoidal profile, with an aspect ratio (width:height) of 3.6. The examples illustrated are for a transverse linear obstacle of height

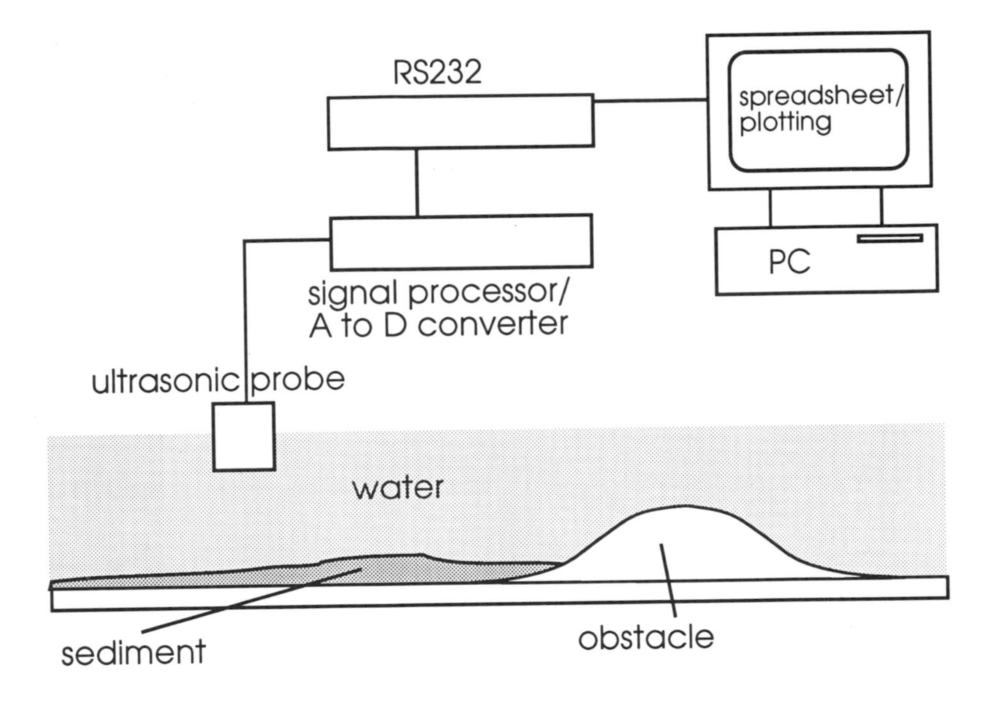

Fig. 16. Schematic diagram of the bed profiling system used to measure sediment thicknesses in the following experiments. The sediment thickness is calculated from the difference in bed profiles before and after deposition from an experimental flow.

31.3 mm (Figs 18–20) and circular obstacles of height 31.3 and 62.5 mm (Figs 21–23).

The transverse obstacle in Fig. 18 was placed with its centreline 303 mm from the channel mouth. It simulates a confined linear sub-basin with a lateral entry point opposite a frontal slope. The diagram represents the spatial distribution of sediment thickness, T^{o}_{xy}, in mm; the obstacle itself is not represented on the plot. The flow is partially blocked by the obstruction (head height was *c.* 1.5 times the obstacle height) reducing deposition downstream of it; maximum thickness occurs adjacent to the upstream side of the obstacle opposite the entry point, and is almost double the maximum thickness of the unobstructed flow. However, much of the flow is diverted laterally so that deposition is shifted 'axially'. The diverting effect of the obstruction can be represented more effectively by displaying the difference between this deposit thickness, T^{o}_{xy}, and that of an unobstructed flow T^{u}_{xy} (Fig. 19); positive values indicate thicknesses greater (values in mm) than those produced by an unobstructed flow; negative values indicate reduced deposition relative to unobstructed flows. Deposition is relatively enhanced over most of the 'basin' upstream of the obstruction, and is over twice that of the unobstructed flow, as illustrated by Fig. 20, in which the values given in Fig. 19 are ratioed to T^{u}_{xy}. Detail which is not revealed in the contour plots includes two narrow (few millimetre) linear zones of non-deposition, one immediately upstream and the other immediately downstream of the obstacle, small-scale hummocks (wavelength a few millimetres, amplitude a few hundred micrometres) on the deposit surface upstream of the obstacle, and sediment ridges of similar scale parallel to flow, downstream of the obstacle.

Figure 21 illustrates the sediment thickness distribution immediately adjacent to a small circular obstruction of similar profile, height and position to that shown in Fig. 20 (note scale difference), analogous to a small diapiric feature on the seafloor. The flow is partially blocked, and deposition downstream is reduced compared to an unobstructed flow, but the most marked effects are lateral deflection of the flow, and the development of non-depositional areas in separation zones on the downstream quarters of the obstacle.

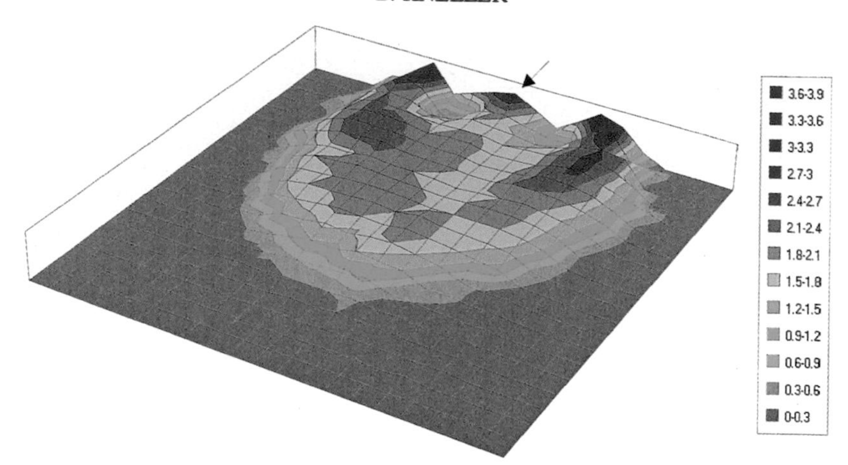

Fig. 17. Sediment thickness distribution (T^u_{xy}) of a standard unobstructed flow (see text for flow parameters) in a 1 × 1 m tank, measured on a grid spacing of 40 × 40 mm; the entry point of the flow is marked by an arrow; the sediment thickness contour interval is 0.3 mm.

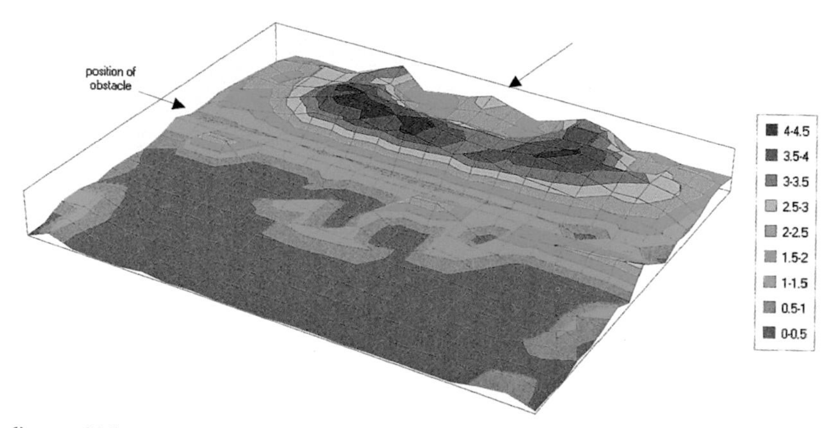

Fig. 18. Sediment thickness distribution produced by a standard flow around a transverse linear obstacle of height 62.5 mm, width 117 mm, contoured on a 40 mm grid; contour interval is 0.5 mm. The entry point of the flow is the mid-point of the upper edge.

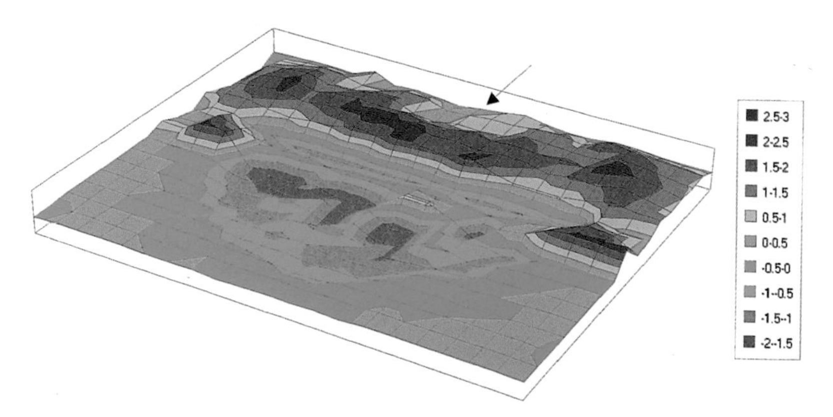

Fig. 19. Plot of the difference between T^u_{xy} (Fig. 17) and T^o_{xy} (Fig. 18), indicating increased or decreased deposition compared to the unobstructed flow; the contour interval is 0.5 mm.

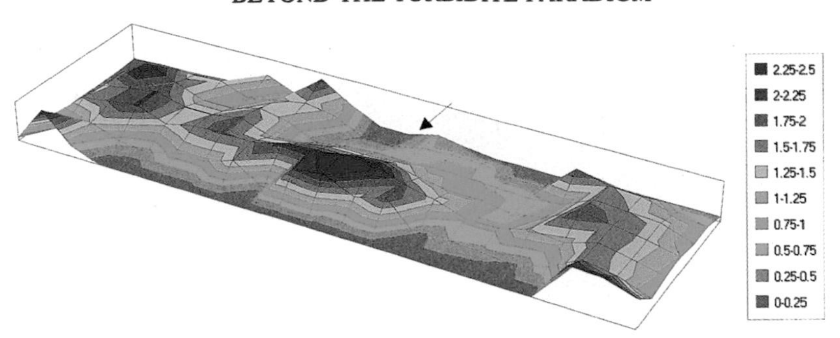

Fig. 20. Non-dimensionalized values for sediment thickness shown in Fig. 18 for the area upstream of the obstacle, showing the difference from the unobstructed flow deposit expressed as multiples of T^u_{xy}.

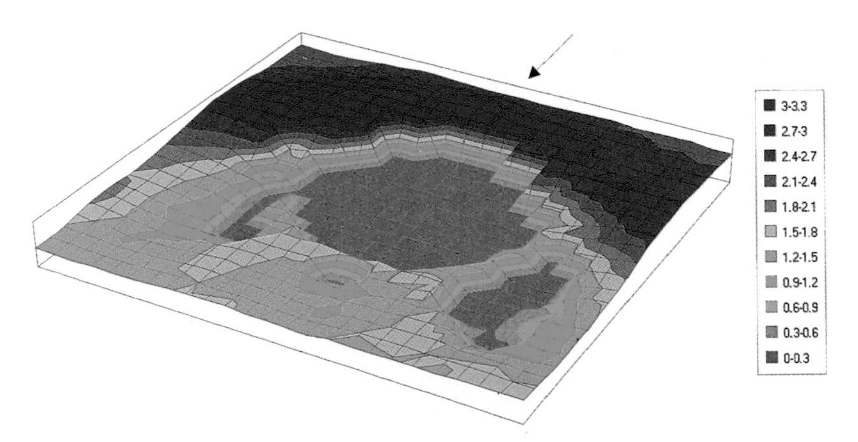

Fig. 21. Sediment thickness distribution (central part of tank only) produced by a standard flow around a circular obstacle of height 31.3 mm, width 58.5 mm, contoured on a 10 mm grid; contour interval is 0.3 mm.

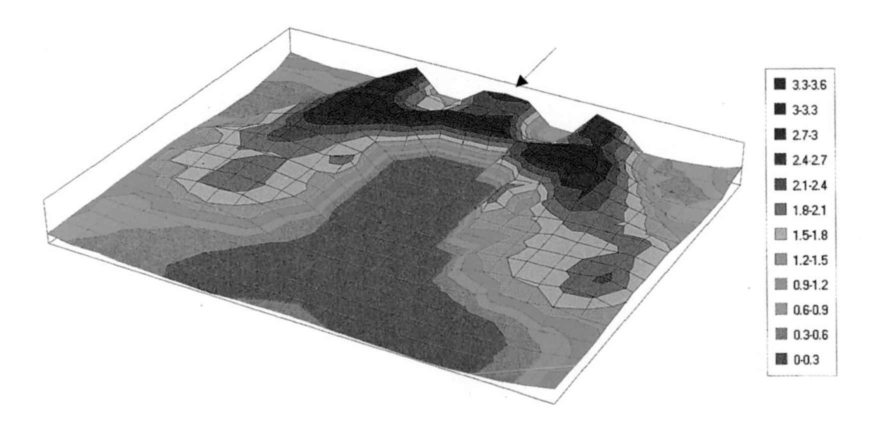

Fig. 22. Sediment thickness distribution produced by a standard flow around a circular obstacle of height 62.5 mm, width 117 mm, contoured on a 40 mm grid; contour interval is 0.3 mm.

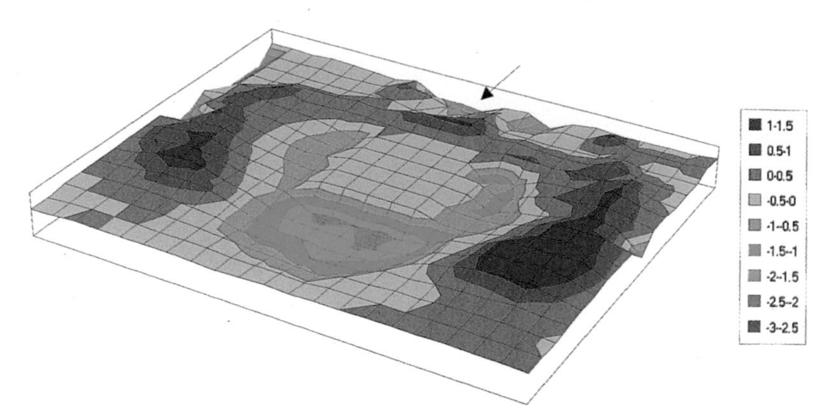

Fig. 23. Plot of the difference between T^u_{xy} (Fig. 17) and T^o_{xy} (Fig. 22), indicating increased or decreased deposition compared to the unobstructed flow; the contour interval is 0.5 mm.

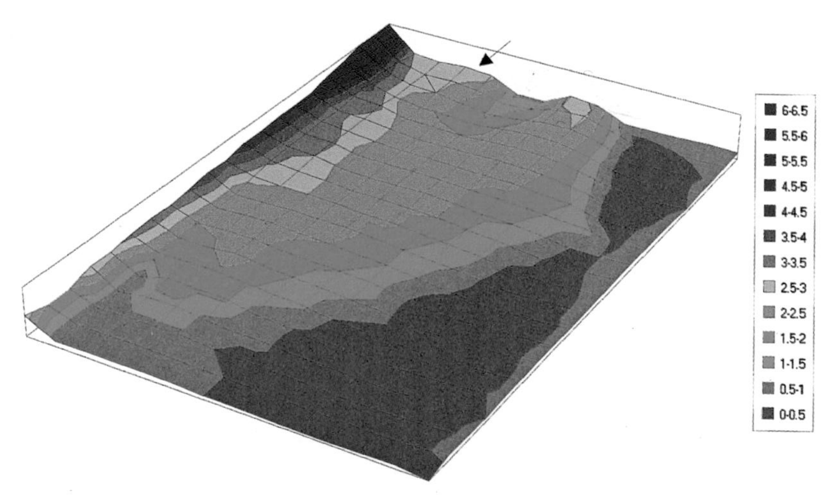

Fig. 24. Sediment thickness distribution produced by a standard flow against a 30° lateral ramp (foot of ramp along left-hand edge of plot); contour interval is 0.5 mm; the entry point of the flow is shown by the arrow.

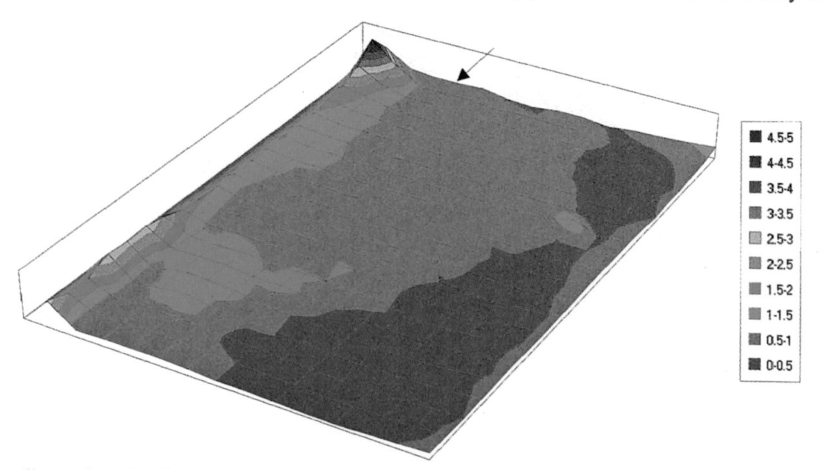

Fig. 25. Non-dimensionalized values for sediment thickness shown in Fig. 24, showing the difference from the unobstructed flow deposit expressed as multiples of T^u_{xy}; values close to zero indicate thicknesses similar to those of the unobstructed flow deposit.

Figure 22 illustrates the sediment thickness distribution around an obstacle 62.5 mm high – *c.* 1.5 times head height) with its centre at 303 mm from the channel. Directly downstream of the obstacle the flow is completely blocked and there is no deposition. The deflection of the flow is clear both from the isopach distribution in Fig. 22, and also from Fig. 23, which shows the difference between T^u_{xy} and T^0_{xy}; along the flanks of structure sediment thicknesses are less (by 1–1.5 mm) than the corresponding unobstructed flow deposits.

For circular obstacles, such as diapirs, there are some common features irrespective of their height compared to that of the flow; the thickest sediment accumulations are upstream of the obstruction, and in lobes to each side which are displaced some distance from the obstacle. As with the linear obstacle, there is a zone of non-deposition (and possibly scour) at the foot of the upstream slope.

Figure 24 illustrates the sediment thickness distributions associated with a 30° lateral slope. The overall sediment distribution is rather similar to that for the unobstructed flow except close to the foot of the slope (represented by the left edge of the plot); here the sediment is several times thicker than that of the unobstructed flow. This effect is highlighted in Fig. 25, which shows the dimensionless sediment thickness $[(T^u_{xy} - T^0_{xy})/T^0_{xy}]$. Sediment thickness at the foot of the slope is three to four times that of an unobstructed flow deposit; this effect is probably enhanced by slumping of sediment deposited on the slope. Distribution is shifted slightly 'down-fan' close to the slope, but over much of the area there is little difference in the form or thickness of the deposit from that of the unobstructed flow.

Conclusions and implications for reservoir modelling

The character and distribution of a turbidite sandstone is influenced by the nature of the accelerations which operate upon it during transport and deposition. The temporal acceleration is a function largely of the initiation mechanism of the flow, which determines whether it has a waxing phase and/or is maintained in a quasi-steady state for a significant period. The spatial acceleration is largely a function of the topography over (or through) which the flow moves. Both these terms influence the character of an individual bed – its local vertical sequence, grain-size sorting and lateral variation (and thus values and gradients of 'net to gross', and of porosity and permeability), and also its length:thickness ratio; whereas a bed's spatial distribution is controlled largely by flow non-uniformity and therefore by topography. Experimental analogues can be useful in constraining the shape of sandbodies deposited around seafloor topography where this topography is known, but the internal character of these bodies and their lateral variability can be modelled only by deducing the temporal evolution of the flows which produced them.

Single vertical sequences (individual wells, for example) may be inadequate to establish the conditions under which the deposits were formed, since the regime can be defined uniquely only with a knowledge of the streamwise variations in properties such as vertical grading (Fig. 6). None the less, individual vertical sequences may contain subtle pointers to the broader sandstone distribution. For example, a well drilled in the centre of the basin modelled in Fig. 25 would contain little or no more sand than if the basin were unconfined. It might, however, include beds which contain palaeocurrent indications of reflection from the bounding surface (Figs 13 & 14), which thus point to a significant (but otherwise unknown) sand accumulation at the foot of the slope, or (perhaps no less importantly) simply indicate the existence and orientation of such a slope.

Many of the clues that individual turbidite beds contain concerning their overall shape and properties have been overlooked. Better numerical and experimental models may yet prove more valuable than outcrop analogues.

Mike Leeder, Mike Branney and Henry Pantin all contributed substantially to the ideas contained in this paper. Mike Leeder and Bill McCaffrey helped with the experimental work. Chris Clayton and Fabiano Gamberi helped in the field. The work was funded by BP International, Conoco UK and Arco British, and I am particularly grateful to Richard Dixon, Mike Bowman, Ian Sweetman and Colin Taylor for their support. Many thanks to John Cater and Gilbert Kelling for helpful reviews.

References

ALEXANDER, J. & MORRIS, S. 1994. Observations on experimental non-channelized turbidites: thickness variations around obstacles. *Journal of Sedimentary Research*, **A64**, 899-909.

ALLEN, J. R. L. 1985. *Principles of Physical Sedimentology*. George Allen & Unwin, London.

BEST, J. L. & ASHWORTH, P. 1994. A high resolution ultrasonic bed profiler for use in laboratory flumes. *Journal of Sedimentary Research,* **A64,** 674-675.

BOUMA, A. H. 1962. *Sedimentology of Some Flysch Deposits: A Graphic Approach to Facies Interpretation*. Elsevier, Amserdam.

BRANNEY, M. J., KNELLER, B.C. & KOKELAAR, B. P. 1990. Disordered turbidite facies (DTF): a product of continuous surging density flows. *International Sedimetological Congress, Nottingham, Abstracts*, 38.

CAZZOLA, C., MUTTI, E. & VIGNA, B. 1985. Cengio turbidite system, Italy. *In:* BOUMA, A. H., NORMARK, W. R. & BARNES, N. E. (eds) *Submarine Fans and Related Systems*. Springer-Verlag, Berlin, 179–183.

DIETRICH, W. E. 1982. Settling velocities of natural particles. *Water Resources Research,* **18**, 1615–1626.

DRUITT, T. H. 1992. Emplacement of the May 18, 1980 lateral blast deposit ENE of Mount St. Helens, Washington. *Bulletin of Volcanology,* **54**, 554–572.

EDWARDS, D. A. 1991. *Turbidity Currents: dynamics, deposits and reversals*. PhD thesis, University of Leeds, UK.

——, LEEDER, M. R. & BEST, J. L. 1994. On experimental reflected density currents and interpretation of certain turbidites. *Sedimentology,* **41**, 347-461.

ELLIS, D. 1982. *Palaeohydraulics and computer simulation of turbidites in the Marnoso-Arenacea, Northern Apennines, Italy*. PhD thesis, University of St Andrews, UK.

GARCIA, M. O. & HULL, D. M. 1994. Turbidites from giant Hawaiian landslides: results from Ocean Drilling Program site 842. *Geology,* **22**, 159–162.

HEEZEN, B. C. & EWING, M. 1952. Turbidity currents and submarine slumps, and the 1929 Grand Banks earthquake. *American Journal of Science;* **250**, 849-873.

HISCOTT, R. N. 1994. Traction-carpet stratification in turbidites – fact or fiction? *Journal of Sedimentary Research,* **64,** 204–208.

KNELLER, B. C. & BRANNEY, M. J. 1995. Sustained high density turbidity currents and the deposition of thick massive sands. *Sedimentology,* (in press).

——, EDWARDS, E., MCCAFFREY, W. & MOORE, R. 1991. Oblique reflection of turbidity currents. *Geology,* **19** 250–252.

——, MCCAFFREY, W. D. & BENNET, S. J. In press. Velocity and turbulence structure of gravity currents and internal solitary waves: potential sediment transport and the formation of wave ripples in deep water. *Sedimentary Geology.*

KOKELAAR, B. P. 1992. Ordovician marine volcanic and sedimentary record of rifting and volcanotectonism: Snowsdon, Wales, United Kingdom. *Geological Society of America Bulletin,* **104**, 1433–1455.

KUENEN, PH. H. & MIGLIORINI, C. I. 1950. Turbidity currents as a cause of graded bedding. *Journal of Geology,* **58**, 91-127.

LIPMAN, P. W. & MULLINEAUX, D. R. 1981. The 1980 eruptions of Mount St Helens, Washington. *United States Geological Survey Professional Paper No. 1250.*

LOWE, D. R. 1982. Sediment gravity flows: II. Depositional models with special reference to the deposits of high-density turbidity currents. *Journals of Sedimentary Petrology,* **52**, 279–297.

—— 1988. Suspended-load fallout rate as an independent variable in the analysis of current structures. *Sedimentology,* **35**, 765–776.

LUTHI, S. 1980. Some new aspects of two-dimensional turbidity currents. *Sedimentology,* **28**, 97–105.

MARJANAC, T. 1990. Reflected sediment gravity flows and their deposits in the flysch of Middle Dalmatia, Yugoslavia. *Sedimentology,* **37**, 921–930.

MORGENSTERN, N. R. 1967. Submarine slumping and the initiation of turbidity currents. *In:* RICHARDS, A. F. (ed.) *Marine Geotechnique.* University of Illinois Press, Urbana, Illinois, 189–220.

MUCK, M. T. & UNDERWOOD, M. B. 1990. Upslope flow of turbidity currents: a comparison among field observations, theory, and laboratory methods. *Geology,* **18**, 54–57.

MUTTI, E. 1992. *Turbidite Sandstones,* Agip, San Donato Milanese.

—— & NORMARK, W. R. 1991. An integrated approach to the study of turbidite systems. *In:* WEIMER, P. & LINK, M. H. (eds) *Seismic Facies and Sedimentary Processes of Submarine Fans and Turbidite Systems.* Springer-Verlag, New York, 75–106.

——, E. & RICCI-LUCCI, F. 1978. Turbidites in the northern Appenines: introduction to facies analysis. *International Geology Review,* **20**, 127–166.

NORMARK, W. R. & PIPER, D. J. W. 1991. Initiation processes and flow evolution of turbidity currents: implications for the depositional record. *In:* OSBORNE, R. H. (ed.) *From Shoreline to Abyss.* SEPM Special Publication, **46**, 207-230.

——, POSAMENTIER, H. & MUTTI, E. 1993. Turbidite systems: state of the art and future directions. *Review of Geophysics,* **B31B**, 91–116.

PANTIN, H. M. & LEEDER, M. R. 1987. Reverse flow in turbidity currents: the role of internal solitons. *Sedimentology,* **34**, 1143–1155.

PICKERING, K. T. & HISCOTT, R. N. 1985. Contained (reflected) turbidity currents from the Middle Ordovician Cloridorme Formation, Quebec, Canada: an alternative to the antidune hypothesis. *Sedimentology,* **32**, 373–394.

——, HISCOTT, R. N. & HEIN, F. J. 1989. *Deep Marine Environments: Clastic Sedimentation and Tectonics.* Unwin Hyman, London.

——, UNDERWOOD, M. B. & TAIRA, A. 1992. Open-ocean to trench turbidity-current flow in the Nankai Trough: flow collapse and reflection. *Geology,* **20**, 1099–1102.

REYNOLDS, S. 1987. A recent turbidity current event, Hueneme Fan, California: reconstruction of flow properties. *Sedimentology,* **34**, 129–137.

ROTTMAN, J. W. & SIMPSON, J. E. 1989. The formation of internal bores in the atmosphere: a laboratory model. *Quarterly Journal of the Royal Meteorological Society*, **115**, 941–963.

SIMPSON, J. E. 1987. *Gravity Currents in the Environment and the Laboratory*. Ellis Horwood, Chichester.

STANLEY, D. J., PALMER, H. D. & DILL, R. F. 1978. Coarse sediment transport by mass flow and turbidity current processes and downslope transformations in Annot sandstone cayon-fan valley systems. *In:* STANLEY, D. J. & KELLING, G. (eds), *Sedimentation in Submarine Canyons, Fans and Trenches*. Dowden, Hutchinson & Ross, Stroudsberg, PA, 85–115.

THORNBURG, T. M., KULM, L. D. & HUSSONG, D. M. 1990. Submarine-fan development in the southern Chile Trench: a dynamic interplay of tectonics and sedimentation. *Geological Society of America Bulletin*, **102**, 1658–1680.

WALKER, R. G. 1967. Turbidite sedimentary structures and their relationship to proximal and distal depositional environments. *Journal of Sedimentary Petrology*, **37**, 25–43.

WEAVER, P. P. E., ROTHERWELL, R. G., EBBING, J., GUNN, D. & HUNTER, P.M. 1992. Correlation, frequency of emplacement and source directions of megaturbidites on the Madeira abyssal plain. *Marine Geology*, **109**, 1–20.

WRIGHT, L. D., WISEMAN, W. J. JR, YANG, Z. S., BORNHOLD, B. D., KELLER, G. H., PRIOR, D. B. & SUHAYDA, J. N. 1990. Processes of marine dispersal and deposition of suspended silts off the modern mouth of the Juanghe (Yellow) River. *Continental Shelf Research*, **10**, 1–40.

Anisotropic grain fabric: volcanic and laboratory analogues for turbidites

SIMON R. HUGHES[1], JAN ALEXANDER[1] & TIM H. DRUITT[2]

[1] *The Marine Geoscience Research Group, Department of Earth Sciences, Cardiff University, PO Box 914, Cardiff CF1 3YE, UK*

[2] *Département des Sciences de la Terre, Université Blaise Pascal, 5 Rue Kessler, 63038 Clermont-Ferrand, France*

Abstract: A variety of grain fabrics are commonly observed in coarse-grained turbidites that form potential reservoir rocks. Such fabrics may impart an anisotropy to permeability; however, their origin is poorly understood. Observations of analogous volcanic and laboratory deposits are used here to infer the origin and character of clast alignments. Experimental gravity currents were produced by releasing suspensions of silt-grade silicon carbide (10, 15 or 20% by volume) with minor volumes of larger, lower-density perspex clasts into a tank of water. Fabric was measured in the experimental deposits and in the Quaternary Upper Laacher See Tephra, Germany. The experimental and pyroclastic deposits both had a clast fabric with grain long-axes predominantly transverse, but not necessarily perpendicular, to the mean flow direction. The perspex clasts in the laboratory currents were initially transported in suspension, and marks on the sediment surface indicated that these clasts rolled along the bed immediately prior to final deposition. The perspex clasts were buried at differing levels and at a range of distances from source in the laboratory deposits, having been deposited at different stages in the flow history. In the volcanic deposits the same fabric is persistent vertically throughout the thickness of the bed at each site, suggesting that progressive aggradation provided a surface on which the clasts rolled prior to deposition. It is proposed here that similar fabrics in high-density turbidites are analogous to those observed in these experimental and pyroclastic deposits, and therefore they may be predicted from palaeocurrent interpretations, or conversely used to assess palaeocurrent directions.

A variety of grain fabrics (patterns of distribution and orientation of clasts) are commonly observed in coarse-grained turbidites that form potential reservoir rocks. Such fabrics impart an anisotropy to permeability, and have been used to infer palaeocurrent direction and mechanisms of deposition (e.g. Tiara & Scholle 1979; Hiscott & Middleton 1980; Middleton 1993). These fabrics are commonly caused by alignment of sand and gravel clasts, organic particles, such as shell fragments or foramanifera tests, and also by rip-up clasts. Fine-grained rip-up clasts may be particularly important during hydrocarbon production as they can complicate the patterns of fluid flow through reservoir rocks. The characteristics and distribution of grain fabrics through individual beds are incompletely documented and their origins are poorly understood. Here, in an attempt to improve understanding of reservoir anisotropy, grain fabrics in deposits of analogous gravity currents are described and the possible implications for coarse-grained turbidites discussed. In both analogues, mixed-grain populations (density and size) are deposited rapidly from turbulent suspension on to an aggrading bed. As rip-up clasts have considerably lower densities than the sand and gravel with which they are associated during transport and deposition, this paper concentrates on patterns of preferred orientation of relatively low-density (compared with the bulk of the sediment clasts) elongate clasts.

It is not possible to observe directly coarse sediment deposition from naturally occurring turbidity currents. Recent coarse-grained turbidites have not been cored (piston and capstan cores do not easily penetrate coarse-grained sediments and deep water vibro-coring techniques are only now being developed), and ancient turbidites observed at outcrop are invariably lithified and have been deformed dur-

From Hartley, A. J. & Prosser, D. J. (eds), 1995, *Characterization of Deep Marine Clastic Systems*, Geological Society Special Publication No. 94, pp. 51–62.

ing uplift. Thus, it is difficult to record three-dimensional fabric patterns accurately or to understand their origin. To overcome some of these problems, an unconsolidated volcanic deposit (deposited from a high-concentration turbulent suspension similar in behaviour to some turbidity currents) and experimental turbidites that are, in part, analogous to turbiditic reservoir rocks have been examined. In the volcanic analogue, a pyroclastic surge deposit in the upper Quaternary Upper Laacher See Tephra (Germany), slate clasts derived from the country rock were incorporated into the pyroclastic surge and form a strong fabric in the deposit. Excavation of the unconsolidated sediment allows accurate measurement of grain-size distribution, individual grain shapes, and three-dimensional grain orientations. In addition, unlike ancient turbidites, the original topography is known and the spatial distribution of the features can be observed directly. In the laboratory experiments, silt-grade silicon carbide and larger elongate perspex prisms were put into suspension in water and the sediment suspensions were released as high-density turbidity currents into tanks of standing water. The plastic particles were deposited with the silicon carbide on the flat floor of the tank where their position and orientation were recorded.

Johansson (1963, 1976) observed a preferred orientation of elongate clasts with long-axes normal to the flow direction in sediment deposited by unidirectional currents. Johansson (1963) suggested that such transverse orientations resulted from clasts rolling on a flat bed. Johansson (1963) recorded flow-parallel long-axis trends where: (1) objects on the bed such as stationary pebbles hindered depositing clasts and caused their rotation; (2) where elongate clasts were deposited directly from suspension with no movement on the bed; (3) where pebbles slipped along the bed. Rust (1972) observed that on fluvial braid bars, rotation of pebbles into flow-parallel orientations (1 above) is more probable at high pebble concentrations, and that when pebbles are isolated on a sandy bed they can roll freely into resting position with long-axes tranverse to flow. Smith (1986) found a bimodal distribution of clast orientations in hyperconcentrated flood deposits (aqueous flows with sediment concentrations 40–80 Wt%; Beverage & Culbertson 1964). Large cobbles and boulders (maximum diameter 0.15–0.50 m) were orientated with long-axes tranverse to mean flow direction, but small cobbles and pebbles (maximum diameter 0.01–0.15 m) were preferentially orientated with long-axes parallel to the mean flow direction. Smith (1986) inferred that this bimodality indicated that the flow was competent to carry small cobbles and pebbles dispersed in suspension above the bed, but the larger clasts were transported by traction. Johansson (1963, 1976) noted that smaller pebbles are more likely to leave the bed and move in suspension, and consequently different grain-size populations will have differing preferred orientations.

In many published studies of ancient turbidites, a predominant long-axes flow-parallel clast fabric has been inferred (e.g. Nilsen & Simoni 1973; Davies & Walker 1974; Walker 1975). In many cases, however, fabric is inferred from two-dimensional exposures and often can only be described qualitatively. Differences between grain populations are not generally recorded. Postma *et al.* (1988) found outsized clasts with long-axes both tranverse and parallel to flow in experimental (and natural) high-density flows.

The volcanic analogue

Cas & Wright (1987), Fisher (1990), Druitt (1992) and others suggest that 'dry' pyroclastic surges (i.e. those with a low initial water content) are analogous to turbidity currents. Both are turbulent, gravity-controlled flows with high particle concentrations moving in a cohesionless manner. Wilson (1980) has shown that the fluidization and sedimentation behaviour of gas–particle mixtures is fundamentally different from those of liquid–particle suspensions, but once initiated, pyroclastic surges and turbidity currents both move as gravity-controlled, turbulent masses. There are three main advantages in studying deposits of 'dry' pyroclastic surges: (1) Recent pyroclastic surge deposits can be studied in exposures distributed over the whole area covered by an individual surge, where the pre-eruption topography is known; (2) In the unconsolidated deposits it is relatively easy to examine bedforms, grain size and fabric, because exposures can be excavated; and (3) The wide range of clast sizes, densities and shapes transported simultaneously in a pyroclastic surge tend to produce strongly developed grain sorting and fabric variations in its deposits, exaggerating trends that appear more subtly in turbidites where the variability in clast type is less.

The Laacher See Tephra was deposited by the eruption of the Laacher See volcano in the Eifel, Germany (Schmincke 1970; Schminke *et al.* 1973; Fisher *et al.* 1983). The Upper Laacher See Tephra represents the final phase of the

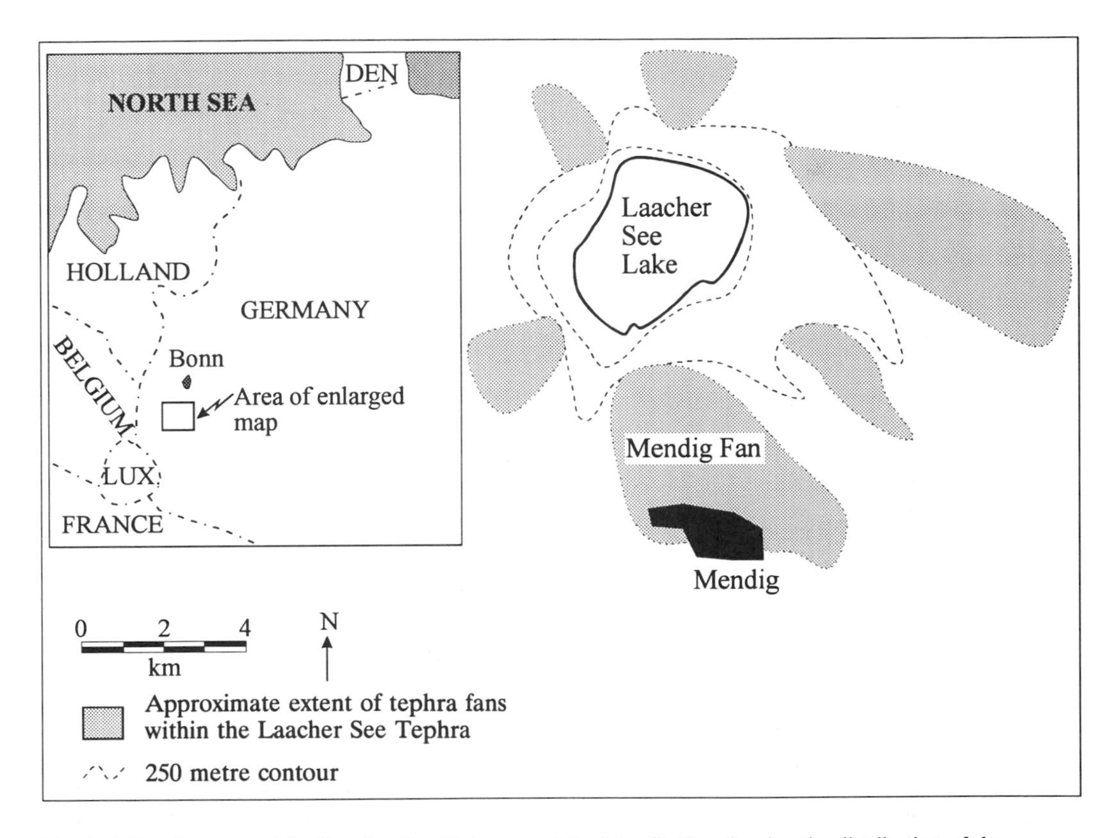

Fig. 1. A location map of the Laacher See Volcano and the Mendig Fan showing the distribution of the tephra around the vent. The insert map shows the location of the study site. The vent of the volcano was situated within the area of the Laacher See lake (Schmincke *et al.* 1973).

eruption (11 ka) and is the product of pyroclastic surges with relatively low water contents. It is confined to six topographically controlled fans (Fig. 1). The Mendig Fan covers an area > 9 km². It is exposed in quarries both near the edge of the crater and up to 6.0 km from the centre of Laacher See (Fig. 1). The fan is composed of couplets of lithic breccia overlain by low-angle, cross-bedded, sand-grade beds (B- and D-layers of Schminke *et al.* 1973, respectively). The couplets are interbedded with massive and laminated ash. There are six couplets in which breccias pass abruptly upwards (> 0.05–0.1 m) into finer grained cross-bedded deposits. The top of each couplet is marked by a thin airfall ash that is partly or wholly eroded by the overlying couplet. Each couplet is believed to be the product of a single pyroclastic surge pulse, with an initial waxing phase (depositing breccia) followed by a prolonged waning phase (depositing cross-laminated sand-grade beds and fallout ash). Fabric data from one representative well-exposed couplet (beds B6/7–D4 of Schmincke *et al.* 1973) are presented here. The fabric patterns recorded from the other couplets are similar to that in B6/7–D4.

Lithic breccia B6/7

The B6/7 breccia generally has an open framework and is fines-deficient. The mean grain size of the breccia is 8 mm at proximal localities and 4 mm at distal localities. The maximum diameters of both juvenile (magma derived) and accidental lithic clasts decrease with distance from vent. The breccia thins quasi-exponentially (half-distance of 1.23 km) with increasing distance from the vent. At 2.5 and 3.3 km from the vent, B6/7 has a low-angle (< 25°) cross-stratified sandy-gravel base (0.10–0.15 m thick) overlain by two or three (depending on location) inversely graded units, each 0.12–0.26 m thick, varying between fine-granule and fine-cobble grade (Fig. 2). In more distal sites, B6/7 is a single, parallel-bedded ungraded to normally graded granule-grade bed (Fig. 2). The breccia pinches and fines over dune crests of the preceding (underlying) flow deposit (named M14

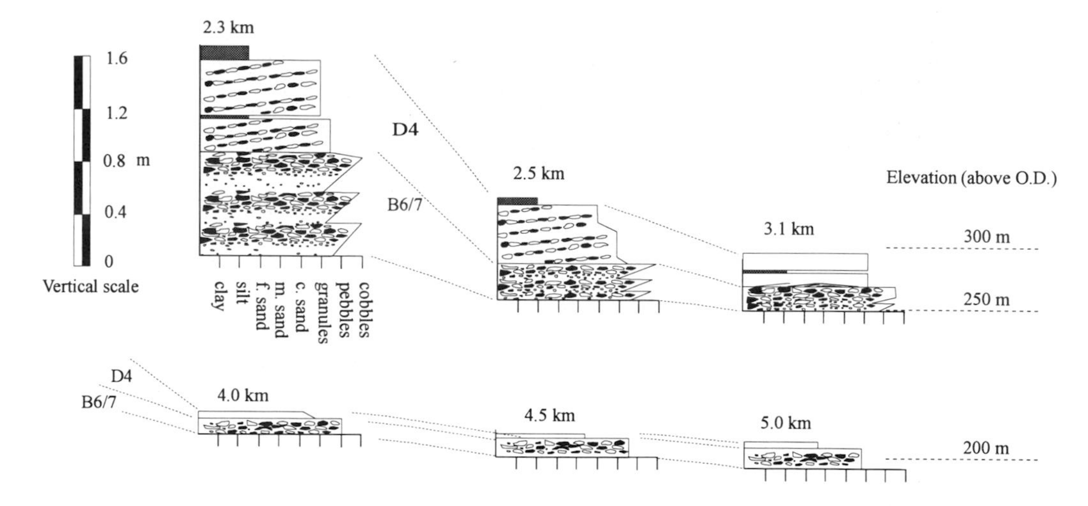

Fig. 2. Representative sedimentary logs of couplet B6/7–D4 (nomenclature after Schmincke *et al.* 1973) on a proximal to distal transect from 2.3 to 5.0 km from vent (taken as the geographical centre of Laacher See lake).

by Schmincke *et al.* 1973), which formed the rough surface over which B6/7 was deposited.

Cross-bedded unit D4

There are seven sand- to gravel-grade, cross-bedded units in the Upper Laacher See Tephra, ranging in thickness from 3 m (proximally) to only a few centimetres distally. Cross-bedded unit D4 has a mean grain size of 0.7 mm, decreasing to 0.4 mm at distal sites. The sand-grade cross-bedded unit tends to show a quasi-exponential decrease in thickness (half-distance of 0.34 km) with increasing distance from vent. At proximal sites D4 is characterized by spectacular bedforms and low-angle cross-bedding (10–20°), formed of low amplitude (< 2 m), long wavelength (10–12 m) bedforms (Fig. 3). Many of the bedforms are similar in appearance to antidunes and chute-pool structures (Schmincke *et al.* 1973). In exposures that are perpendicular to flow, many of these units are seen to be trough cross-stratified. The cross-bedded units are often capped by up to 0.10 m of silt-grade massive ash. Distally D4 becomes more massive until all evidence of cross-bedding is lost.

Fabric in couplet B6/7–D4

The fabric in the B6/7 unit was determined at five localities, of which four are presented here, distributed over the area of the fan. Data were collected by excavating material from quarry faces that were orientated either approximately parallel or transverse to the mean flow direction. At each locality, between 100 and 120 clasts were examined from an area *c.* 2 × 0.5 m. The long-axis plunge (magnitude and orientation) and the strike and dip of the a–b plane were measured *in situ* using a compass-clinometer. This was possible as the clasts protruded from the face on the wind-etched surfaces of the quarries. Each clast was then removed from the sediment using tweezers, in order to measure the lengths of its long, intermediate, and short axes (a, b and c, respectively). The measurements were limited to basement-derived slate fragments, with short to intermediate axial ratios of 2.5:1 or greater and with a minimum long-axis length of 5 mm. The long-axes data from four sites are plotted as poles on the equal area stereographic projections in Fig. 4. Throughout the area, the clasts have a preferred long-axis orientation transverse, but not necessarily perpendicular, to the mean palaeocurrent direction (interpreted from each site's position relative to the site of the vent).

The cross-bedded unit D4 was examined at the same sites as breccia B6/7 using the same procedure. The mean grain size of the cross-bedded unit is less than that of the breccia and this made it more difficult for reliable fabric measurements to be obtained. Most of the clasts were less than the imposed threshold size (5 mm) for this data analysis. At proximal sites, where the clasts were larger, reliable fabric data were obtained.

Fig. 3. (*a*) Flow-parallel and (*b*) flow transverse views of the Upper Laacher See Tephra deposits at a distance of 2.3 km from vent showing bedforms similar to those described by Schminke *et al.* (1973) as antidunes and chute-pool structures. The vertical bars represent a scale of 5 m.

Interpretation of the couplets

The origin of couplets forming within the one eruption period that deposited the whole of the Mendig Fan has still to be investigated, but it is assumed that each couplet represents a pulse in a quasi-continuous eruption. In sites proximal to the vent, the nature of the lithic breccia suggest that they were emplaced by a depositional regime characterized by particle movement with a strong horizontal component (i.e. flow dominated). Bedforms and cross-bedding in the basal parts of some breccias (e.g. Fig. 2) suggest that, in some parts of the Mendig Fan, initial deposition of pyroclastic density current was from turbulent suspension. The near-bed turbulence was later suppressed by higher sediment concentrations associated with the emplacement of traction carpets. The stacked, inversely graded nature of the breccias may be the result

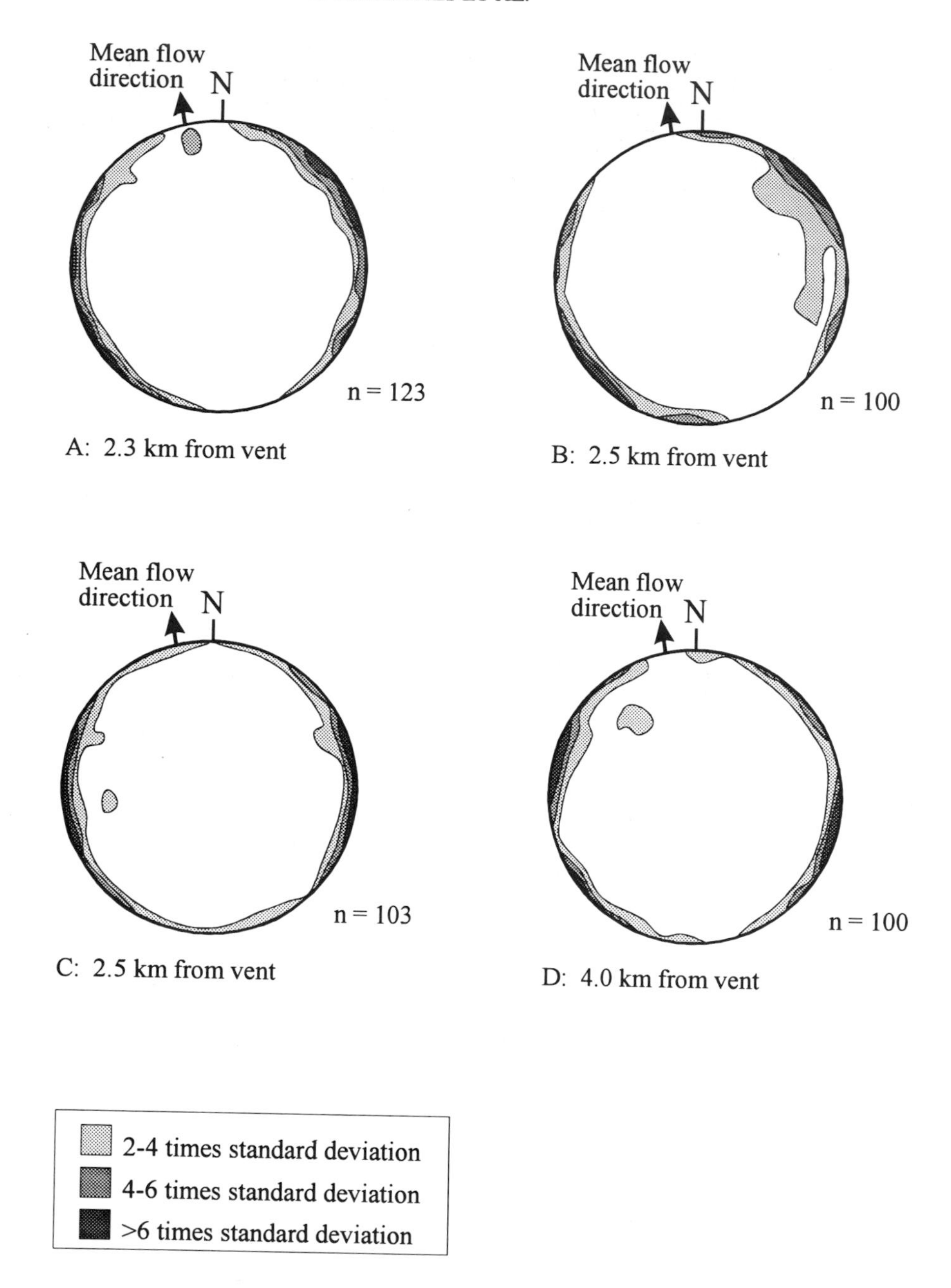

Fig. 4. Stereographic projections showing the long-axis orientations of elongate slate clasts in B6/7 at distances of 2.3, 2.5 and 4.0 km from vent (taken as the geographical centre of the lake).

of emplacement of a series of traction carpets formed due to bedload shear and dispersive pressure (cf. Bagnold 1954), and kinetic filtering (cf. Middleton 1970) influencing final stages of deposition from high concentrations of sediment near to the bed. Continued sediment flux towards the bed loads the traction carpet causing it to freeze in place, and further traction carpets accrete above, forming a stack (Lowe 1982). Todd (1989) found clasts had a flow-parallel orientation in high-density traction carpets, where grain inertia effects were prominent.

In B6/7 the long-axis fabric transverse to flow suggest that the clasts rolled into place during their final phase of deposition. This rolling would be unlikely to happen in a shearing granular mass and therefore the inverse grade units may in fact be associated with fluctuations in flow and sediment flux due to a pulsating flow. Distally, the breccias show sedimentary features which suggest a depositional regime that had a weak horizontal component of sediment emplacement (i.e. normal grading, few bedforms and mantling of topography).

The bedforms and cross-bedding in the upper part of the couplets indicate that, proximally at least, they were emplaced by a turbulent flow that had relatively low sediment concentration (compared with the breccias). The cosets with thin airfall ashes separating sets (Fig. 2) suggest that the cross-bedded units of each couplet were emplaced in a series of pulses. Thus, it seems that cross-beds illustrate both a temporal (at a given location) and a spatial radial waning of a pulsatory pyroclastic density current. The thin ashes on the top of individual bedsets represent fallout from the dilute wake of the surge, perhaps equivalent to the dilute mixing cloud of a turbidity current.

The laboratory analogue

Laboratory-generated turbidity currents have been used by a number of workers to investigate the behaviour of turbidity currents and the nature of their deposits (see the reviews by Edwards 1993 and Middleton 1993). Here the results are presented from a short series of experiments that were designed to assess the feasibility of measuring outsized grain fabrics and investigating the mechanisms of their formation in laboratory-generated turbidity currents. In these experiments high-concentration particle suspensions were released from a lock into a large tank of clear standing water. The resulting gravity current spread over the horizontal floor of the tank, depositing sediment and becoming more dilute, until eventually the flow deflated (the experimental procedure is described in more detail in Alexander & Morris 1994).

Silicon carbide grains (grain characteristics in Table 1) were used in these feasibility experiments to form sediment suspensions in water, which had initial sediment concentrations of 10, 15 or 20% by volume. Fifty-two outsized clasts (plastic prisms) of various sizes and shapes (Table 2) were added to the turbulent suspension before it was released axially into a long narrow tank (4.76 × 0.21 m, filled with water to a depth of 0.17 m) representing a laterally restricted (channelled) flow (Table 3). Sediment was deposited from the moment that the suspension was released into the tank. Once the silicon carbide had settled from suspension, and all appreciable current activity had stopped, the nature of the deposits, the position and orientation of each outsized clast and bed surface features were recorded.

In all of the experiments the perspex clasts were transported and deposited with the silicon carbide grains. Perspex clasts were incorporated at varying depths within the deposit (although rarely at the base). Most perspex clasts had at least a veneer of silicon carbide grains over them; some were relatively deeply buried, although in all cases their position was obvious from topographic expression on the surface of the deposits.

In the narrow tank experiments, the perspex clasts were deposited at a range of distances from the lock (Table 3). For example, the perspex clasts transported by a gravity current that had an initial silicon carbide concentration of 10% by volume (Run 1), were deposited between 2 and 3 m from the lock gate. The areas over which the perspex clasts were deposited were coincident with the areas where the gravity current showed pronounced deceleration (observed in video recordings).

The orientations of perspex clasts deposited in the straight tank experiments are shown in the stereographic projections in Fig. 5. Unlike the

Table 1. *Physical properties of silicon carbide particles used in the experiments*

d (μm)	Md_f	s_f	r (g cm^{-3})	U_0 (cm s^{-1})	Re	Grain shape
57	4.13	*C.* 0.4	3.22	0.52 (±.0.18)	0.29	Very angular, variable sphericity

d is the mean equivalent spherical diameter; Md_f and s_f are the median and sorting parameters of Inman (1952); r is particle density as supplied by the manufacturers; U_0 is the mean settling velocity; Re is the particle Reynolds Number (Re = $U_0 ßd/\mu$) where ß is the density of water (0.998 g cm^{-3}) and μ the viscosity of water (10^{-3} N sm^{-2}) at 20°C.

Table 2. *Clast characteristics of perspex prisms*

Clast class	Long-axis length (mm)	Intermediate-axis length (mm)	Short-axis length (mm)	Clast density (g cm^{-3})
A	30	9	3	2.33
B	30	4	3	2.33
C	15	9	3	2.33
D	15	4	3	2.33
E	3	4	4	2.33

Table 3. *Experimental conditions of runs*

Run number	SiC concentration in initial suspension (% of total volume)	Number of clasts found at varying distance from lock gate (m)				Number of clasts deposited outside lockgate	Total number of clasts
		0–0.1	0.1–0.2	0.2–0.30	> 0.3		
1	10	7	4	35	0	46	52
2	15	12	11	8	9	40	52
3	20	16	18	0	0	34	34

Tank dimensions: 4.76 × 0.21 × 0.19 m (volume 0.19 m^3); lock dimensions: 0.31 × 0.21 × 0.19 m (volume 0.01 m^3); water depth: 0.17 m; water temperature: 20°C.

field observations described above, the perspex clasts lay on a thin film of silicon carbide and the long-axes were generally horizontal. The long-axes of perspex clasts were generally transverse, but not necessarily perpendicular, to the mean flow direction (Fig. 5). The clasts near to the tank walls were orientated with their long-axes parallel to the mean flow direction.

Marks were observed on the sediment surface (Fig. 6). These marks consisted of depressed grooves in the sediment surface that had lengths approximating that of the long-axis length of the perspex clast they were near. Series of parallel marks formed a trail behind (upstream) some of the clasts. The grooves were generally transverse to flow and were spaced at distances a little greater than the associated perspex clast's intermediate axis length. The grooves suggest that some of the outsized clasts rolled or skipped along the bed prior to final deposition and burial by the silicon carbide. In low-concentration suspension experiments (not presented here), this rolling behaviour was observed directly within the body of the flow. At the upstream end of the trail, the first mark was sometimes asymmetrical (deeper at one end than the other) and skewed at a slight angle (*c.* 10°) to other marks in the trail.

Discussion of laboratory experiments

A common problem with the comparison of laboratory analogues and natural turbidites is that of scaling (see discussion by Middleton 1966; Edwards 1993). The experimental gravity currents are hydrodynamically similar to small, high-sediment concentration turbidity currents transporting sand and gravel (as for the experiments described by Alexander & Morris 1994). These experiments should not be compared with other types of turbidites except with extreme caution. The purpose of the experiment was to investigate only the depositional mechanisms and resulting fabrics. Therefore, the lower-density outsized clasts were chosen so that they behaved differently to the silicon carbide grains, in that they could be transported easily as a traction load following a period of turbulent suspension, while the silicon carbide was deposited largely by fallout from turbulent suspension. The perspex clasts are large relative to the size of the current and, therefore, comparison with slate fragments in the pyroclastic deposit or shale and organic fragments in turbidites should be made with care. The perspex

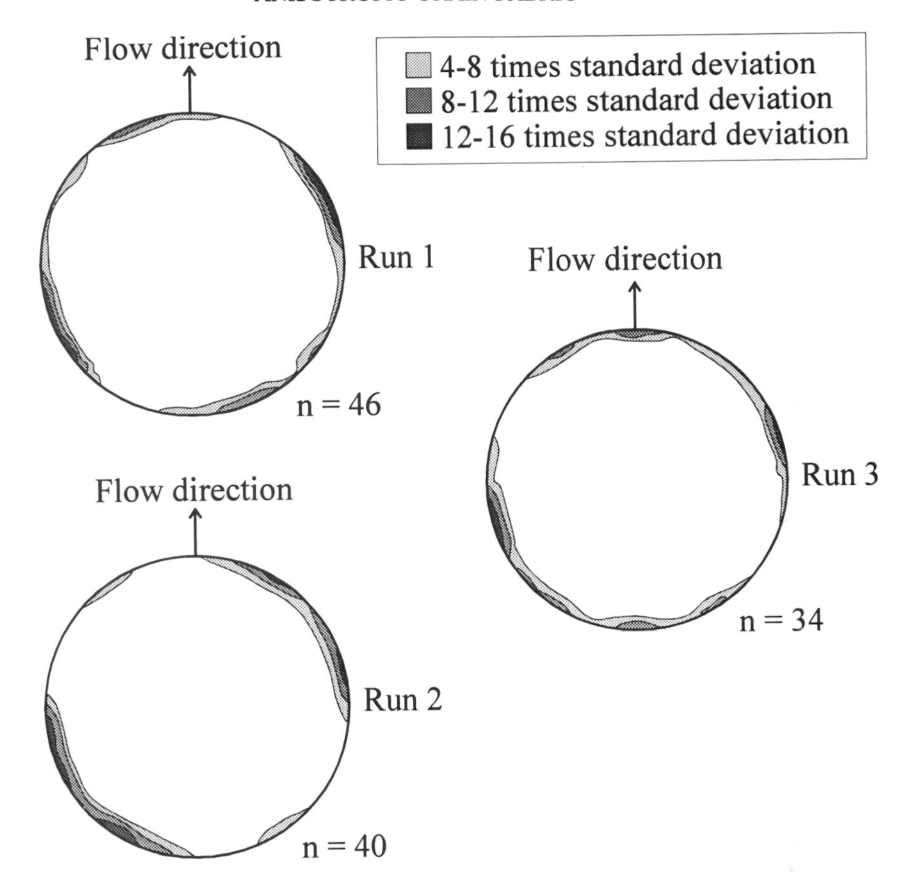

Fig. 5. Stereographic projections showing the long-axis orientations of outsized clasts in experiments using initial suspended sediment concentrations of 10% (Run 1), 15% (Run 2) and 20% (Run 3) silicon carbide concentrations (see Table 3 for experimental conditions).

clasts used in the experiments, if directly scaled with the size of the current, are large in comparison with those found in the pyroclastic deposits. The experiments used a single grain size for the bulk of the sediment suspension that produced a well-sorted deposit; however, in natural currents there is more variety in size, shape and density and it is uncertain how these factors will influence the processes of deposition and therefore resulting grain fabrics.

The data presented here suggest, that following initial transport from the lock in turbulent suspension, most of the outsized clasts rolled into position. Generally, the perspex clasts show a range of orientations transverse to flow direction; however, none of the runs produced a mode exactly perpendicular to flow. The parallel to flow orientation of some perspex clasts near to the edge of the tank was produced by interference with the tank walls, in a similar way to the effect of stationary pebbles on depositing clasts observed by Johansson (1963, 1976). Away from the edges of the tank, a flow-parallel orientation was not observed as either the perspex clasts themselves were not of a size to interfere with each other or, more likely, the concentration of perspex clasts was insufficient to allow for such interference. These feasibility experiments represent a first approximation to the volcanic and turbidity current deposits, and further research is in progress.

The implications of the analogue studies for fabrics in turbidites

Low-density particles, such as those considered in this paper, are often coarser grained than the bulk of the sediment because of the effects of hydraulic equivalence. When low-density particles are present in relatively low concentrations, they behave differently to finer, denser grains deposited simultaneously. The low-density clasts

(a)

(b)

Fig. 6. Outsized clasts deposited from experimental gravity current. (**a**) Single class A clast (see Table 2) and (**b**) two class C clasts. Note the transverse to flow (mean flow from right and left) grooves suggesting that the outsized clasts rolled or saltated into place.

reach the sediment interface with the finer material, but low-density grains tend to be more prone to movement along the sediment interface. In both the volcanic and the experimental analogues considered in this paper, the low-density clasts were generally orientated transverse, but not necessarily perpendicular, to mean current directions. These transverse fabrics are similar to pebble fabrics in mixed sand and gravel alluvial deposits (e.g. Johansson 1963, 1976; Rust 1972, 1975; Smith 1986). In natural turbidity currents, it is envisaged that low-density clasts may behave in a similar fashion, being transported initially in suspension and then in traction immediately prior to final deposition. Postma *et al.* (1988) considered the orientation to be dependent on where the clast was deposited in the system; clasts orientated transverse to flow were deposited from a fast moving turbulent flow layer undergoing traction with clasts rolling into place, while clasts with a parallel to flow orientation were deposited from an inertia flow phase with clasts gliding into place. In a turbiditic reservoir rock, outsized mudstone clasts are likely to have long-axes preferentially transverse to the mean flow direction giving locally pronounced permeability anisotropy. This study highlights the need for more research into the nature and causes of fabric heterogeneities and anisotropy in turbiditic reservoirs.

Conclusions

(1) In pyroclastic and experimental gravity current deposits, where differing density particles are present, low-density large clasts tend to be orientated preferentially with their long-axes transverse to mean flow direction. The preferred clast orientation is produced by outsized clasts rolling and saltating after a period of turbulent suspension, prior to coming to rest on the agrading bed of the flow.

(2) Locally flow-parallel long-axes trends are observed where there is restriction of movement on the bed or where flow patterns are disturbed by obstructions.

(3) Low-density outsized clasts are expected to produce similar fabrics in turbidites. Local variations between flow-transverse and flow-parallel fabrics may aid interpretation and produce variable barriers to permeability anisotropy in reservoir rocks.

Simon Hughes acknowledges receipt of a NERC Studentship no. GT4/91/GS/129. Thanks to Stephen Morris, Melissa Johansson and members of the Cardiff undergraduate school for assistance with experiments. Thanks also to Norman Fry, Ben Kneller and an anonymous reviewer for constructive criticism of early versions of this paper.

References

ALEXANDER, J. & MORRIS, M. 1994. Observations on experimental, nonchannelized, high-concentration turbidity currents and variations in deposits around obstacles. *Journal of Sedimentary Research*, **A64**, 899–909.

BAGNOLD, R. A. 1954. Experiments on a gravity-free dispersion of large solid spheres in a Newtonian fluid under shear. *Proceedings of the Royal Society*, **A225**, 49–63.

BEVERAGE, J. P. & CULBERTSON, J. K. 1964. Hyperconcentrations of suspended sediment. *American Society of Civil Engineers Proceedings, Hydraulics Division Journal*, **90**, 117–128.

CAS, R. A. F. & WRIGHT, J. V. 1987. *Volcanic Successions Modern and Ancient.* Chapman & Hall, London, 528.

DAVIES, L. C. & WALKER, R. G. 1974. Transport and deposition of resedimented conglomerates: the Cap Enragé Formation, Gaspé, Quebec. *Journal of Sedimentary Petrology*, **44**, 1200–1216.

DRUITT, T. H. 1992. Emplacement of the May 18, 1980 lateral blast deposit ENE of Mount St. Helens, Washington. *Bulletin of Volcanology*, **54**, 554–572.

EDWARDS, D. A. 1993. *Turbidity Currents: Dynamics, Deposits and Reversals.* Springer-Verlag, Lecture Notes in Earth Sciences, **44**.

FISHER, R. V. 1990. Transport and deposition of a pyroclastic surge across an area of high relief: the May 18, 1980 eruption of Mount St. Helens, Washington. *Geological Society of America Bulletin*, **102**, 1048–1054.

——, SCHMINCKE, H.-U. & BOGAARD P.VD. 1983. Origin and emplacement of a pyroclastic flow and surge unit at Laacher See Germany. *Journal of Volcanology and Geothermal Research*, **6**, 305–318.

HISCOTT, R. N. & MIDDLETON, G. V. 1980. Fabric of coarse deep-water sandstones, Tourelle Formation, Quebec, Canada. *Journal of Sedimentary Petrology*, **50**, 703–721.

INMAN, D. L. 1952. Measures for describing the size distribution of sediments. *Journal of Sedimentary Petrology*, **22**, 125–145.

JOHANSSON, C. E. 1963. Orientation of pebbles in running water. A laboratory study. *Geografiska Annaler*, **45**, 85–112.

—— 1976. Structural studies of frictional sediments. *Geografiska Annaler*, **58A**, 201–301.

LOWE, D. R. 1982. Sediment gravity flows II: depositional models with special reference to the deposits of high density turbidity currents. *Journal of Sedimentary Petrology*, **52**, 279–297.

MIDDLETON, G. V. 1966. Small-scale models of turbidity currents and the criterion for auto-suspension. *Journal of Sedimetary Petrology*, **36**, 202–208.

—— 1970. Experimental studies related to problems of flysch sedimentation. *Geological Association of Canada Special Paper*, **7**, 253–272.

—— 1993. Sediment deposition from turbidity currents. *Annual Review of Earth and Planetary Science*, **21**, 89–114.

NILSEN, T. H. & SIMONI, T. R. 1973. Deep-sea fan palaeocurrent patterns of the Eocene Butano Sandstone, Santa Cruz Mountains, California. *US Geological Survey Journal of Research*, **1**, 439–452.

POSTMA, G., NEMEC, W. & KLEINSPEHN, K. L. 1988. Large floating clasts in turbidites: a mechanism for their emplacement. *Sedimentary Geology*, **58**, 47–61.

RUST, B. R. 1972. Pebble orientations in fluvial sediments. *Journal of Sedimentary Petrology*, **42**, 384–388.

—— 1975. Fabric and structure in glaciofluvial gravels. *In*: JOPLING, A. V. & MACDONALD, B. C. (eds) *Glaciofluvial and Glaciolacustrine Sediments.* SEPM, Special Publication, **23**, 238–248.

SCHMINCKE, H.-U. 1970. "Base surge" Ablagerungen des Laacher-See-Vulkans. *Aufschluß*, **21**, 350–364.

——, FISHER, R. V. & WATERS, A. C. 1973. Antidune and chute and pool structures in base surge of the Laacher See area (Germany). *Sedimentology*, **20**, 1–24.

SMITH, G. A. 1986. Coarse grained nonmarine volcaniclastic sediment: terminology and depositional process. *Bulletin of the Geological Society of America*, **97**, 1–10.

TIARA, A. & SCHOLLE, P. A. 1979. Deposition of resedimented sandstone beds in the Pico Formation, Ventura Basin, California, as interpreted from magnetic fabric measurements. *Geological Society of America Bulletin*, **90**, 952–962.

TODD, S. P. 1989. Stream driven gravelly traction carpets: possible deposits in the Trabeg Conglomerate Formation, S.W. Ireland and some theoretical considerations of their origin. *Sedimentology*, **36**, 513–530.

WALKER, R. G. 1975. Generalised facies models for resedimented conglomerates of turbidite association. *Geological Society of America Bulletin*, **86**, 737–774.

WILSON, C. J. N. 1980. The role of fluidization in the emplacement of pyroclastic flows: an experimental approach. *Journal of Volcanology and Geothermal Research*, **8**, 231–249.

The liquification and remobilization of sandy sediments

ROBERT J. NICHOLS

Department of Geology, University of Bristol, Wills Memorial Building, Queens Road, Bristol, UK; Present address: Department of Earth Sciences, University of Leeds, Leeds, UK

Abstract: A number of large, unusually shaped sandbodies have been interpreted from three-dimensional seismic data of hydrocarbon-bearing Tertiary submarine fan deposits of the North Sea. The unusual sandbody shape is considered to have resulted from post-depositional liquification of turbidite deposits. An understanding of liquification processes may therefore be important in delineating reservoir body geometry. The unusual shapes take the form of sheet-like intrusions, either along faults or as dyke and sill complexes, and domes with oversteepened sides. Three processes can cause liquification: (1) fluidization, which results from pore fluid movement; (2) liquefaction, caused by the agitation of grains during cyclic shear stress; and (3) shear liquification which results from the movement of grains during the application of a shear stress across the sandbody. In laboratory experiments each liquification process produces its own style of deformation. Fluid escape and dish structures are produced during fluidization, load structures form during liquefaction, and mass flow structures result from shear liquification. The three liquification processes can interact to create an even greater diversity in deformation style. The active liquification process, or combination of processes, can change in both space and time. In the natural environment it is unlikely that one liquification process will occur independently and therefore most natural liquification must be considered in terms of two or more interactive processes. The large-scale deformation of sandbodies and hence their geometry may vary according to which liquification process or processes are active. In natural systems the sandbody will not deform in isolation from the surrounding material. The rheology of the surrounding material and the nature of the stress field active at the time of deformation will play an important part in controlling the behaviour of the sandbody during remobilization.

A number of workers have recently postulated that large scale remobilization of sandbodies up to many cubic kilometres in volume may occur (e.g. Barriga *et al.* 1992; Brooke *et al.* 1995; Dixon *et al.* 1995). Remobilization may cause sandbodies to take on unusual geometries (e.g. domes with sides that slope more steeply than would be stable during deposition; Dixon *et al.* 1995) and to move to new levels within the surrounding strata causing an unexpected juxtaposition of lithologic units (e.g. deeply buried sands of high metamorphic grade rising up into material at shallower depths with lower metamorphic grade; Barriga *et al.* 1992). This remobilization behaviour is of particular interest to petroleum geologists who, with the aid of three-dimensional seismic imaging, have recognized unusual geometries in a number of reservoir sandbodies from the Tertiary of the North Sea (Brooke *et al.* 1995; Dixon *et al.* 1995).

Deformation of clay-rich sediments is dominant in most cases of remobilization of large pre-lithified sediment bodies (see Maltman 1994 for recent review). In the remobilization described here, however, deformation of sand-rich (clay-poor) sediments dominate. For sand-rich sediments to deform more than adjacent clay-rich sediment the sand-rich sediment must be liquified. The term liquification describes any process that transforms a sediment into a liquid-like state (Allen 1982). Two other important elements that control the deformation of sandbodies are: (1) the rheology of the surrounding material, and (2) the stress field in which the deformation is taking place (Maltman 1994). The purpose of the present paper is to emphasize liquification processes and the production of unusual sandbody geometries.

Previous workers consider liquification to be a single process which is characterized by pore-pressure changes within a sediment body (e.g. Seed & Lee 1966; Youd & Perkins 1978; Lambe & Whitman 1979; Seed 1979). However, there is evidence that liquification is not simply a single

From Hartley, A. J. & Prosser, D. J. (eds), 1995, *Characterization of Deep Marine Clastic Systems*, Geological Society Special Publication No. 94, pp. 63–76.

process but comprises at least three processes (e.g. Davidson & Harrison 1971; Melosh 1979; Campbell & Brennan 1985; Barker & Mehta 1993). Evidence is presented here which suggests that: (1) these processes can interact to a greater or lesser degree; (2) the style of deformation is different for each of the three processes; and (3) the interaction of processes produces even greater diversity in behaviour and style of deformation. The implications of these observations in the study of the remobilization of large sandbodies are discussed.

Definitions of liquification

Liquification of sand

The term sand is used here to describe any granular non-cohesive material with a grain size (d) in the range $63 < d < 2000\ \mu m$. Liquification of a sand occurs when the grains of sand are no longer supported by static intergranular contact. As the shear strength of the sand is transferred by friction through these static contacts, liquification causes a reduction in shear strength to the extent that the sand is able to flow as though it were a liquid. Three common liquification processes can be identified.

Fluidization (seepage liquification). As pore fluids seep past the grains within a body of sand, force is transferred from the pore fluid to the grains by fluid drag. The transfer of force is expressed as a pressure drop across the sandbody (Richardson 1971). If the pore fluid flow is upward the vertical component of the drag will counteract the downward force due to the weight of the grains. As the fluid flow increases, the amount of drag force acting on the grains will also increase until it equals the downward force due to the weight of the grains. The grains will no longer be supported by static grain-to-grain contact but instead will be supported within the fluid flow and therefore the sand will be free to liquify (Davidson & Harrison 1971). If the grains are homogeneous, having the same size, shape, density and packing arrangement, the whole body of sand will began to fluidize at a single flow velocity. This flow velocity is called the minimum fluidization velocity, U_{mF}, (Davidson & Harrison 1971). When the grains are heterogeneous, and the fluidizing fluid is a liquid, portions of the body will remain static while other portions fluidize during the initial stages of fluidization (Nichols *et al.* 1994).

When the sand is fluidized by a gas, typical behaviour is for the gas to pass through the sand in an irregular manner causing the formation of pipes, bubbles and small bodies of particles or 'slugs' (Zenz 1971). Although fluidization due to the passage of a liquid is described here, there is evidence that some remobilization of gas-filled sandbodies occurs (Brooke *et al.* 1995). On a large scale (metres and greater), fluidization by both liquid and gas produces the same effects (loss of shear strength in the material and deformation by liquid-like flow). On a small scale (metres and less), fluidization by a gas is more likely to produce pipe structures and bodies of unfluidized material (e.g. Wilson 1980, 1984); fluidization by a liquid is more likely to produce an even reorganization of particles, normal grading (i.e. grain size decreases upward) and a loss of original structures (Nichols *et al.* 1994). For simplicity, the term fluidization is used in the rest of the article to mean fluidization by a liquid.

Liquefaction (vibration liquification). When sand is subjected to a cyclic shear stress, either as individual shocks or as a continuous vibration, the grains can become agitated moving side to side in response to the shear stress. If the agitation is great enough the movement of the grains can cause them to collide with their neighbours. This movement causes the grains to become momentarily suspended within the pore fluid. While in suspension the weight of the sand is no longer supported by static intergranular contacts and instead the weight of the sand is transferred by grain-to-grain collisions. The sediment loses its shear strength and is free to flow as a liquid (e.g. Allen 1982, 1985).

Shear liquification (liquification by body shear). The term body shear is used here to describe a unidirectional shear force applied across a sandbody. This shear is applied to the body either due to the downslope component of the gravitational force when the body is lying on a slope, or by the lateral motion of material over the body. Models have been produced which compare the shear of granular materials with the transfer of heat between atoms (e.g. Campbell & Brennen 1985; Walton & Braun 1986; Nadkarni & Peters 1987). From these models it can be seen that shear causes grains to collide. Kinetic energy is transferred from grain to grain during these collisions. The energy transfer can be great enough to balance the gravitational potential energy (due to the weight of the grains) so that force is no longer transferred by static grain-to-grain contacts. In this state the intergranular friction is greatly reduced from that seen in the static state; therefore, the shear strength of the granular material is also greatly reduced and the material becomes liquified. Shear liquification is very similar to liquefaction, but in liquefaction the stress and resulting grain movements are cyclic with rapid alternations in direction. In shear liquification the stress is unidirectional resulting in a gross displacement of the sand-body in the direction of the stress.

The shear strength of a cohesionless material is proportional to the effective stress (Lambe & Whitman 1979). A high pore pressure will greatly reduce the shear strength of a material. Therefore, if the pore pressure is high, shear deformation is more likely. The pore fluid acts as a lubricant and allows much of the shear strain to be taken up in deformation of the pore fluid. The deformation may be concentrated at horizons where the permeability decreases rapidly and the pore pressure drop changes rapidly (Maltman 1994). This is a similar process to the gliding processes envisaged as being active in some debris flows (e.g. Ineson 1985). Shear liquification may occur in these circumstances, depending upon how the sediment reacts to the applied shear stress.

A previous model

The emphasis on pore-pressure measurement in many studies of liquefaction has led to the belief that liquefaction can result from an increase in absolute pore pressure. The description of the processes involved is either unclear (e.g. Owen 1987) or discussed as one of a number of ways of causing liquefaction (Allen 1985). In the 'liquefaction by increase in pore pressure' model (Allen 1985, 183–184) a granular sediment is trapped between layers of impermeable material (e.g. clay) allowing the pore pressure within the sand to be greatly elevated if extra fluid is added. As the pore pressure is increased some of the weight of the grains is supported by the high pore fluid pressure. Eventually a point is reached where the entire weight of the grains is supported by the pore fluid and the granular sediment liquifies. However, in an unconsolidated sandbody, all the pore spaces are interconnected, and an increase in pressure will be transferred, via adjacent pore spaces, evenly throughout the sediment body. No mechanism has been described which would enable a pressure force, evenly distributed within a static pore fluid, to be transferred as lift to the sand grains. Lift will only be transferred to the grains if the pressure gradient is increased, but the increase in the pressure described in the 'liquefaction by increase in pore pressure' model only considers the absolute pressures. If pressure gradients increase there will be a flow of fluid and if the flow is great enough fluidization can occur.

Therefore, in this model, where sand is trapped within layers of clay and pore pressures are elevated, liquification will occur if the seal is broken to release pore fluid. The result is fluid flow through the sediment body and fluidization results. Ascribing liquefaction simply to an increase in the absolute pore pressures cannot adequately explain the process. Increases in pore pressure at the base of a sand layer can cause a pressure drop across the layer and the resulting increase in the pressure gradient can cause liquification. However, the resulting process is fluidization, which is already adequately described. It is therefore unnecessary, and may even cause confusion, to describe the processes as liquefaction by increase in pore pressure.

Other similar processes

Liquefaction of very fine-grained sediments

Some clay or clay-rich sediments (more accurately, very fine grain-sized cohesive sediments) liquify during vibration. The shear strength rapidly returns once vibration ceases. This is described as thixotropic behaviour. The change in behaviour during vibration results from the realignment of the platey clay minerals and their dispersion within the pore fluid, causing a loss in shear strength (Owen 1987). The long-range bonding of the clay minerals allows the shear strength to be recovered rapidly once vibration ceases (Smalley 1976).

In general, the presence of clay bonds within clay-rich sediments reduces the susceptibility of a sediment to liquification. However, clay can naturally deform plastically without alteration, i.e. without liquification. The plastic deformation of clay can produce very similar structures to those produced during liquification of otherwise solid materials (Rettger 1935; Dzulynski 1966) and care must be taken not to confuse the two behaviours.

The liquid behaviour of dense suspensions immediately prior to deposition

In the final stages of sedimentation, granular sediments can behave as a liquid-like layer immediately above the already deposited sediment. Deformation within this layer or between this layer and an underlying plastic layer (e.g. a previously deposited layer of clay) can produce liquification structures (summarized in Anketell *et al.* 1970). However, as the granular material is only liquified during sedimentation the structures are restricted to within a single sedimentary unit, or at the boundary between a single sedimentary unit and the material immediately underlying it (a sedimentary unit being the material sedimented in a single event, e.g. a turbidite). This mechanism cannot account for a liquification event which results in the deformation of more than two sedimentary units.

Evidence for the variation in liquification behaviour

Differentiation between liquification processes would be purely academic if it were not for one important observation: the style of deformation is dependent upon the active liquification process. For example, in experiments it has been found that fluid escape structures and dish structures are formed during fluidization (Tsuji & Mivata 1987). In experiments where load structures have been produced by the liquification of a previously solid sediment the process was driven by a shock or vibration, i.e. liquefaction (Kuenen 1958; Anketell *et al.* 1970). Also, fluidization of sand typically produces a fining-up arrangement of grains (Nichols *et al.* 1994), whereas liquefaction often produces a coarsening-up arrangement of grains (see discussion by Barker & Mehta 1993).

The differences in behaviour can be demonstrated by the comparison of a liquefaction experiment and a fluidization experiment described below. The latter experiment was specifically designed to reproduce the starting conditions of the first but to fluidize the sediment instead of vibrating the sediment.

The liquefaction experiment.

A box 600 mm tall, 600 mm wide, and 600 mm high, was constructed from 5 mm thick Perspex sheets welded together using a solution of perspex dissolved in chloroform. The box was placed on a 7 mm thick mild steel plate, and the plate was attached to a 3 kW Derritron exciter via a 1.5 m long driving rod. The exciter was similar to a loudspeaker, but instead of transferring movement to a conical diaphragm (the loudspeaker cone), the movement was transferred to a steel 'driving rod'. The length of the driving rod was tuned so that its resonant vibration frequency was close to the required frequency, which greatly increased the efficiency of the system. A long driving rod would be used when a lower frequency was required.

Table 1. *The grain size distribution and composition of two sand samples (a fine sand sample and a medium sand sample) used in a series of vibration experiments*

Parameter	Fine sand*	Medium sand*
Sizes (mm)		
> 1400	1 Wt%	4 Wt%
1400–1000	1 Wt%	6 Wt%
1000–500	8 Wt%	25 Wt%
500–250	20 Wt%	50 Wt%
< 250	70 Wt%	15 Wt%
Composition of clasts		
Quartz	95%	55%
Lithics	5%	30%
Shell fragments	0%	15%

* Wt% refers to the percentage of the total weight of the sample trapped between the relevant sieves, % composition refers to the percentage of the sample belonging to each type of composition.

Two types of sand were used, one a fine sand and the other a medium sand (Table 1). The box was filled with water and fine sand was deposited through the water to a thickness of 240 mm. Thin layers of medium sand were periodically deposited to produce a roughly horizontal lamination through the fine sand. A hollow was dug in one corner 150 mm deep and *c.* 400 mm across, this was filled to a depth of 110 mm with medium grain sand. Thin layers of fine sand were periodically deposited to produce a roughly horizontal lamination through the medium sand. The last 40 mm of the hollow were filled with fine sand (again with horizontal laminations of medium sand). In this way a lens-shaped structure of medium sand enclosed within fine sand was produced (Fig. 1a).

The box was then shaken at 10 Hz. Downward lobes in the base of the coarse sand lens spread sideways and down, producing flat-bottomed lobes similar to the load structures described by Anketell *et al.* (1970). There was no evidence of water escape structures at the height of the load structures (Fig. 1b). The laminae in the top 75 mm of the lens structure deformed into a series of antiforms and synforms. A small diapiric body (*c.* 30 mm across) of fine sand rose from the apex of one of the antiforms. The formation of these convolute

a)

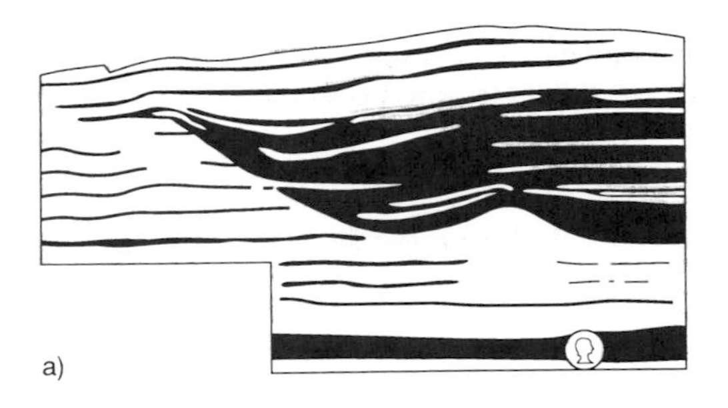

b)

Fig. 1. The liquification structures formed during a vibration experiment. The light coloured sediment is a very fine sand and the dark coloured sediment is a medium sand (see Table 1). **(a)** The structure prior to the vibration experiment. (b) The structures formed during the vibration experiment. The lower third of the container shows little sign of deformation; the middle third shows load structure deformation with little evidence of fluidization structures; and the top third shows a distinct upward deformation and convolution of structures.

structures was associated with the escape of water during vibration. No deformation occurred in the bottom 150 mm of the box.

The fluidization experiment

Two sand samples were used in these experiments (the sand was from the same source as that used in the liquefaction experiments). Three layers of sand were deposited in a three-dimensional fluidization rig (see Nichols *et al.* 1994, for details of this apparatus) (Fig. 2a). The bottom layer of medium sand was 100 mm thick. Three thin marker bands of fine sand were deposited within the base layer, the lowest of which (26 mm above the base of the layer) was 6 mm thick. The middle layer of fine sand was 55 mm thick and contained one marker band of medium sand. The top layer of medium sand was 67 mm thick and contained two marker bands of fine sand. Water flow was initiated and separations developed within the fine sand. In the middle layer the separation spread across the Perspex container and the material above the separation lifted up as a solid plug. At the thick marker band within the base layer a triangular void formed (Fig. 2b).

(a)

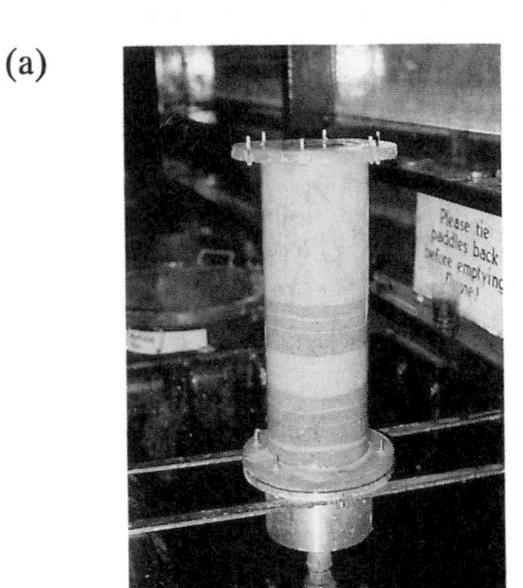

(b)

(c)

Fig. 2(a). The sample in the three-dimensional rig prior to the fluidization experiment. **(b)** A separation formed within a thick marker horizon near the base of the sample. **(c)** Burst out from the lower void resulted in the formation of a fluid escape structure.

The overlying material was deformed to form an antiform over the void. The deformation caused localized weakening of the material, especially near the axial plane of the antiform. A burst out occurred through this weakened material and a fluid escape structure was produced (Fig. 2c). The development of a void, rupture of overlying material, burst out and formation of a fluid-escape structure was as observed in other experiments (Nichols *et al.* 1994). At no point was there any sign of the plastic interfingering of lobes of material seen in the liquefaction experiment.

Liquefaction and fluidization structures

These experiments suggest that load structures are produced during liquefaction and water-escape structures during fluidization. However, liquefaction of water saturated sand causes the upward movement of water (e.g. Seed 1979; Allen 1985). It has been observed by other authors that liquefaction of sediment can result in fluid-escape structures being formed at the surface (e.g. Wills & Manson 1990). In the liquefaction experiment the structures in the upper part of the sample had a different form (convolute lamination) to those at a lower level (load structures). These observations indicate that if the vertical pore-fluid flow is great enough fluidization structures will overprint liquefaction structures.

The increase in dominance of fluidization with increased height can be explained by considering how the discharge of water passing through the sand changes with height. The water driven out of the pore spaces during vibration moves upward. At any horizon water will flow out from the sand in the horizon and water will flow through from the material below the horizon. The higher the horizon, the more sand underlies it and the more water discharges through it – water discharge increases with height. At the base of the system vibration is not enough to liquify the sand. Further up, the increase in water flow is enough to reduce the shear strength of the sand to allow liquefaction to occur. With increased height the water flow is great enough to modify the liquification behaviour so that liquification is by both liquefaction and fluidization. As the height increases still further fluidization may dominate if fluid flow is great enough.

The production of convolute laminations in the upper parts of the sample in the liquefaction experiment suggest that convolute structures are formed when neither liquefaction nor fluidization dominate. Therefore, convolute structures form in a liquefaction environment with a high water throughput.

Experiments on the interaction of fluidization and liquefaction

Experiments were carried out on two samples of glass ballotini in a three-dimensional fluidization rig (described in Nichols *et al.* 1994). The samples were fluidized by passing water up through them. An electric motor (240 V, 50 Hz) was attached to the frame that held the three-dimensional rig. An off-centre weight was fitted to the motor so that when the motor was run a vibration occurred which shook the whole apparatus (Fig. 3). A dimmer switch was wired into the power supply to the motor. The speed of the motor could be controlled by adjusting the dimmer switch and hence the intensity of vibration could be modified. An accelerometer was fitted to the frame to measure the vibration. The off-balance was the result of a 22 g mass being positioned 72.5 mm away from the centre of a spinning piece of steel that was otherwise balanced about the centre. The maximum speed of the motor was 12000 rev min^{-1} and at this speed a force of *c.* 800 N would be applied to the apparatus at a frequency of 200 Hz. Some of the force was lost in driving the apparatus and therefore the force applied to the sample would be < 800 N. It was not possible to distinguish the vibrations which were exciting the samples in the fluidization rig from those exciting the

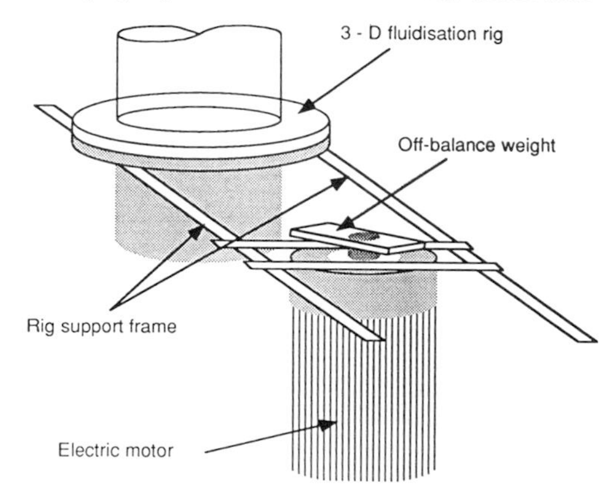

Fig. 3. The three-dimensional rig with the electric motor attached to the support frame. When the electric motor was run the off-balance weight caused a vibration which was transferred to the three-dimensional rig via the support frame.

frame and the rig itself. This would require detailed vibration studies of the apparatus. Such detailed vibration analyses were beyond the scope of these qualitative experiments. Change in vibration, therefore, was measured as a change in intensity from low to high. High intensity referred to the vibration produced when the motor was at full speed and 800 N was being produced at 200 Hz. Low intensity was characterized by accelerations which had displacements an order of magnitude smaller than those characterizing the high intensity vibration (as measured with the accelerometer) and was produced by reducing the rotational speed of the motor. The vibration experiments were confined to the two levels of vibration, i.e. high and low.

Initial experiments involved setting the superficial flow velocity at various levels both above and below the minimum fluidization velocity and then turning on the motor to investigate how the vibration affected the behaviour of the sample. Whilst the rig was being vibrated shock waves travelled up the pressure tube causing the pressure readings to oscillate. These oscillations centred about the normal static pressure causing a maximum variation of ± 0.91 mbar around the normal mean when 1.8 kg of ballotini was vibrated (the pressure required to fluidize the material would be in excess of 7 mbar). At first it was thought that these fluctuations were caused by the vibration of the apparatus, rather than the sample. However, when the apparatus was vibrated with no sample present (only water being present in the rig) no oscillations were seen. The valves in the bleed tubes positioned either side of the pressure transducer were opened, and this slowed down the response time of the transducer and effectively averaged the pressure. Small-bore glass pipes positioned in the bleed pipes allowed them to be used as manometers and visual observation of these manometers confirmed the nature of the pressure oscillation as cyclic around a mean. It was initially proposed that the change in pressure drop across the sample would be used to assess the effect of the vibration, as liquefaction is usually described in terms of pressure fluctuations (e.g. Seed & Lee 1966; Lowe 1976; Lambe & Whitman 1979; Allen 1982, 1985; Wills & Manson 1990). However, the pressure fluctuations were too rapid to measure accurately with the equipment available. The pressure drop measured before vibration and the average pressure drop measured during vibration were the same.

The experimental apparatus was modified to allow a 25.4 mm diameter steel ball, weighing 67 g, to be lowered on to the top of the sample (which had been fluidized and allowed to settle to produce a flat top). The static sediment would support the ball, but once liquification started the loss of shear strength would allow the ball to sink and therefore the onset of liquification could be assessed. Fluid flow was initiated and the water flow was slowly increased until the ball sank progressively through the sample. The ball was counterbalanced to slow the descent of the ball so that it could be measured more easily (the effective weight of the ball was reduced to 22 g by the counterweight). The flow velocity was recorded when the ball sank to the points where: (1) only half of the ball was visible above the top of the sample; (2) only the ring, by which the ball was attached (via a cord) to the counterbalance, was visible; and (3) the ball reached the bottom of the sample. Water flow was then stopped and the ball was pulled out of the sediment. The sample was fluidized and then allowed to settle to give a new flat top. The motor was switched on at a velocity which produced a low-intensity vibration, and the experiment was repeated. The experiment was then repeated with the motor set to give a high-intensity vibration.

It was found that the vibration caused a drop in the fluid flow velocity, U, required for liquification to occur (Fig. 4). For 250–355 μm ballotini the reduction was greatest when the vibration intensity was greatest. For 710–850 μm ballotini the increased vibration intensity resulted in less reduction in the fluid flow velocity required to cause liquification. This unexpected result in the behaviour of the coarse-grained material, could be due to the nature of the source of the vibration. To increase the intensity of the vibration the motor speed was increased so that an increase in vibration amplitude was coupled to an increase in the frequency of the vibration. It is thought that the coarse-grained material was more susceptible to liquefaction at low frequencies than at high frequencies and that the increase of amplitude was not enough to off-set this effect. This suggests that granular grains exhibit harmonic behaviour where a material has natural frequencies at which it vibrates easily, separated by frequency regions where vibration is more difficult. This frequency-dependent behaviour is less important where the driving cyclic shear stress is from an instantaneous shock, as the nature of such a shock is to vibrate a body at all frequencies. However, where a vibration or shock wave has travelled through surrounding strata the damping effect and resonant behaviour of this material will result in the

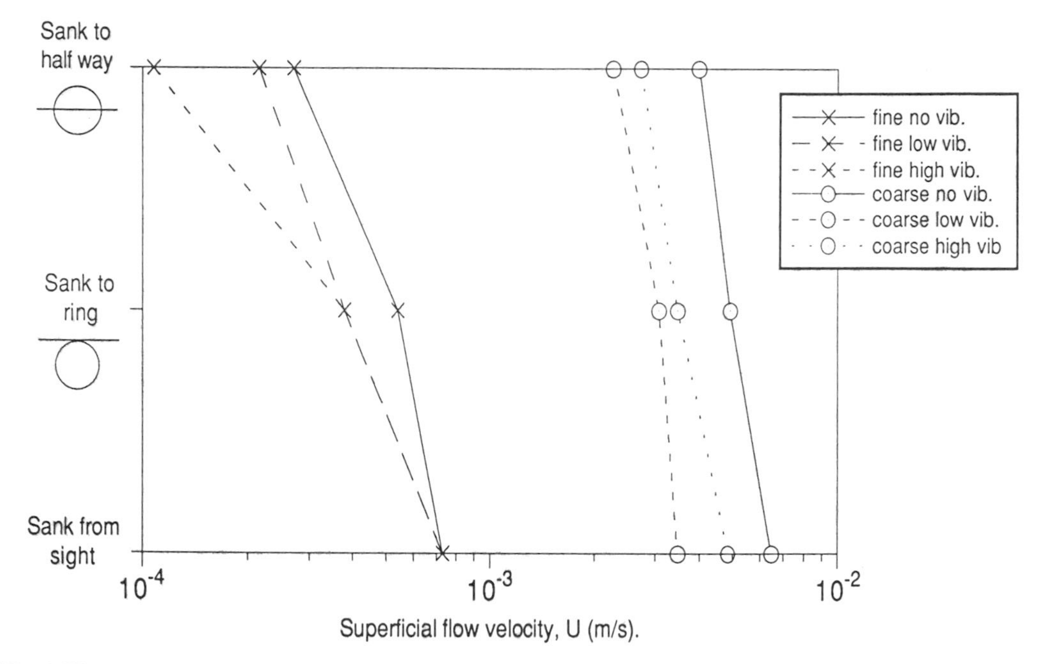

Fig. 4. The sinking of a steel ball through fine (250–355 μm) and coarse (710–850 μm) ballotini at different flow velocities and vibration intensities.

frequency range of the vibration being greatly modified. Thus, during an earthquake the seismic shock may be transmitted many kilometres through various rock types, taking various paths, before it reaches a point at the surface. In such a case the resonant behaviour of a sand may alter the way it responds to a far travelled cyclic shear stress, i.e. fine-grained sand may be more or less susceptible to liquefaction than coarse-grained sand because of the difference in resonant behaviour.

Interpretation of the results from the vibration experiments

These experiments show that vibration can cause liquefaction without producing a change in the mean pressure drop across the layer of granular material; i.e. that liquefaction is not driven by an increase in the mean pressure drop. This conclusion is supported by the vibration behaviour of dry granular material (e.g. Rosato *et al.* 1991).

A fluctuation in pore pressure during vibration liquification is widely recognised where the pore fluid is a liquid (e.g. Seed & Lee 1966; Allen 1982, 296–298) and was also observed in these experiments. These observations must be accounted for in any model of liquefaction. One possible explanation for the fluctuation in pore pressure is as follows. During vibration enough kinetic energy is transferred by intergranular collision to keep the grains in suspension, but as soon as the vibration stops the grains sink back through the pore fluid. As the grains sink their weight is transferred to the pore fluid by drag, resulting in an increase in the pressure drop across the layer. During a shock the grains will be liquified initially, after which the grains will sink to the floor of their container and therefore the shock will be associated with a subsequent temporary increase in pore pressure. If the frequency of the vibration cycles is low there could be enough time for the grains to start sinking during the low points in each cycle. Therefore, liquefaction by vibration would be associated with cyclic increases in pore pressure as seen in experiments (e.g. Seed & Lee 1966).

The experiments described above demonstrate that fluidization occurs more easily when liquefaction is present and *vice versa.* An expulsion of pore fluid is usually associated with an increase in packing density after liquefaction (e.g. Selley & Shearman 1962; Seed 1979; Allen 1985) and therefore it is envisaged that in most natural systems, where the pore fluid is a liquid, the two processes will occur together and complement one another. In the geological environment it is likely that liquefaction will occur more easily in thixotropic muds and quick soils. The liquefaction of granular noncohesive material is more difficult, but the added action of fluidization by escaping pore fluids will increase the likelihood of liquification. The experiments also indicate that liquefaction is a complicated process that is not only effected by the amplitude of the vibration, but also the frequency of the vibration.

Experiments on the interaction of fluidization and shear liquification

A set of experiments were carried out in a two-dimensional fluidization rig (described in Nichols *et al.* 1994). The samples were fluidized, the water flow was stopped and the particles were allowed to settle, forming an even layer with a horizontal top. The two-dimensional rig was then raised at one end (care was taken to ensure that the sample was not disturbed) so that the sample rested at an angle and therefore a gravitational shear component was applied to the sample; the angle was recorded. Fluid flow was initiated and the fluid flow slowly increased until the top of the sample could be seen to liquify. When a slope was present the grains at the top of the layer would start to flow downslope. At very low angles of slope the grains at the top of the layer would be seen to move randomly. The velocity at which liquification first occurred was recorded. The end of the two-dimensional rig was then lowered and the sample was fluidized once again. The experiment was then repeated at a new angle of rest.

Figure 5 shows the results from these experiments. When the sample was horizontal liquification was seen at a velocity slightly lower than the calculated minimum fluidization velocity, U_{mf}, i.e. *c.* 0.8 U_{mf}. This apparent anomaly can be explained as follows. U_{mf} was calculated for the mean grain size and as preparation of each sample was by sieving, a range of grain sizes would be present. The test samples were fluidized before each experiment and as fluidization produces normal grading (Nichols *et al.* 1994) the grains at the top of each sample would be finer than the mean grain size. The top grains would have a lower actual grain size to that calculated for the whole sample, and this could explain the reduction in the observed liquification point in respect to the calculated U_{mf}. The consistency of the results over a range of grain sizes implies that the first liquification criteria used here is valid. The use of pressure gauges, which is a more accurate way of determining U_{mf}, was not used due to the problems in assessing accurately the pressure drop across the sample when it was tilted at an angle other than zero.

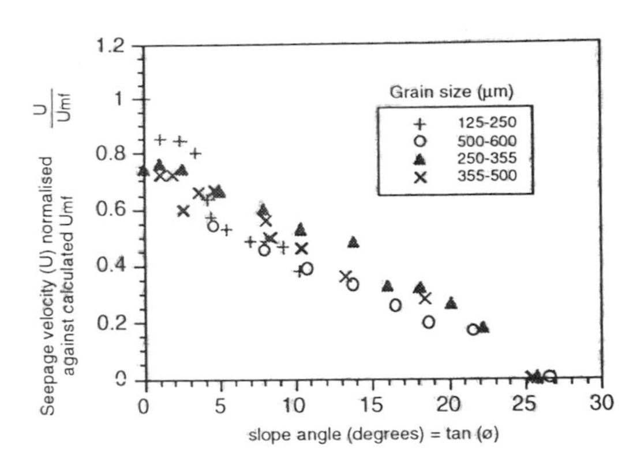

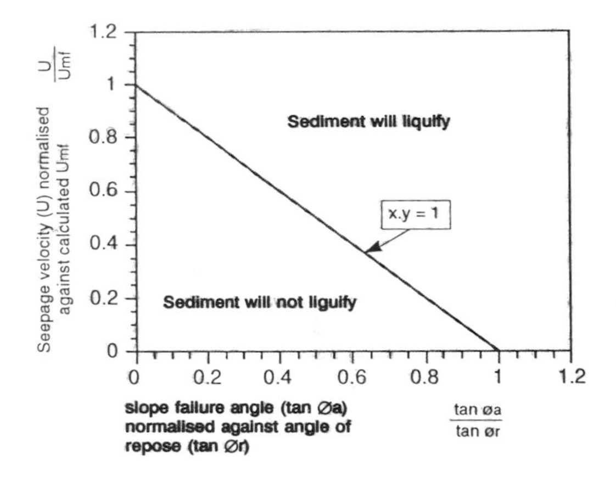

Fig. 5. The relationship between shear liquification and seepage liquification: a comparison between experimental results and a theoretical model. **(a)** Results from a series of experiments carried out on ballotini in a range of grain sizes. **(b)** A graphical representation of equation 2.

As the angle of the layer was increased, progressively less water flow was required to cause failure, until the angle of repose (or angle of shearing resistance; 26–30° for uniform fine to medium sand, Lambe & Whitman 1979, 149) was reached, at which point failure occurred without water flow. The change in behaviour is linear and suggests that a relationship can be used to determine the onset of liquification when both fluidization and shear liquification are active. The onset of liquification will occur when the following relationship occurs:

$$\frac{U}{U_{mf}} + \frac{\text{applied shear stress}}{\text{critical shear stress}} = 1 \quad (1)$$

Where U is the superficial fluid flow velocity and the critical shear stress is the minimum shear stress required to cause liquification when no other liquification process is active. For these experiments, where the sample is homogeneous and at the free surface the change in shear stress is due to a change in the angle of rest of the material:

$$\frac{U}{U_{mf}} + \frac{\tan \varnothing}{\tan \varnothing_r} = 1 \qquad (2)$$

Where: $\varnothing$ is the angle at which the sample is resting when it fails, and $\varnothing_r$ the angle of repose of the sample material in the absence of pore-fluid movement (e.g. Lambe & Whitman 1979). Figure 5a is a plot of the measured results and can be compared with Fig. 5b which is a plot of Equation 2. It can be seen that there is a reasonable correlation between the actual plot (Fig. 5a) and the theoretical plot (Fig. 5b), except that the actual plot has consistently lower values for U/U_{mf} (as discussed above).

The use of the failure criterion used here is not an accurate way of determining the onset of liquification and a better way of determining this onset is required. However, these experiments demonstrate qualitatively that fluidization is easier if a shear is applied and liquification by shear is easier when a vertical fluid flow is present. They also demonstrate that there is a continuous change in behaviour from fluidization to shear liquification and that these two styles form end members in a continuum of fluidization–shear liquification behaviour.

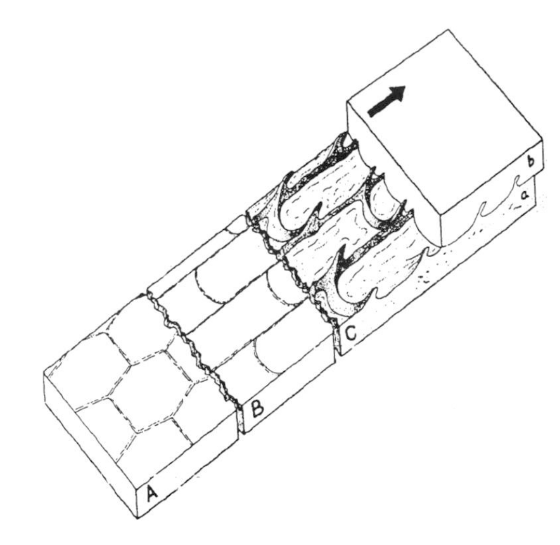

Fig. 6. The change in the pattern of liquefaction structures from hexahedra (A) to elongate 'scales'(B) with the application of shear (in the direction of the arrow). C shows the three dimensional pattern of the structures. Layer 'a' is less dense than layer 'b' (after Anketell *et al.* 1970).

Variation in liquification behaviour

In the experiments described above liquification occurred not only due to the processes of fluidization, liquefaction and shear, but also due to the interaction of fluidization with liquefaction, and fluidization and shear. There is also evidence that liquification can occur as a result of the interaction of liquefaction and shear.

Anketell *et al.* (1970) gave a detailed account of the change in style of load structures with increasing amounts of shear. With no shear the load structures formed a polygonal network of hexahedral columns in plan view (Fig. 6). With small amounts of shear these hexahedra were stretched in the direction of shear, into what were described as 'scales' (Fig. 6). Large amounts of shear resulted in the loss of the structures perpendicular to the shear direction, with the polygonal network being replaced by a series of ridges running sub-parallel to the shear direction (Fig. 7a). This model was complicated if the internal friction of the material involved was high (Anketell *et al.* 1970), as in such cases the lobes of the load structures were found to rotate around an axis perpendicular to the shear direction (Fig. 7b). These results support the notion that there is a continuum of liquification behaviour between liquefaction and shear liquification.

A general scheme for liquification

By combining all of the above observations it is possible to develop a general scheme to describe liquification behaviour. There are three end member processes: (1) fluid flow driven fluidization; (2) vibration driven liquefaction; and (3) liquification by body shear. Between these three end members lies a range of liquification styles where any combination of these three processes may occur (Fig. 8). The fluidization–shear liquification and fluidization–liquefaction experiments indicate that the changes of behaviour are gradual. The liquification behaviour can be characterized by regimes where either one process dominates or two processes act equally together. For example: the fluidization regime where fluid-escape structures are the dominant structure; the liquefaction regime where load structures are dominant; and the shear liquification regime where mass

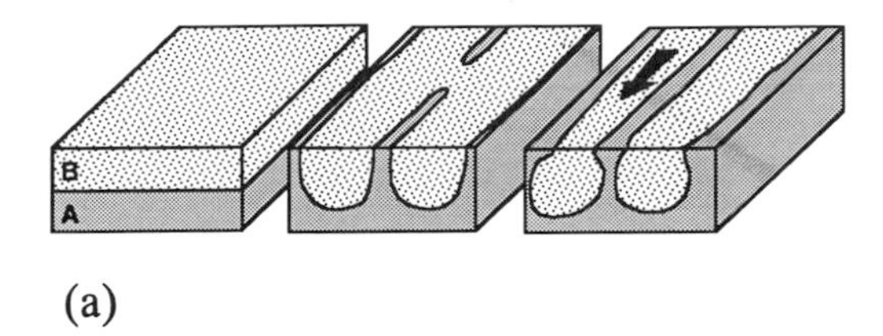

(a)

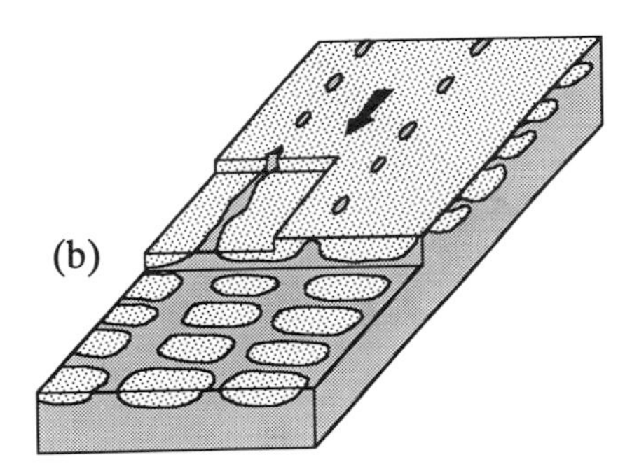

(b)

Fig. 7. The deformation pattern of load structures, formed whilst a relatively large shear force is applied (in the direction of the arrow). Layer B is more dense than layer A and the kinematic viscosity of A is much greater than layer B. (after Anketell *et al.* 1970). **(a)** The structures formed when the materials have low internal friction. **(b)** The structures formed when the materials have high internal friction.

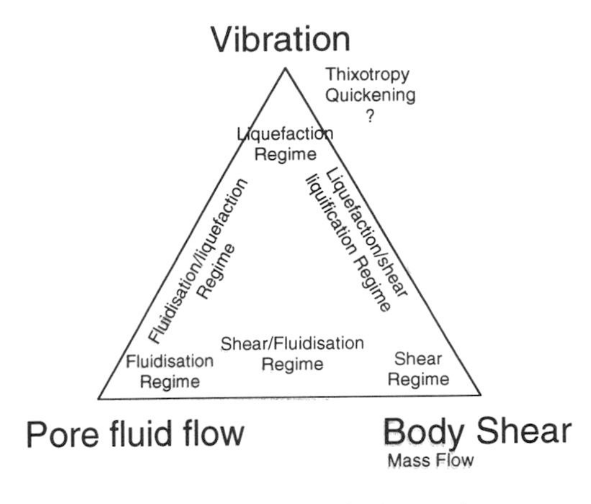

Fig. 8. A graphical representation of the interaction of the three main end-member liquification processes.

flow structures are dominant. There are also transitional regimes between two end members, where each has a roughly equal effect. At present, there are not enough laboratory investigations of these transitional regimes to describe them accurately, but a tentative scheme is suggested here: liquefaction–shear where elongate ridges and rotated (around a horizontal axis perpendicular to the shear direction) load blocks are produced; liquefaction–fluidization where convolute lamination occurs; and fluidization–shear where elongate or stretched fluid escape structures develop.

Large-scale remobilization

As small-scale deformation structures vary depending upon the active liquification process, it is reasonable to suggest that the liquification behaviour will affect the deformation of a large sandbody during remobilization. Dixon *et al.* (1995) document three styles of deformation: (1) doming and oversteepening of the sides of sandbodies; (2) injection of material into dykes and sills; and (3) movement of sand along fault planes. Variation in the active liquification processes may account for the diversity in deformation styles. There is insufficient data at present to test this hypothesis and therefore no attempt has been made here to assign a dominant liquification process to a particular deformation style. However, with more data from both field observation and laboratory modelling, assigning liquification process to deformation style may be possible in the future.

In studying large-scale remobilization of sandbodies it should be noted that the liquification process can change in time and space, causing further complication of the deformation history. This will be more evident in the remobilization of large sandbodies as larger distances and greater time scales are involved. As with facies models, the change in structural styles through a body may give more information than could be gained from studying the specific structural style at one point. For example, if the water fluidising a sandbody is derived from within the sandbody the amount of water passing through a point should increase with height (as explained above). Therefore, the effects of fluidization should also be greater with height. Conversely, if the water is derived from outside the sandbody the fluidization effects should be similar throughout the sandbody.

Liquification of a layer of sand will result in the sand no longer supporting the overlying strata. The overlying strata may deform in response to this loss of support. The style of deformation will depend upon the rheology of the overlying strata, and the nature of the stress field present during deformation. In fluidization experiments it was found that the form of the fluid-escape structures produced, altered depending upon the rheology of the material overlying the fluidized layer (Nichols *et al.* 1994). In natural large-scale deformations, where the stresses are more complicated and the overlying material is stronger and thicker, the rheology of the overlying material and the nature of the stress field may play a larger role than is apparent from laboratory models.

Two models for the type of deformation above a sandbody encased within clay-rich sediments can be envisaged. Firstly, irregular compaction of clay-rich sediment around a mound of sand could cause localized weakening of the clay-rich sediment around the top of the mound. If the overlying clay-rich sediment ruptured, pore waters would be released and the flow-through of pore fluid could fluidize the sand, allowing fluidized sand to flow into the rupture. The removal of sand from within the mound and into the rupture would further weaken the overlying clay-rich sediments, increasing the rupture, and so on. This process has been used to explain the injection structures seen in Tertiary turbidites of the Bruce-Beryl embayment (located in the South Viking Graben, UKCS) (Dixon *et al.* 1995). Secondly, movement along faults associated with uneven compaction around a mound of sand or older tectonic structures, could also cause a release of pore fluids up the fault. The seepage effect combined with the shear associated with the fault movement could cause fluidization-shear liquification of sand around the fault plane and injection of that sand along the fault plane. Dixon *et al.* (1995) show evidence for this type of injection of sand along fault planes in the Forth-Gryphon area (in the southern part of the Bruce-Beryl embayment).

The injection of sediment into overlying strata

The geometry of injection features in fluidizing systems is controlled by the mechanical behaviour of the material into which the fluidized material is injected. In fluidization experiments, injections into loose sand were cylindrical with splayed ends, and into clay-rich unconsolidated sediment formed smaller pipes commonly with angular bends (Nichols *et al.* 1994). Where the clay content of a layer was > 15% the passage of fluidized material was unable to break up balls of this clay-rich material, after the initial layer had broken up (Nichols *et al.* 1994). These results suggest that injection into consolidated clays will not be associated with large amounts of erosion and the geometry of the injected bodies will be controlled by the manner in which the ruptures in the static material were formed. The injection of sediment is analogous to the injection of igneous dykes and sills.

The liquifying processes will cause the break-up of sections of consolidated clay-rich sediments above a liquifying layer. Claystones overlying the main liquified body of the Forth Field contained a series of compaction lineations, indicating that this material had ruptured along these lineations. The liquified sands infilled these ruptures to form elongate sheets. Near the top of the main body a number of blocks of overlying material were seen, which suggests that stoping may also be found, again analogous with some igneous intrusions.

Conclusions

Liquification of a sandbody results in most of the deformation occurring within the sand rather than in surrounding mud-rich sediments. The unusual deformation of the sandbody can produce a change in shape, often causing doming, oversteepening of sides, and/or partial injection along faults or fractures to form sheet-like intrusive bodies. Liquification describes a number of processes that cause a sediment to be transformed from a solid to a liquid-like state. Once a sediment body is liquified most of its shear strength is lost making it very susceptible to remobilization and deformation. Sand can be liquified by a number of processes: fluidization, liquefaction and shear liquification. The style of small-scale structures (up to metres) is dependent upon the liquification process active at the time of deformation. There is some variation in the style of deformation seen in large-scale remobilized sandbodies and the active liquification process may control this to some extent. However, at present there is no clear evidence to test this hypothesis.

The three liquification processes are each driven by a different driving force. Fluidization is caused by the action of pore fluids passing through a sandbody, liquefaction is caused by cyclic shear stresses and shear liquification results from the application of a body shear across a sandbody. In the natural environment it is unlikely that these processes will act independently and instead, in most natural liquification, two or more liquification processes will act together. The experiments described here demonstrate that: (1) the combination of liquification processes results in a greater diversity of liquification behaviour; (2) liquification by one process becomes easier as a second process becomes active; and (3) the liquification behaviour will change gradually as the part played by a second process increases. When studying the remobilization of sandbodies in the natural environment it is important to consider the effect of each of the liquification processes and also the interaction between processes. It is important not to limit the study to just one process. For example, if a sandbody is subject to a body shear, due to the body lying on a slope, it will be more susceptible to liquefaction during an earthquake than it would be if no body shear were present. If pore waters are seeping up through the sandbody at the same time the susceptibility of the sandbody to liquefaction will be greater still.

This paper has concentrated on the liquification behaviour of sand and how liquification can allow the sand to behave in an unusual fashion during remobilization. In natural systems, however, it must be borne in mind that the sandbody will not deform in isolation to the surrounding material. The rheology of the surrounding material and the nature of the stress field active at the time of deformation will also play an important part in controlling the behaviour of the sandbody during remobilization.

The author was funded by NERC during most of this work, and further funding was also provided by DSS. Much of the experimental equipment was purchased using a BP venture fund. The vibration liquification experiments were carried out under the supervision of Jan Alexander, and the other experiments under the supervision of Steve Sparks and Colin Wilson. I would like to thank them for their help and encouragement during this project. Thanks also to Gordon Kenneway for his help with experiments. I would like to thank those at Rolls-Royce (Bristol) who allowed me to use their vibration equipment, especially Phil Christie and Ian Jones. I am grateful to Candace Brook and Richard Dixon for their encouragement and assistance with this work. I would also like to thank Adrian Hartley, Jeremy Prosser, Mike Leeder and Ken Glennie whose comments helped turn what was a rambling mess into a publishable piece of work.

References

ALLEN, J. R. L. 1982. *Sedimentary structures: Their Character and Physical Basis, Vol II.*, Elsevier, Amsterdam.

—— 1985. *Principles of Physical Sedimentology.* George, Allen & Unwin.

ANKETELL, J. M., CEGLA, J. & DZULYNSKI, S. 1970. On the deformational structures in systems with reversed density gradients. *Rocznik Polskiego Towarzystwa Geologicznego Annales de la Societe Geologique de Pologne,* **40,** 3–30.

BARKER, G. C. & MEHTA, A. 1993. Size segregation mechanisms. *Nature,* **364,** 486–487.

BARRIGA, F. J. A. S., FYFE, W. S., LANDEFELD, L. A., MUNHA, J. & RIBEIRO, A. 1992. Mantle eduction: tectonic fluidization at depth. *Earth-Science Reviews.* **32,** 123–129.

BROOKE, C., TRIMBLE, T. J., MACKAY, T. A. 1995. Mounded shallow gas sands from the Quaternary of the North Sea-Analogues for the formation of sand mounds in deep water Tertiary sediments? *This volume.*

CAMPBELL, C. S. & BRENNEN, C. E. 1985. Computer simulations of granular shear flows. *Journal of Fluid Mechanics,* **151,** 167–188.

DAVIDSON, J. F. & HARRISON, D. 1971. *Fluidization.* Academic Press, London.

DIXON, R. J., SCHOFIELD, K., ANDERTON, R., REYNOLDS, A. D., ALEXANDER, R. W. S., WILLIAMS, M.C. & DAVIES, K. G. (1995) Sandstone diapirism and clastic intrusion in the Tertiary fans of the Bruce-Beryl Embayment, Quadrant 9, UKCS. *This volume.*

DZULYNSKI, S. 1966. Sedimentary structures resulting from convection-like pattern of motion. *Rocznik Polskiego Towarzystwa Geologicznego Annales de la Societe Geologique de Pologne,* **36,** 4–21.

INESON, J. R. 1985. Submarine glide blocks from the Lower Cretaceous of the Antartic Peninsula. *Sedimentology,* **32,** 659–670.

KUENEN, P. H. 1958. Experiments in geology. *Geological Society of Glasgow,* **23,** 1–28.

LAMBE, T. W. & WHITMAN, R. V. 1979. *Soil Mechanics, SI Version.* John Wiley & Sons, New York.

LOWE, D.R. 1976. Subaqueous liquefied and fluidized sediment flows and their deposits. *Sedimentology,* **23,** 285–308.

MALTMAN, A. 1994. Prelithification deformation. *In:* Hancock, P. (ed.) *Continental Deformation, Pergamon Press, Oxford. 143–158.*

MELOSH, H. J. 1979. Acoustic fluidization: A new Geologic process? *Journal of Geophysical Research,* **84,** 7513–7520.

NADKARNI, A. R. & PETERS, M. H. 1987. Analytical solutions to the granular temperature profiles and effective particle-phase viscosities for single-bubble motion in a fluidized-bed. *International Journal of Multiphase Flow,* **13** 493–510.

NICHOLS, R. J., SPARKS, R. S. J. & WILSON, C. J. N. 1994. Experimental studies of the fluidization of layered sediments and formation of fluid escape structures. *Sedimentology,* **41,** 233–253.

OWEN, G. 1987. Deformation processes in unconsolidated sands. *In:* JONES, M. E. & PRESTON, R.M.F., Deformation of Sediments and Sedimentary Rocks. *Geological Society, London, Special Publication*, **29**, 11–24.

RETTGER, R. E. 1935. Experiments on soft-rock deformation. *American Association of Petroleum Geologists Bulletin*, **19**, 271–292.

RICHARDSON, J. F. 1971. Incipient fluidization and particulate systems. *In:* DAVIDSON, J. F. & HARRISON, D. (eds.) *Fluidization*, Academic Press, London.

ROSATO, A. D., LAN, Y. & WANG, D. T. 1991. Vibratory particle size sorting in multi-component systems. *Powder Technology*, **66**, 149–160.

SEED, H. B. 1979. Soil liquefaction and cyclic modility evaluation for level ground during earthquakes. *American Society of Civil Engineers GT2*, **105**, 201–225.

—— & LEE, K. L. 1966. Liquefaction of saturated sands during cyclic loading. *Proceedings of the American Society of Civil Engineers, Journal of Soil Mechanics and Foundations Division*, **92**, 105–134.

SELLEY, R. C. & SHEARMAN, D. J. 1962. The experimental production of sedimentary structures in quick-sands. *Proceedings of the Geological Society, London*, **1599**, 101–102

SMALLEY, I. 1976. Factors relating to the landslide process in Canadian quickclays. *Earth Surface Processes*, **1**, 163–172.

TSUJI, T. & MIVATA, Y. 1987. Fluidization and liquefaction of sand beds – Experimental study and examples from Nich inan Group. *Journal of Geological Society of Japan*, **11**, 791–808.

WALTON, O. R. & BRAUN, R. L. 1986. Viscosity, granular-temperature, and stress calculations for shearing assemblies of inelastic, frictional disks. *Journal of Rheology*, **30**, 949–980.

WILLS, C. J. & MANSON, M. W. 1990. Liquefaction at Soda Lake: Effects of the Chittenden earthquake swarm of April 18, 1990. Santa Cruz County, California. *California Geology*, **October**, 225–232

WILSON, C. J. N. 1980. The role of fluidization in the emplacement of pyroclastic flows: An experimental approach. *Journal of Volcanology and Geothermal Research*, **8**, 231–249.

—— 1984. The role of fluidization in the emplacement of pyroclastic flows, 2: Experimental results and interpretation. *Journal of Volcanology and Geothermal Research*, **20**, 55–84.

YOUD, T. L. & PERKINS, D. M. 1978. Mapping liquefaction induced ground failure potential. *Journal of Geotechnical Engineering Division, American Society of Civil Engineers GT4, 13659*, **104**, 433.

ZENZ, F. A. 1971. Regimes of fluidization behaviour. *In:* DAVIDSON, J. F. & HARRISON, D. (eds.) *Fluidization*, Academic Press, London, 1–23.

Sandstone diapirism and clastic intrusion in the Tertiary Submarine fans of the Bruce–Beryl Embayment, Quadrant 9, UKCS

R. J. DIXON,[1] K. SCHOFIELD,[2] R. ANDERTON,[1] A. D. REYNOLDS,[3] R. W. S. ALEXANDER,[1] M. C. WILLIAMS[4] & K. G. DAVIES[1]

[1] *BP Exploration, Farburn Industrial Estate, Aberdeen, AB2 0PB, UK*

[2] *BP Exploration, Sage Plaza One, 5151 San Felipe, PO Box 4587, Houston, Texas 77210, USA*

[3] *BP Exploration, 4/5 Long Walk, Stockley Business Park, Uxbridge, Middlesex, UB11 1BP, UK*

[4] *BP Exploration Company (Columbia) Ltd, Carrera 9a 22-09, A.A. 59824, Santafe de Bogota, Columbia*

All correspondence to: Dr. R. J. Dixon, BP Exploration, Farburn Industrial Estate, Aberdeen, AB2 0PB

Abstract: Submarine fans of Late Palaeocene and Early Eocene age form important hydrocarbon reservoirs in the Bruce–Beryl Embayment, northern North Sea. The Early Eocene fans are the main reservoirs in the Forth–Gryphon oilfields and in the giant Frigg gasfield. Significant oil discoveries have also been made in Late Palaeocene fans. Forth and Gryphon lie on the flanks of the Crawford Anticline, a drape structure that developed during the Palaeocene above the crest of a Mesozoic tilted fault block. The Early Eocene fans pinchout against the flanks of the anticline implying continued growth of the structure throughout the Eocene. Growth was accompanied by the development of major gravity slides that detached in a sequence of altered, basaltic tephras at the base of the Eocene sequence. Seismic-scale, post-depositional deformation (sandstone diapirism and the intrusion of clastic sills and dykes) connected with this sliding dramatically modified the original depositional geometries of the fans. A detailed account of the deformation features, illustrated with core, wireline log and three-dimensional seismic data is presented together with a discussion of their exploration/appraisal significance.

The Bruce–Beryl Embayment is a major North Sea hydrocarbon province located on the UKCS between the North Viking Graben and the East Shetland Platform (Fig. 1). Exploration in the Bruce–Beryl Embayment began in the late 1960s and by the mid 1970s a cluster of giant petroleum fields had been discovered, including the Frigg gasfield (Brewster 1991), the Beryl oilfield (Knutson & Munroe 1991) and the Bruce gasfield (Beckly *et al.* 1993). The Bruce and Beryl reservoirs (Fig. 2) are in deeply buried (3000–4000 m) Triassic and Jurassic strata and trapped in complex tilted fault blocks. This petroleum play formed the main exploration focus in the Bruce–Beryl Embayment throughout the late 1970s and early 1980s. Although major discoveries such as the Frigg gasfield (7Tcf Reserves) had been made in the Tertiary successions, it was the discovery of the Forth (BP) and Gryphon (Kerr Mcgee) oilfields in the late 1980s that shifted the main focus of exploration to the shallower (1600–2000 m) Tertiary play. Forth and Gryphon are located in the southern part of the Bruce–Beryl Embayment (Fig. 2) and comprise Late Palaeocene and Early Eocene sandstone reservoirs with complex structural and stratigraphic trapping geometries. Since the discovery of Forth and Gryphon a number of other Tertiary discoveries have been made in the Bruce–Beryl Embayment, e.g. 9/19–8 (Timbrell 1993) and 9/14b–T1 (Dixon & Pearce 1995; Fig. 2).

The Forth and Gryphon oilfields comprise a number of discrete oil pools within predominantly massive, fine-grained sandstones. Where penetrated the sands have excellent reservoir qualities (35% porosity, 12 D permeability) and are locally over 100 m thick with net to gross values of 100%. Laterally, however, sand thickness varies dramatically. Analysis of regional

From Hartley, A. J. & Prosser, D. J. (eds), 1995, *Characterization of Deep Marine Clastic Systems*, Geological Society Special Publication No. 94, pp. 77–94.

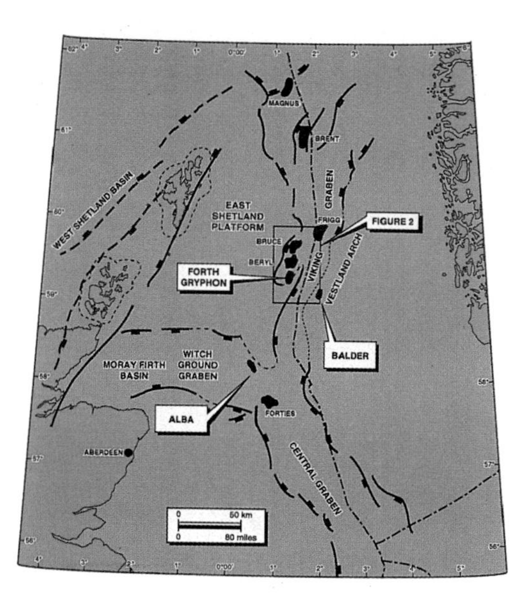

Fig. 1. Location map, showing Bruce–Beryl Embayment (inset, Fig. 2), Alba Field and Balder Field.

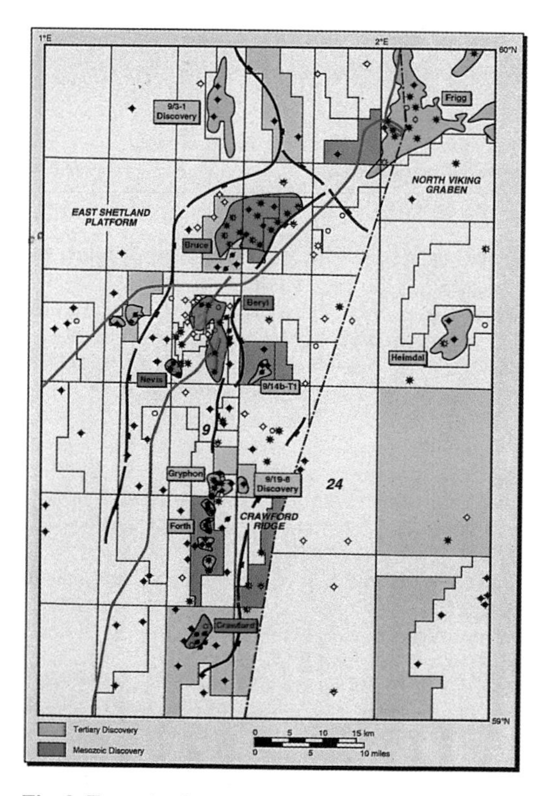

Fig. 2. Tectonic elements and major petroleum fields of the Bruce–Beryl Embayment.

seismic data suggests that the Early Eocene sandstones were deposited seaward of a Late Palaeocene offlap break suggesting that deposition took place in relatively deep water. A number of depositional models have been proposed for the Tertiary sequences (Alexander *et al.* 1992; Newman *et al.* 1993; Timbrell 1993; Dixon & Pearce 1995). All of these models are in agreement with regard to the deep marine setting of the strata, however, there is some debate as to the range of sedimentary processes involved and also as to the extent to which post-depositional processes have modified the original (depositional) sandbody geometries.

In order to optimize appraisal drilling it was important to understand and predict lateral thickness variations within the sandstone sequences. It was also important (for reserves and reservoir management purposes) to elucidate sandbody geometries in a zone of sandstone and mudstone which overlies many of the massive sandstones. In addition, certain features of the sandbodies, specifically their dramatically mounded geometry, steeply dipping (up to 18°) flanks (observed on seismic data), and association with numerous sandstone dykes and sills (observed in core) were not easily explained by reference to the depositional characteristics of turbidite systems. It is our contention that post-depositional processes (detachment faulting, sandstone diapirism and clastic intrusion) have locally substantially modified the original depositional geometries of the reservoir sandstones.

In the following sections the stratigraphy of the North Sea Tertiary and the subregional evolution of the Bruce–Beryl Embayment are discussed. This allows a depositional model for the reservoir sandstones to be outlined and the stage to be set for a detailed account of the deformation features, illustrated with core, log and three-dimensional seismic data. Geological models for the deformation are developed (with reference to other published North Sea examples), and the exploration and appraisal significance of post-depositional deformation is discussed.

Regional setting and sequence stratigraphy

BP have divided the Early Tertiary of the North Sea Basin into a number of basinwide sequences. These are broadly equivalent to the lithostratigraphic units defined by Deegan & Scull (1977) and are from oldest to youngest, T10 (Ekofisk), T20 (Maureen), T30 (Andrew),

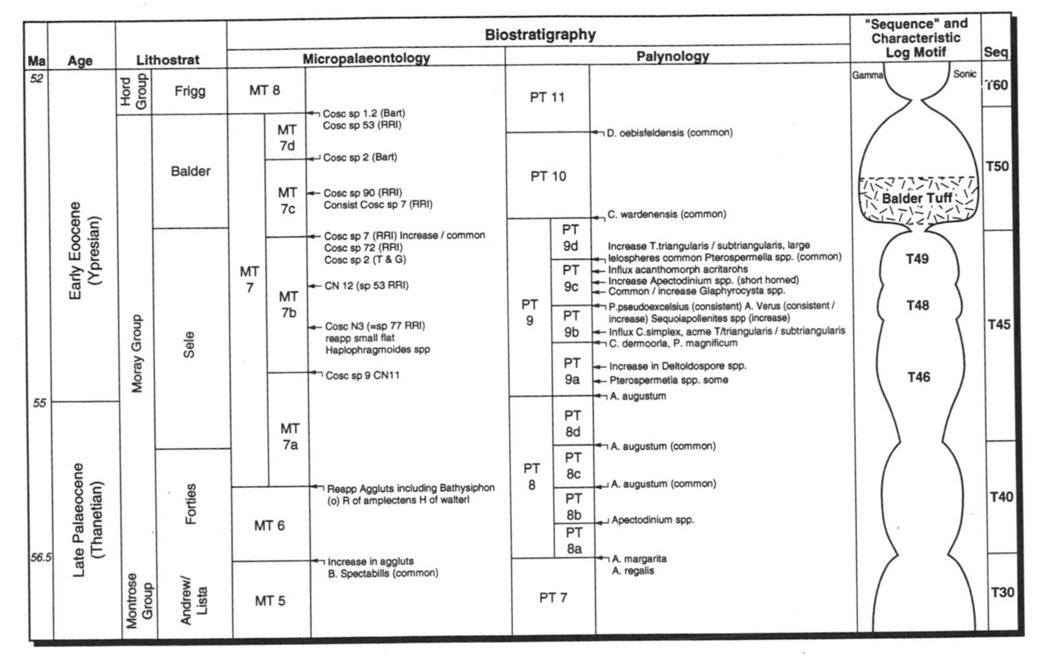

Fig. 3. Late Palaeocene to Early Eocene sequence-stratigraphic framework.

T40 (Forties), T45 (Sele), T50 (Balder) and T60 (Frigg). These are all genetic stratigraphic sequences and their boundaries are picked on geophysical logs as gamma-ray maxima (believed to correspond to times of maximum flooding or transgression). The sequence-stratigraphic framework used in the Forth–Gryphon area is given in Fig. 3 and was developed using good quality core, log and three-dimensional seismic data (Dixon & Pearce 1995). A distinctive horizon composed of numerous basaltic, ash-fall tephra layers was also identified (the 'Balder Tuff', Phase 2 Pyroclastic's of Knox & Morton 1988). Sedimentological and petrophysical analysis of the tephra sequence revealed that it had a consistent internal stratigraphy across the Forth–Gryphon area (Hatton *et al.* 1992) and thus represented a significant marker horizon.

The T50 fans were deposited at or near the toe of a complex slope that developed during the Late Palaeocene (T40–45; Fig. 3). The T40 and T45 sequences were fed by a sand-prone, shelf-edge delta system that built eastwards into the Bruce–Beryl Embayment. Delta progradation was punctuated by at least two downward shifts in coastal onlap, that produced significant basin margin relief in the form of incised valleys and lowstand terraces (Dixon & Pearce 1995; Fig. 4a & b). By the end of T45 times, however, the delta had started to backstep westwards in response to a second-order rise in sea-level (Milton *et al.* 1990). The flooded basin margin was subsequently draped by hemipelagic muds and ash-fall tephras of the overlying T50 sequence (Fig. 3). Despite this drape, relief on the drowned delta complex was still significant, so that during progradational phases, T50 shelf-deltas locally fed directly into the heads of the relict incised valleys (Fig. 4c). These valleys (now submarine canyons) funnelled sediment into the basin to accumulate as fans at the base of the old T45 delta-slope (Fig. 4c). In addition to these major canyon systems sediment was also supplied to the basin where retrogressive gravity slides initiated on the old T45 delta-slope tapped back into the sandy T50 shelf-deltas.

As a result of extensive winnowing by shelfal processes (storms, waves, tides) and by recycling of previous T45 topsets the T50 shelf-deltas were relatively sand-prone and the fans correspondingly sand-rich. With continued sea-level rise the T50 shelf-deltas moved further west and the fans retreated, first on to the lowstand terraces and eventually back along their feeder canyons (Dixon & Pearce 1995). The Forth and Gryphon fans were subsequently blanketed by dark, laminated mudstones deposited relatively quickly from plumes of fine, suspended detritus sourced from the deltas further west. This is supported by the general westward thickening of the T50 sequence (Timbrell 1993). The top of T50 is represented by a condensed section

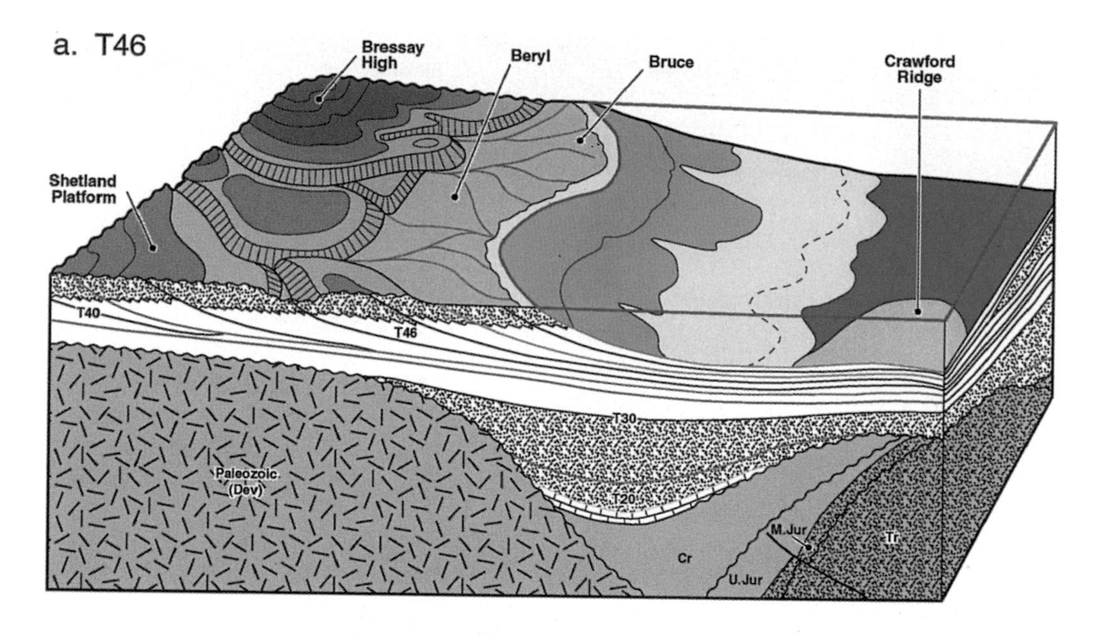

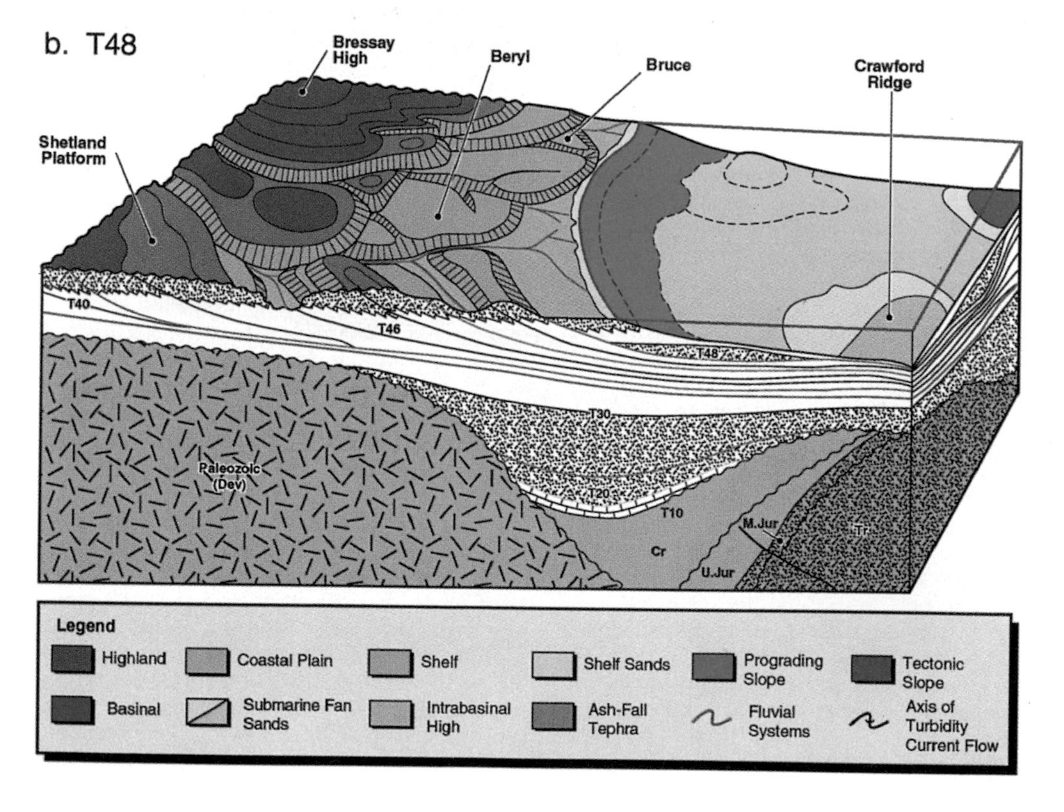

Fig. 4. Geological models illustrating the sub-regional setting and stratigraphic evolution of the Balder (T50) submarine fans. **(a)** T46 sequence; **(b)** T48 sequence; **(c)** T50 sequence; **(d)** T60 sequence.

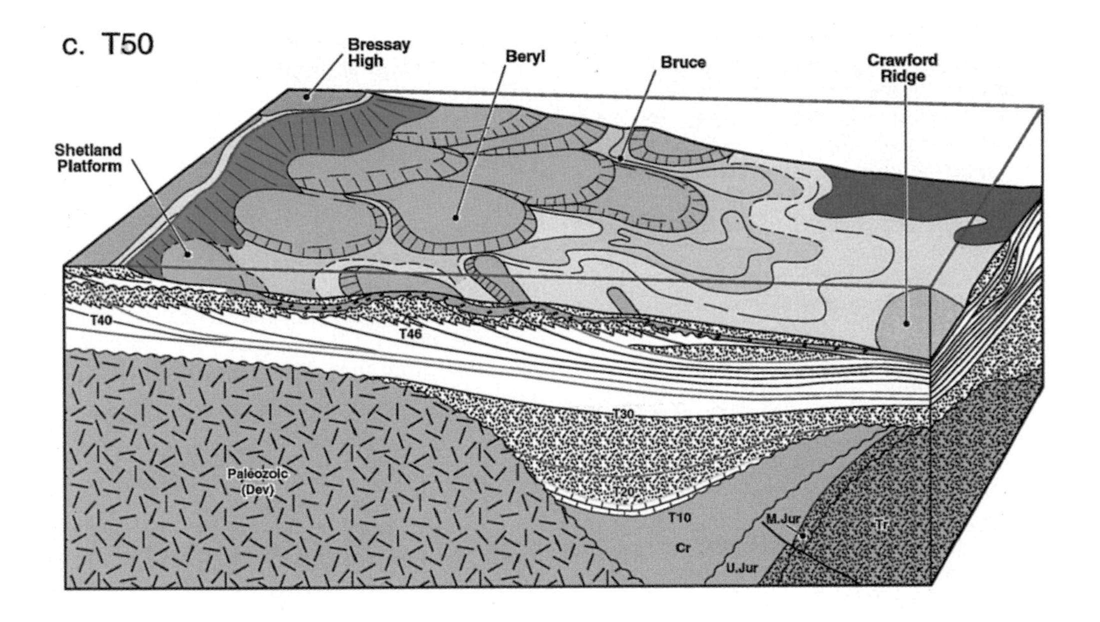

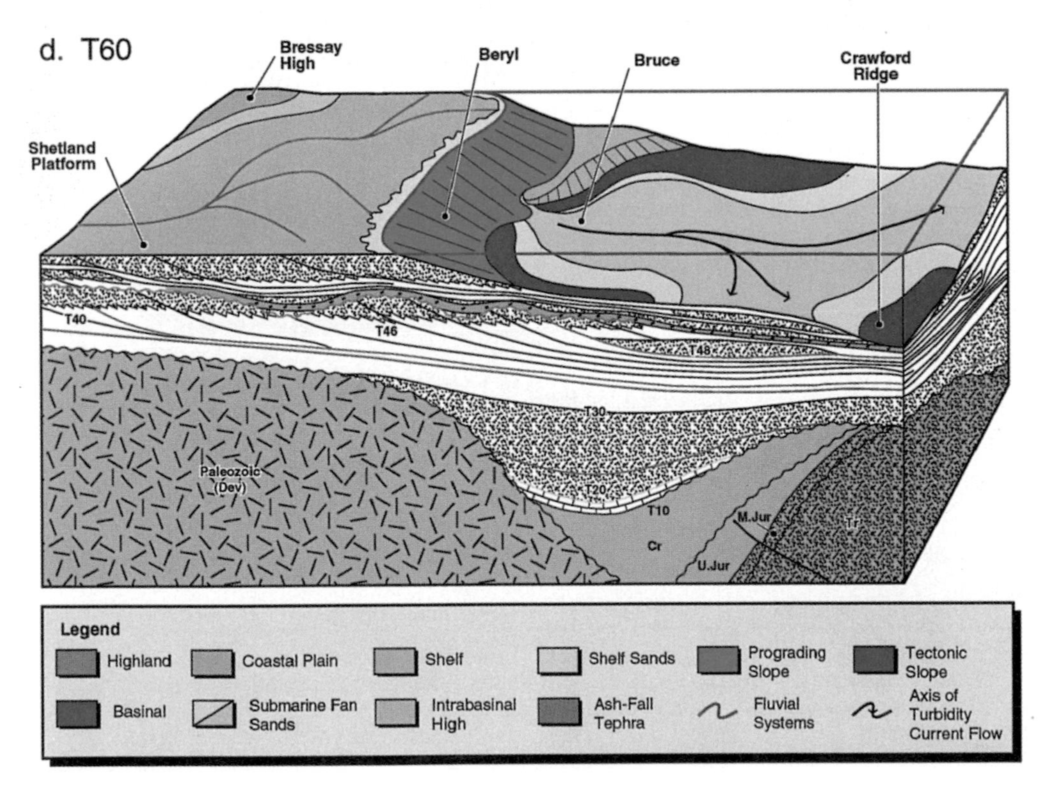

across the whole of the North Sea Basin and has been interpreted as a (second order) maximum flooding surface. Subregional analysis has thus shown that the T50 fans were deposited at or near the base of an east-facing slope and that they form part of a second-order transgressive systems tract.

The overlying T60 sequence was marked by the progradation of a sandy, shelf-delta system into the Bruce–Beryl Embayment from the southwest. The distribution of coeval submarine fan sandstones suggests that the old T50 submarine canyon systems were still actively supplying sediment into the basin early in T60 times (Fig. 4d), but not during the later part of T60 when most appear to have been infilled. A major base-level fall can be inferred at or near top T60 on well and seismic data (Dixon & Pearce 1995). This led to widespread incision of the shelf-delta and the deposition of a major, sand-rich submarine fan system in the basin. This submarine fan forms the main reservoir of the giant Frigg gasfield (Brewster 1991).

Regional studies also suggest the presence of a NE–SW-trending bathymetric feature located above the Mesozoic Crawford Ridge (Fig. 2). The deep structure of the Crawford Ridge is a westerly tilted, NE–SW-trending fault block, downthrown to the east and plunging to the north. A number of NW-SE-trending faults that offset the main NE-SW trend are also mapped (Newman *et al.* 1993). At shallower (Tertiary) levels, however, the structure is a broad, northerly plunging anticline that developed as a result of compactional drape over the deeper structure. This anticline acted as a barrier to the basinward transport of sediment, so that in the Forth area turbiditic sandstones were ponded between it and the base of the T45 delta slope (Fig. 4c). Further north in the Gryphon area the anticline was more subdued and turbidity currents were able to flow eastward into the Viking Graben (Dixon & Pearce 1995).

Lithofacies and depositional models

Three main lithofacies are represented in the T50 strata of the Forth–Gryphon area: sandstone (lithofacies L1); tuff (lithofacies L2); and mudstone (lithofacies L3).

Lithofacies L1

Lithofacies L1 comprises poorly lithified, mostly structureless, moderately well to well sorted, fine-grained, locally micaceous sandstone that occasionally contains small, subrounded mudstone intraclasts. In places, dish structures are present, as are bedding planes, though the latter are rare. Interbeds of lithofacies L2 suggest that beds in lithofacies L1 are mostly >1 m in thickness. In the context of regional seismic evidence for a basinal setting, Lithofacies L1 is considered to reflect deposition from high concentration (high density) turbidity currents. The dish structures are thus interpreted as reflecting syn-depositional dewatering, while the structureless sands may be the result of very rapid sediment deposition.

Lithofacies L2

Lithofacies L2 comprises graded, basaltic, lithic and vitric tuffs. Individual tuff beds vary from 2 to 25 cm in thickness and typically have sharp, planar bases and tops. Occasionally tuff beds have irregular, bulbous (loaded) bases and bioturbated tops (*Helminthopsis* and *Chondrites*), but more often than not display well-developed grading; from grey, fine-grained lithic or vitric tuff upward into extremely fine, blue grey tuffaceous claystone. Slumped tuff beds have also been noted (Alexander *et al.* 1992). The tuffs are interpreted as waterlain, ash-fall tephras and occur in a number of discrete 'packages' (each containing between 20 and 30 tephra layers >2 cm in thickness) separated by laminated mudstones with a lower density of tuff layers. The bulk of the tuffaceous material is contained in the lower part of the T50 sequence though tuffs are present throughout T50 and also occur sporadically in both the T45 and T60 sequences. Petrophysical analysis (Hatton *et al.* 1992) has revealed that the T50 tephra sequence has a consistent internal stratigraphy over most of the eastern Bruce–Beryl Embayment, implying extremely good lateral continuity of the individual tuff 'packages'. The latter and the well-developed grading shown by individual tephra layers suggest that ash settled slowly through the water column to accumulate on a quiet, stagnant basin floor. The existence of local slopes on the basin floor is suggested by the occasional presence of slumped tephra layers. The general lack of bioturbation indicates that the basinal sediments were essentially anaerobic or dysaerobic, an interpretation that is consistent with the nature of the lithofacies L3 mudstones.

Lithofacies L3

Lithofacies L3 comprises blocky, poorly laminated, though occasionally well laminated, grey to dark bluish grey, mudstones. The blocky mudstones are often characterized by an extremely fine mottling that probably reflects bioturbation, an interpretation supported by the local occurrence of pyritized ?*Chondrites* within this facies. Slump structures are also present (*cf.* lithofacies L2). Lithofacies L3 locally contains numerous sub-rounded aggregates of granular, bioclastic debris (mainly comprising echinoid

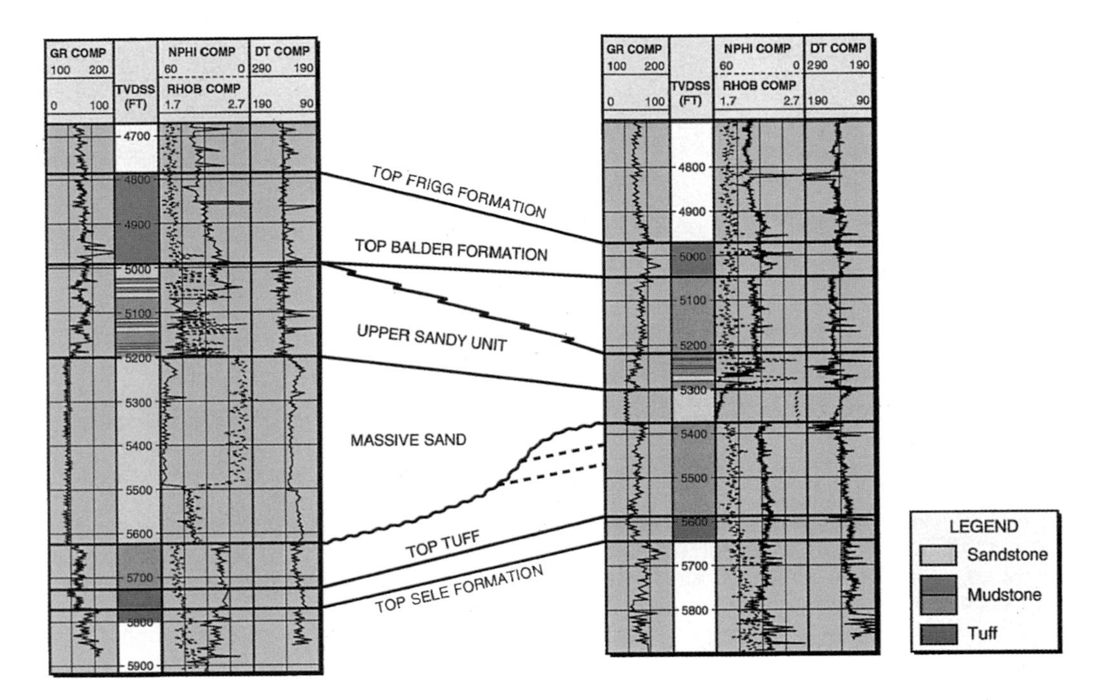

Fig. 5. Well correlation panel between closely-spaced wells in the Forth Field Area.

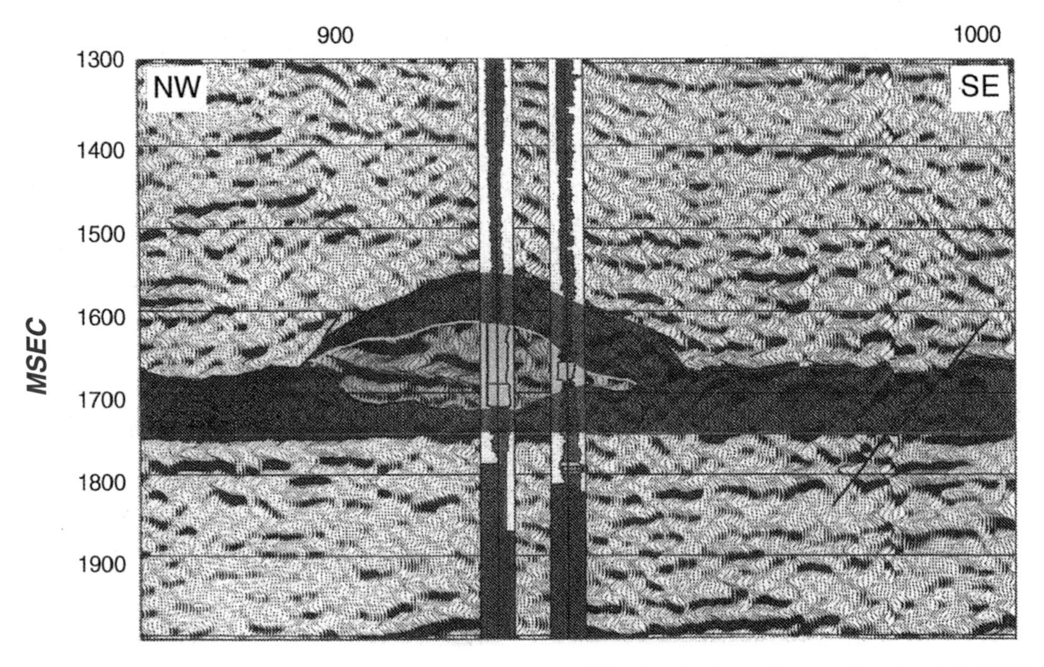

Fig. 6. Well traverse from three-dimensional seismic survey. The data has been 'flattened' on top Sele (T45) to illustrate the discordant nature of the basal contact of the sandbody.

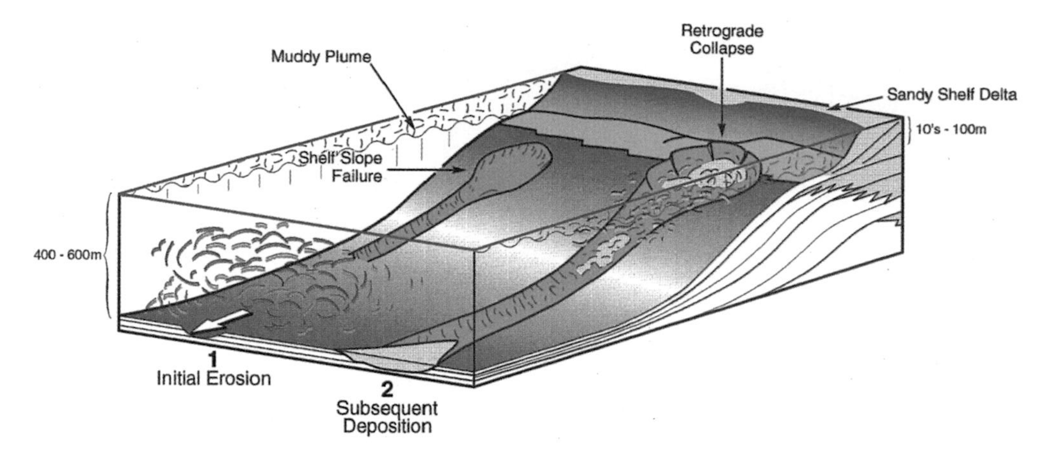

Fig. 7. A depositional model for the Balder (T50) sandstones.

plates and spines) that vary in size from 2 to 15 cm in diameter and are interpreted here as coprolites. Deposition of fines from delta-fed, suspended mud plumes into a current-free, stagnant, basin-floor setting is suggested by the low density and low diversity of the bioturbation and by the preservation of coprolites (implying the lack of a scavenging benthos).

Depositional model

In the Forth area, sandstone deposition took place between the base of the T45 delta-slope and the western flank of the Crawford Anticline. Sandstones occur predominantly as laterally restricted bodies with high net to gross (0.95 – 1.00). Both seismic and well data suggest that many (but not all) of the sandbodies have strongly discordant bases. An example is illustrated in Fig. 5, where a dramatic thickness change occurs between two closely spaced wells and seismic data suggests that the sandbody was deposited within a steep-sided, seafloor depression (Fig. 6). Missing section at the base of the sandstone sequence has been confirmed by a careful analysis of the T50 tephra sequence using the technique outlined by Hatton *et al.* (1992), though whether section has been removed by erosion or by syn-sedimentary faulting is not always clear.

In addition to demonstrating missing section, analysis of the tephra sequence has also allowed a high degree of stratigraphic resolution and revealed that the major sandbodies are all of slightly different ages, an observation also supported by biostratigraphic data (Alexander *et al.* 1992). The absence of interbedded fines (lithofacies L2 and L3) within the sandbodies suggests that seafloor relief was probably rapidly filled by a series of sandy turbidity currents. A model whereby initial erosion and/or mass wasting of the old T45 delta-slope generated scours and/or slide scars that were subsequently filled when ensuing retrogressive slides tapped into the sandy shelf-delta (Fig. 7) is consistent with the available data.

The T50 sandstones are thus interpreted as the deposits of high-density turbidity currents that were introduced into a tranquil, stagnant basin via submarine canyons and retrogressive slide scars. The uniform nature and fine grain size of the turbidites (Dixon & Pearce 1995) suggests that the canyons and scars were fed directly from a large, well-sorted sediment source, that is most likely to have been a coeval wave or storm-dominated, sandy shelf-delta. There is no seismic evidence that the turbidity currents were constrained within leveed channels (or that sands were subsequently deposited within such channels; cf. Timbrell 1993). No wells in the area have cored a recognizable levee facies.

Post-depositional deformation

T50 sandstone reservoirs were deposited as a series of high net to gross, uniformly fine-grained sandstone bodies that filled a suite of seafloor depressions. The dramatic mounded geometries of the T50 sequence and the presence of sandstone intrusions suggest that the depositional geometries were subsequently affected by a variety of post-depositional processes. In the following section a detailed account of the deformation features is presented (based on well data and a three-dimensional seismic survey) and geological models are proposed for their development.

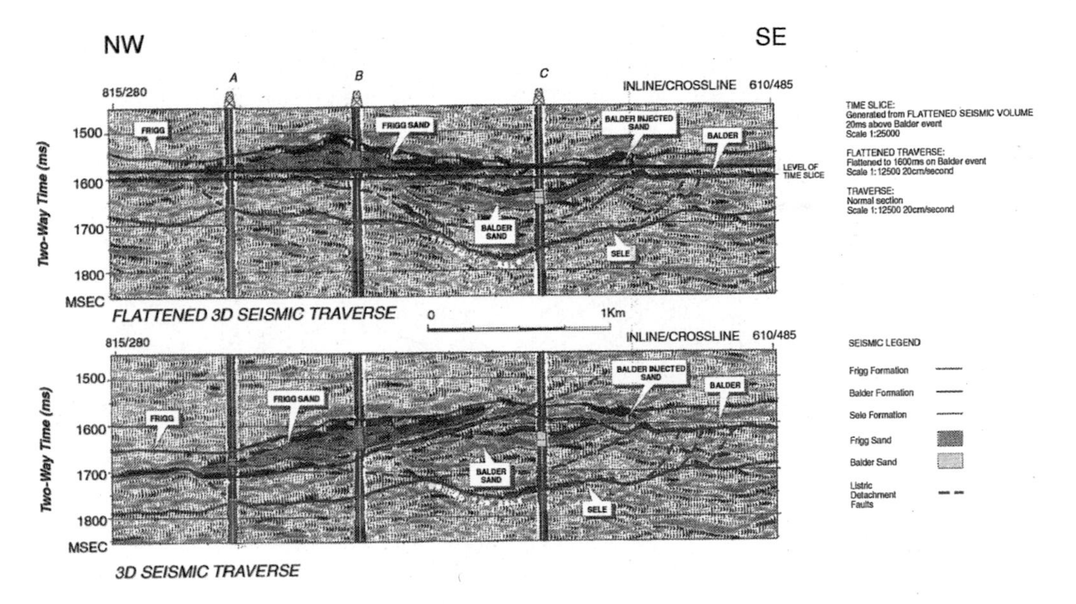

Fig. 8. Three-dimensional seismic well traverse illustrating listric faults and dykes. See Fig. 9 for location.

Seismic evidence

Listric normal faults that detach at the level of the Balder Tuff are the most obvious seismic manifestation of the deformation (Fig. 8). Figure 9 shows a time-slice through the three-dimensional dataset that illustrates the arcuate, plan view of a listric detachment and suggests that they have a cuspate geometry. The slide block appears to be being shed off the western flank of the Crawford Anticline attesting to its importance as a positive feature at this time. That this particular slide scar formed a seafloor depression is suggested by its partial infilling by a T60 lithofacies L1 sandbody. The 'toes' and lateral ramp elements of the cuspate detachments have also been mapped and display evidence of compression in the form of folds and (in places) reverse faults.

Similar relationships have also been illustrated from the Gryphon oilfield where seismic mapping suggests that the steep flanks of the main reservoir sandbodies are formed by major listric normal faults (Newman *et al.* 1993). In core, the upper contacts of lithofacies L1 sandbodies are more often than not markedly discordant (faulted; cf. Newman *et al.* 1993) and are often accompanied by the development of braided or anastomozing arrays of shear fractures within the sandstone. In thin section such fractures are characterized by grain-size reduction (granulation) offering strong support for a faulting model.

Well evidence

Large-scale sandstone intrusion may be inferred on some seismic sections when steep, high-amplitude (bright) reflectors cut across the top Balder (T50) reflector (Fig. 8). These high-amplitude reflectors are discontinuous, but often arcuate in plan (Fig. 9), suggesting that they are intruded along cuspate detachment faults. Intrusion of sandstone dykes along fault planes has been recorded in the literature (Taylor 1982). Sandstone intrusions, however, are best seen in core samples (the role of post-depositional deformation in the Forth–Gryphon area was first realized when oil-stained, sandstone dykes were described from core).

Sandstone dykes

Dykes have been observed in core from vertical wells, though their relationships are best displayed in high-angle wells. Figure 10 illustrates three short, cored sections from such a well. Structural dip in the area penetrated by this well is low (2°), the deviated nature (drilled at 45° from vertical) of the well bore allowing the clear demonstration of the dyke's, discordant contacts. All of the dykes are composed of well sorted, fine-grained sandstone petrographically similar to lithofacies L1. The host mudstones are typically only slightly deformed by the intrusions and fragments of the host sediment are only rarely present within the dykes.

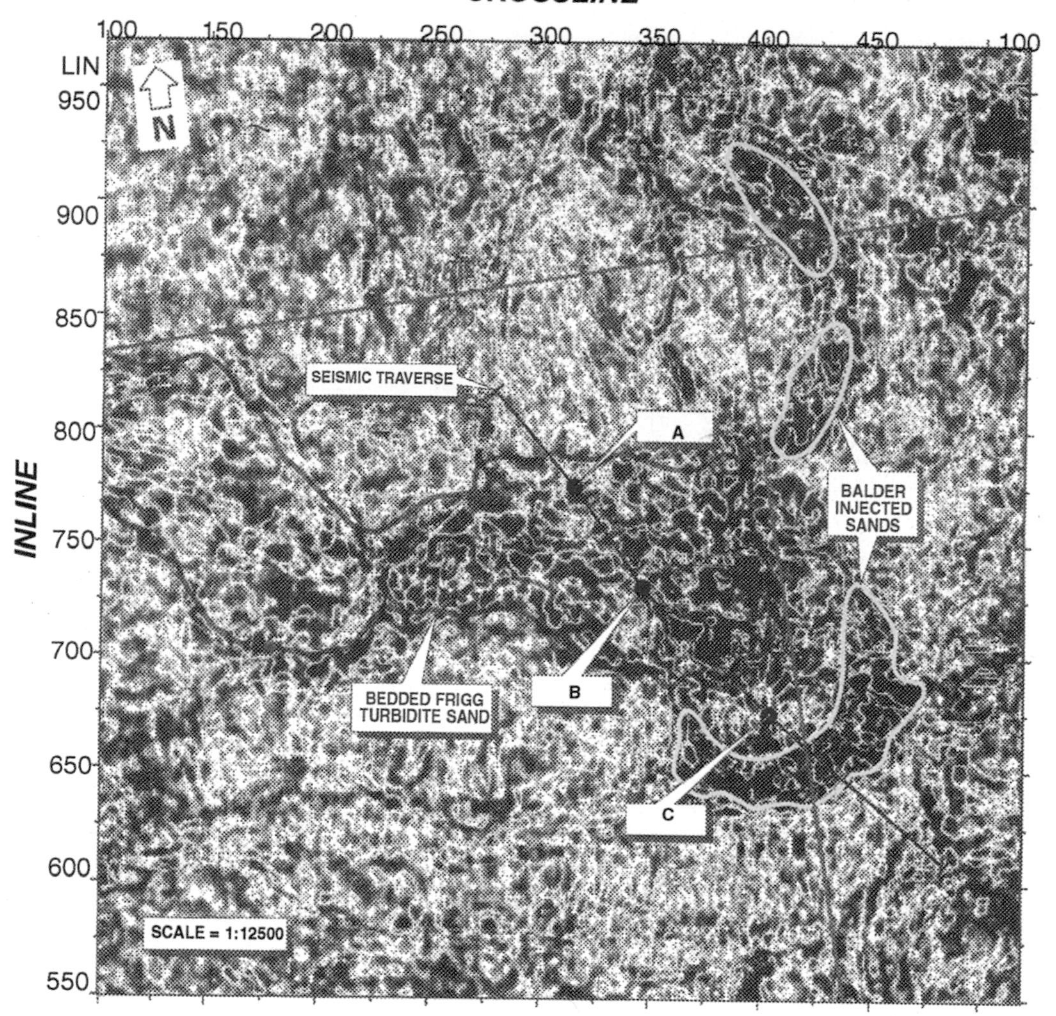

Fig. 9. Timeslice from the three-dimensional seismic data illustrating interpreted listric faults (cuspate features) and dykes.

Where such inclusions are present they are mostly angular to subangular in shape and occur in the centre of the dykes; rounded mudstone inclusions have also been observed. Dykes emanating from lithofacies L1 sandbodies have only rarely been observed in core, but from the examples known both upward and downward injection can be demonstrated.

The dykes illustrated in Fig. 10a & c are approximately normal to bedding, whereas the dyke system in Fig. 10b cuts bedding at a more oblique angle and individual dyke elements are arranged en-echelon. All of the dykes observed in the Forth–Gryphon area have smooth, planar walls characterized by asymmetric 'crenulations'. The crenulations are formed by the intersection of a conjugate set of shear fractures with the dyke wall (Fig. 10c). In the dyke illustrated in Fig. 10a one set of shear fractures is dominant, the conjugate set being only weakly developed. Mudstone inclusions are also dissected by the conjugate shear fractures and clear displacement along the latter may be demonstrated at the margins of the inclusions.

Although dykes have only rarely been observed emanating from Lithofacies L1 sandbodies, the

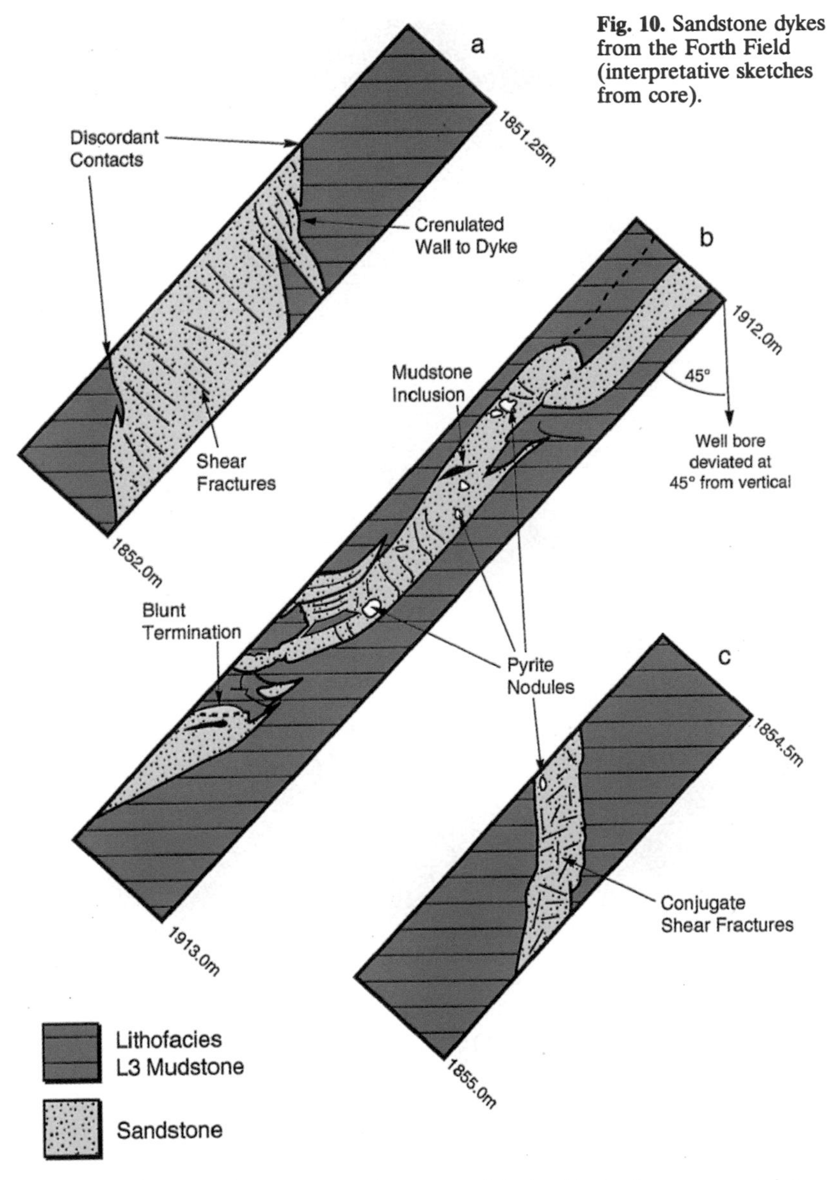

Fig. 10. Sandstone dykes from the Forth Field (interpretative sketches from core).

petrographic similarity of the sandstones strongly suggests that such sandstone beds are the source of the sandstone dykes. The smooth nature of the dyke walls and the general angularity of mudstone inclusions suggest that the host sediments were lithified (or at least cohesive) at the time of intrusion. The general absence of inclusions suggests that little erosion of the host sediment took place during emplacement and the concentration of such inclusions (when present) in the central portions of the dykes suggests that emplacement was by laminar flow.

The smooth nature of the dyke walls, the lack of disturbance in the host sediments and the general lack of inclusions suggests intrusion into a pre-existing set of open or dilated joints or fractures. Because conjugate shear fractures displace the dyke walls and dissect mudstone inclusions it seems likely that they are relatively late features, developed after consolidation of the dykes. The oriention of the fractures suggests they are compressional features that developed during burial as differential compaction took place between the host mudstones and the less compressible dyke. Slight disturbance of the host sediments in Fig. 10b is probably also attributable to compaction around the blunt terminations of en-echelon dyke segments.

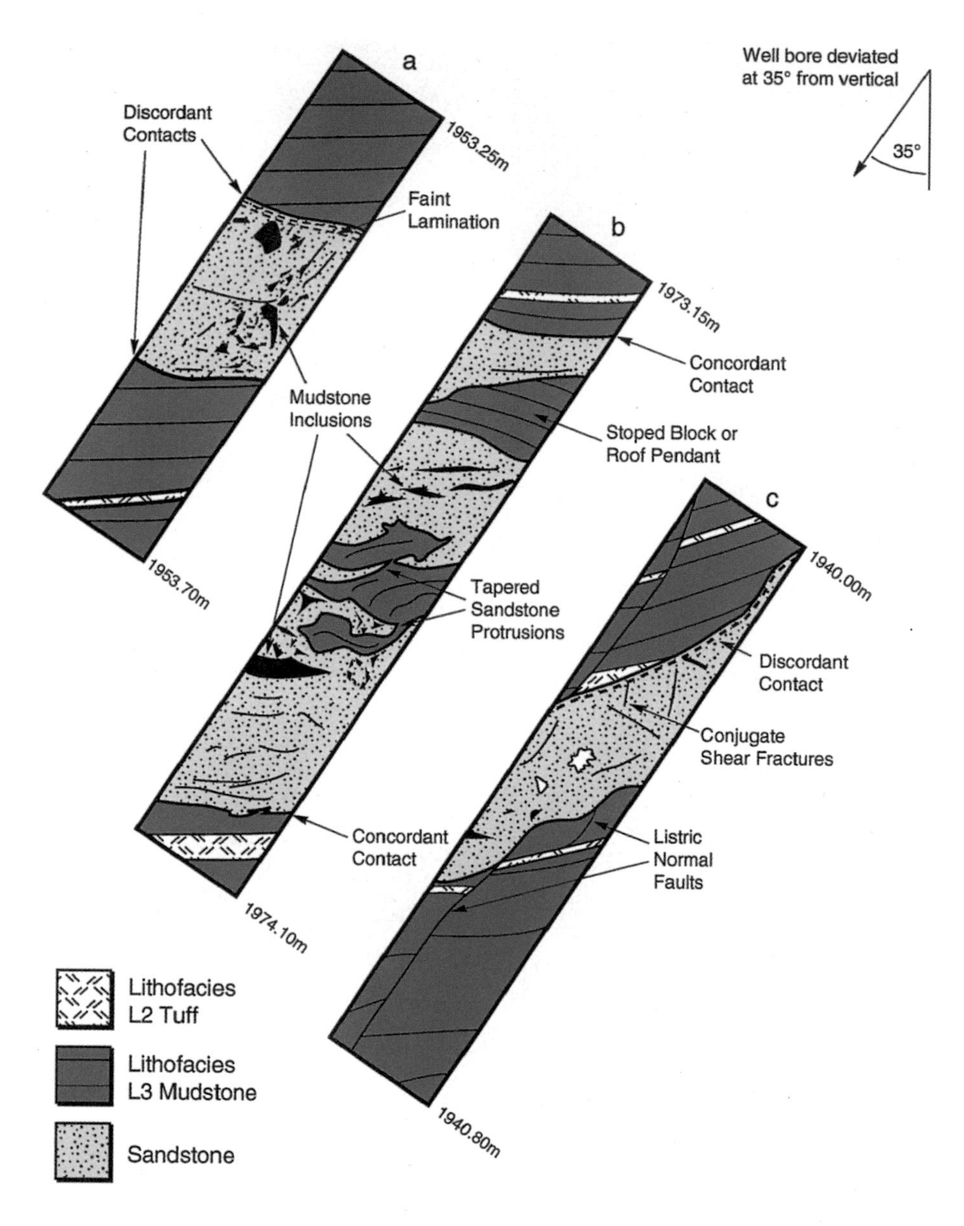

Fig. 11. Sandstone sills from the Forth Field (interpretative sketches from core).

Sandstone sills

Although unambiguous sandstone sills have been identified in core (e.g. Fig. 11a), it is often difficult to confidently discriminate between them and thick-bedded turbidite sandstones, particularly when the latter have undergone *in situ*, post-depositional liquefaction. The upper and lower contacts of the sandstone 'bed' shown in Fig. 11a are sharp, markedly discordant to bedding and the sandstone is clearly a transgressive sill. Likewise, the sandstone illustrated in Fig. 11c, though in this example the situation is complicated by faulting and the sill 'steps' up-stratigraphy across the fault plane. Both of these sills display internal structure, notably a slightly undulose lamination that is essentially parallel to the sill walls. Angular mudstone inclusions are also present. These are apparently oriented parallel to the sill walls in Fig. 11c, but are approximately normal to the sill walls in Fig. 11a. Conjugate shear fractures similar to those described from the sandstone dykes (Fig. 11c) are also present in some sills (Fig. 11c).

Figure 11b shows a more ambiguous sandstone 'bed', where the upper contact of the sandstone is concordant with bedding in the overlying mudstones and only slight discordance is observed at its base. A large slab of laminated mudstone occurs in the uppermost part of the 'bed' apparently 'floating' within the sandstone; similar large mudstone inclusions occur in the centre of the sandbody presenting an 'imbricated' appearance. A number of tapered, sandstone protrusions cut these inclusions and are clearly discordant. Smaller tabular mudstone inclusions are also present and are mostly oriented parallel to the sandbody contacts. Large, angular mudstone clasts have not been observed in any of the cored lithofacies L1 sandbodies, but small rounded mudstone clasts are sometimes present. A much more likely origin for the large slabs illustrated in Fig. 11b is that they are stoped blocks or pendants derived from the overlying mudstone. This hypothesis is supported by the apparent rotation of lamination within the large block at the top of the sandbody (Fig. 11b). Similar detached slabs have been described from outcropping examples of sandstone sills (Hiscott 1979; Archer 1984). Tabular 'offshoots' or dykes that propagate (at angles of *c.* 45°) and link sandstone sills at different stratigraphic levels have also been described (Hiscott 1979).

The intervening host sediment is thus broken-up into a series of lozenge-shaped slabs separated by discordant, intrusive sheets, a relationship that resembles that shown in Fig. 11b. In the outcrop examples illustrated by Hiscott (1979) and Archer (1984) the sills and their 'offshoots' are also characterized by aligned host sediment inclusions and blunt terminations, features also characteristic of the sandbody in Fig. 11b, it seems likely therefore that the latter is in fact a composite intrusion, comprising two sills linked by tabular 'offshoots' or dykes.

As in the sandstone dykes, the angular nature of the mudstone inclusions and the sharp nature of the sill contacts suggest that the host sediments were lithified or cohesive at the time of intrusion. Lamination within the sills and the local oriention of tabular mudstone inclusions suggest that the emplacement mechanism of both sills and dykes was similar, i.e. laminar flow into a pre-existing set of dilated joints (bedding planes in the case of the sills). The presence of dyke-like 'offshoots' from some sills suggests that the sills and dykes formed at roughly the same time.

Complex intrusions

Further useful information concerning the emplacement mechanism of the sandstone intrusions may be deduced from so-called complex intrusions. In the Forth–Gryphon area these are thick (tens of metres) zones of strongly disturbed sandstone and mudstone that often overlie the thick lithofacies L1 sandbodies (Alexander *et al.* 1992). Sandstone sills and dykes form major elements of such zones, but are typically accompanied by more complex intrusive geometries. Unlike the sills and dykes (which do not greatly disturb the host sediments) the intrusive complexes often contain strongly deformed (folded and faulted) rafts of host sediment (Fig. 12).

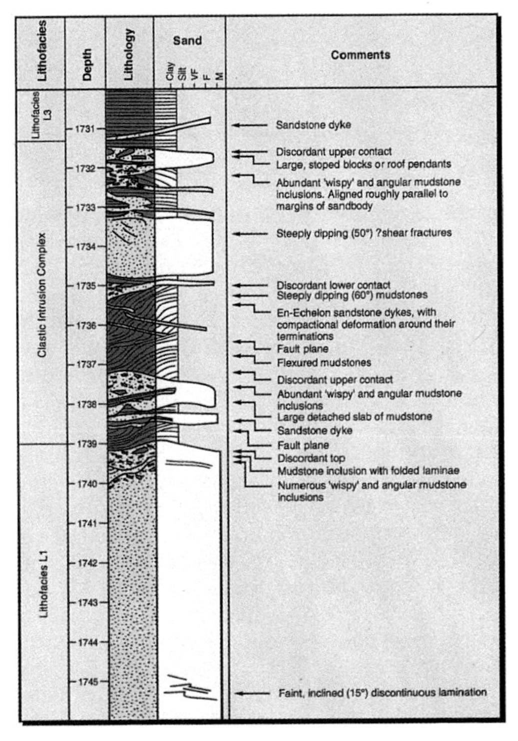

Fig. 12. Graphic sedimentological log through a lithofacies L1 sandbody and associated intrusion complex.

A graphic sedimentological log through an intrusive complex is shown in Fig. 12; the intrusive complex is 8 m thick and overlies a 10 m thick lithofacies L1 sandbody. The upper contact of this sandbody is discordant and its topmost metre contains locally abundant, angular mudstone inclusions. The majority of inclusions are roughly tabular in shape and are

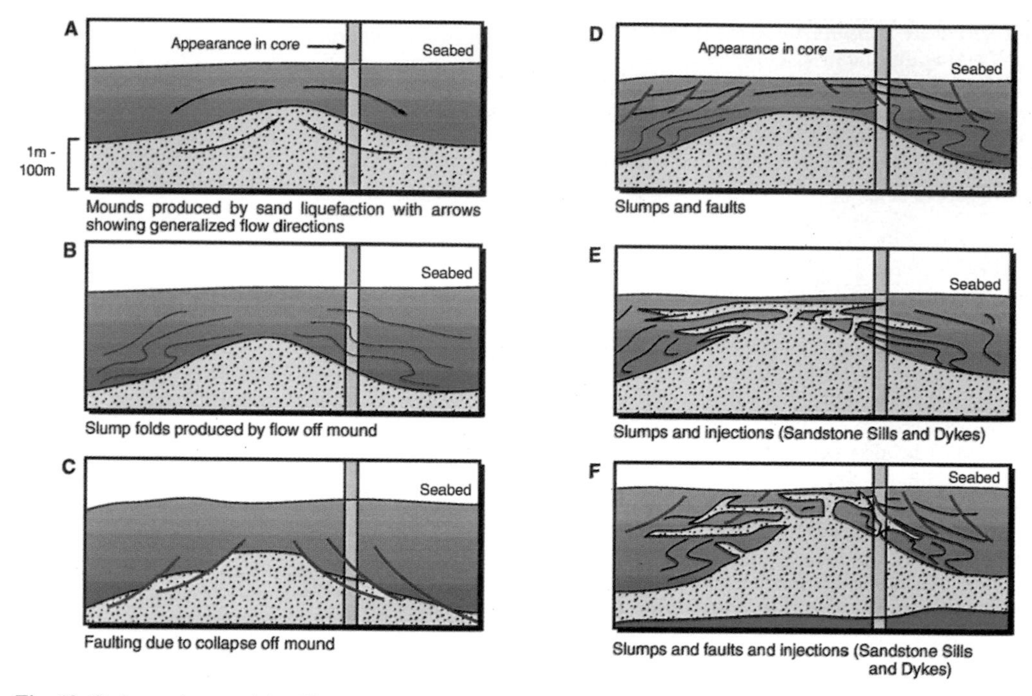

Fig. 13. Deformation model 1 illustrating the evolution of an intrusive complex.

characterized by extremely serrate, 'wispy' terminations. Locally, the inclusions are of folded mudstone. The intrusion complex that overlies the sandbody comprises a number of composite sill-like bodies, linked by sandstone dykes and separated by strongly deformed mudstones. The latter are cut by a number of low angle (2–10°), planar and listric fault planes. The sills contain large, stoped blocks or roof pendants in the uppermost part of the intrusive complex (Fig. 12) and are also characterized by abundant tabular, 'wispy' inclusions similar to those seen in the topmost part of the lithofacies L1 sandbody. The base of the lithofacies L1 sandbody (cored, but not shown in Fig. 11) is sharp, planar and apparently concordant with bedding in the underlying mudstones. The features illustrated in Fig. 12 are typical of the intrusive complexes of the Forth–Gryphon area, though some are considerably thicker than the example shown.

Deformation models

Two main deformation models have been developed by BP to account for the relationships described in the previous sections. Firstly, a model developed from the Forth Field, where sand deposited in deep erosional channels is remobilized into a suite of injection features during shallow burial (Fig. 13). Secondly, a model comparable to that developed for the Gryphon Field by Newman *et al.* (1993) where faulting and sandstone diapirism is important (Fig. 15). The models draw on an analogy with salt tectonics to explain the deformation that occurs when sand is overpressured and capable of flowing. Examples of each model are discussed below with reference to other North Sea examples.

Model 1. The initial upper surface of the lithofacies L1 sandbodies was probably irregular, but of low relief. In response to liquefaction subtle irregularities on the upper surface became exaggerated as sand began to 'flow' producing 'mounds' (Fig. 13a). As the 'mounds' became more prominent the overlying mudstone began to slide off the growing structures. Deformation within the mudstone was initially accommodated by the development of slump folds (Fig. 13b) and/or listric faults (Fig. 13c). However, as the sand continued to flow and pore fluid pressures increased, joints and bedding planes were dilated and filled by sand sills and dykes (Fig. 13d).

Locally, wholesale stoping (or net-veining) of the overlying mudstone took place, liberating large volumes of mudstone inclusions (Fig. 12). The presence of slump-folded fabrics within

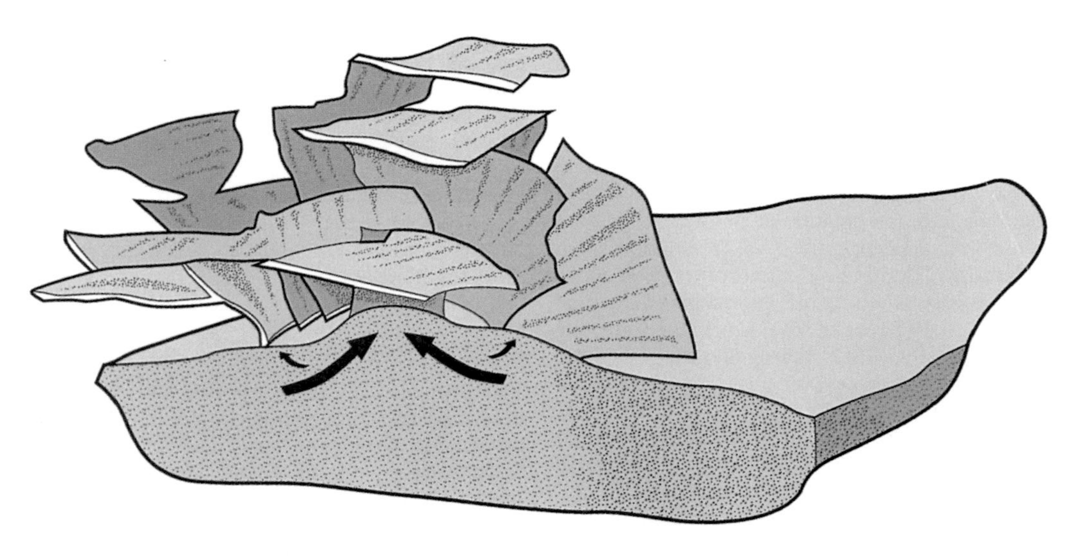

Fig. 14. Schematic three-dimensional geometry of a typical Forth Field intrusive complex.

some of the mudstone inclusions tends to support the hypothesis that intrusion took place relatively late in the sequence of events. The likely three dimensional geometry of the resulting intrusive complex is illustrated schematically in Fig. 14.

Similar features have been described from the Alba Field [located in the Outer Moray Firth in UK Block 16/26 (Fig. 1)]. The field occurs on a terrace at the foot of a SE-facing palaeoslope, is oriented NW–SE and is *c.* 9 km long and 1.5–3 km in width (Newton & Flanagan 1993). The reservoir sandstone of Middle Eocene age is a very fine to fine grained, sub-lithic to subarkosic, quartz arenite, and occurs in two facies: (1) Bedded sandstone – the dominant facies is characterized by discontinuous and continuous laminae, scours, amalgamation surfaces and common dewatering structures. Fining and/or coarsening upward sequences are absent and the sandstones are characterized by 'box-car' wireline log profiles; (2) Injected sandstones – form dykes and sills seen in core. When thickness of injected material exceeds a critical value (3 inches) stoping of the host mudstone is observed.

The sandstones are interpreted to have been deposited in a linear, erosional scour, that was cut and subsequently filled by turbidity currents. Seismic activity resulted in sand remobilization at shallow burial depths and the injection of sand into the overlying mudstone-dominated sequence. An initial interpretation that the sandstones were deposited within a leveed channel is now not favoured for three reasons. Firstly, seismic sections reveal that the sands occur in an erosional scour (up to 500 feet deep) that truncates flat underlying reflectors. Secondly, core and wireline log data show that the sedimentary succession cut by the erosion surface comprises hemipelagic mudstones not thin-bedded turbidites (as would be expected in a levee succession). Thirdly, detailed biostratigraphic correlation suggests that the reservoir sandstones are equivalent to hemipelagic mudstones in an off-field setting (Newton & Flanagan 1993).

Model 2. The T50 sequence comprises a series of discrete lithofacies L1 sandbodies, encased in lithofacies L3 mudstones overlying the T50 tephra sequence (Fig. 15a). A variety of smaller-scale features such as sandstone dykes and sills are well displayed in core (Figs 10-12). Because the T60 sequence thickens into the listric faults (Fig. 8) and as none of the intrusions penetrate formations younger than T60 in age it seems likely that the main phase of deformation took place during T60 times when the T50 sandbodies were at shallow burial depths (50-200 m).

Gravity-driven, sediment instability and seismic activity, together with density contrasts between basinal mudstones and rapidly deposited (and probably overpressured) turbidite sandstones are thought to have triggered the deformation. Extension during the deposition of the T60 sequence rotated sandbodies and triggered liquefaction within the sand. The sand became mobile and intruded along fault planes or generated sills, dykes and intrusion complexes (Fig. 15b). Following the salt analogue sand is also likely to flow into the footwalls of developing faults, producing 'pinched' triangular sandbodies (cf. Newman *et al.* 1993). Once the sand has dewatered, however, 'flowage' will no longer be possible. Sand grains will 'lock' together and are unlikely to deform other than by the self-limiting process of granulation seam generation. Continuing extension is then more likely to be taken-up at the main lithological contrast, i.e. between the sandstone and the mudstone (Fig. 15c), thus generating 'faulted' sandstone–mudstone contacts (Newman *et al.* 1993).

Similar features have been described from the Balder Field [located at the western margin of the Utsira High in Norwegian Blocks 25/10 and 25/11 (Fig. 1)]. Oil is reservoired in submarine fan sandstones of Late Palaeocene (T45) and Lower Eocene (T50) age. The sandstones have reservoir properties and sedimentary facies comparable to those seen in the Forth–Gryphon area. The reservoir sandstones occur in dramatic mounds, with steep margins (15-20°). Initial studies considered that the mounds may have been depositional or erosional features. Analysis of a 1988 three-dimensional seismic survey, however, revealed; (1) that faults commonly cut the top of the reservoir; and (2) that most of the mounds were fault-bounded (Jenssen *et al.* 1992). The seismic survey also revealed that significant sandstone diapirism accompanied the faulting, the sands flowed into the footwalls of the faults resulting in steep-sided triangular sandbodies. As in the Forth–Gryphon area there is no evidence that the sands were constrained within channel–levee complexes or that thinly bedded sand–silt sequences exist as lateral equivalents to the massive sandstones. The working model proposed by Jenssen *et al.* (1992) suggests that extensional faults (which developed at the sandstone–mudstone interface) and sandstone diapirism combined to enhance the thicker sections of an initially extensive sand sheet.

Discussion

Significant soft-sediment deformation is recognized throughout the Tertiary of the Bruce–Beryl Embayment (Alexander *et al.* 1992; Newman *et al.* 1993; Dixon & Pearce 1995). Its importance has also been documented in other North Sea fields with Tertiary reservoirs (e.g. Jenssen *et al.* 1992; Newton & Flanagan 1993). Acceptance of the 'mobility' of sand and the importance of syn-sedimentary deformation in shaping the geometry of submarine fan reservoirs has, in some quarters, been slow. Despite the presence of seismic-scale, soft-sediment deformation features and the lack of supporting evidence, some groups persist with 'established literature' models to explain sand distribution (e.g. channel-levee complexes). Increasingly, however, models are being proposed that invoke sand remobilization.

Exploration significance

The range of post-depositional processes described and discussed in the previous sections have some significant implications with regard to exploration for the Tertiary petroleum play. In the exploration environment, where three-dimensional seismic data is often not available, top reservoir may only be mapped with a limited degree of confidence. Post-depositional deformation processes may give rise to considerable uncertainty in the definition of a trap. For example, closure may not be as mapped (Fig. 16a), but may lie at a higher structural level within an intrusive complex (Fig. 16b). Alternatively, closure at this level may not exist if the intrusion complex is in communication with porous sandstones at a higher stratigraphic level (Fig. 16c).

In exploring for the T50 play it is therefore important to understand the controls on post-depositional deformation processes in order to adequately assess prospect risk. In the Forth–Gryphon area the main controls seem to be the existence of significant slopes (the flanks of the Crawford Anticline) and the presence of a detachment horizon (the T50 tephra sequence). Elsewhere in the Bruce–Beryl Embayment, the old T45 delta-slope (where draped by the T50 tephra sequence) also seems to have focused post-depositional deformation. Tertiary fan prospects in these areas should therefore carry a higher risk on trap definition than those outwith the influence of these unstable slopes.

Appraisal significance

Obviously, the recognition of complex intrusive sandbody geometries like those schematically illustrated in Fig. 14 is of great importance in field appraisal and development. In the absence of core, for example, such features could be interpreted as thinly bedded, turbidite sandstones. This interpretation would result in a completely erroneous assessment of reservoir continuity and connectivity. Where such

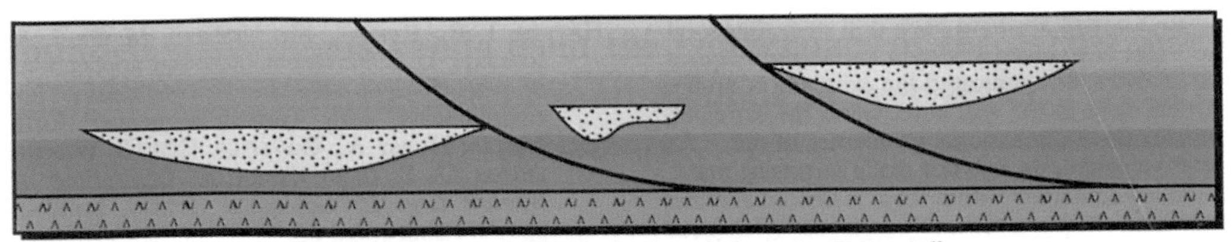

A series of discrete sandstone bodies deposited within mudstone above the Balder tuff.

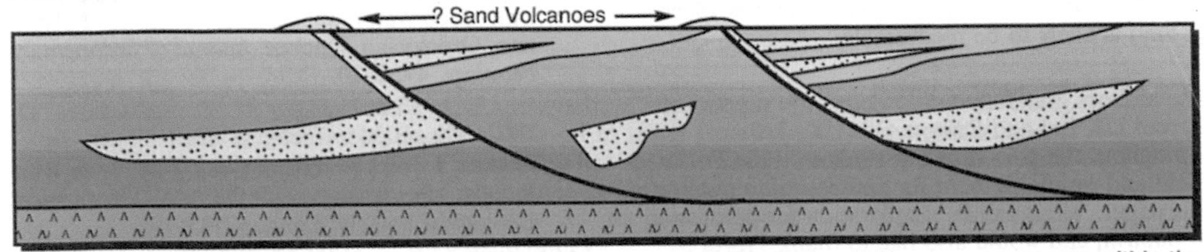

Extension during the deposition of the Frigg Formation rotates sand bodies and triggers overpressure within the sand. The sand becomes mobile and either (i) flows into fault footwalls or (ii) is injected along lines of weakness such as fault planes. The faults generate sea-floor topography, and their hanging walls become sites for preferential sand accumulation.

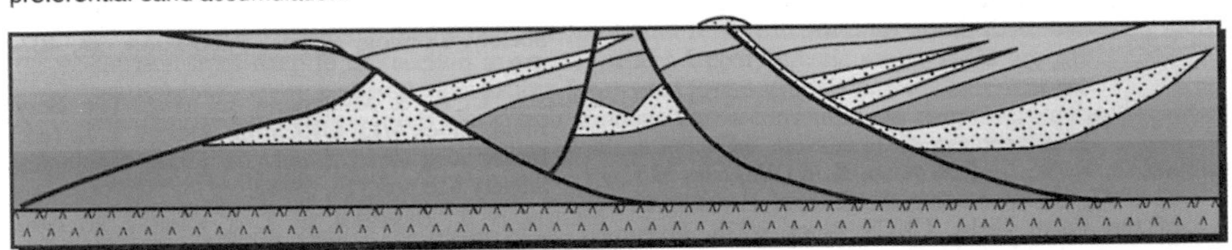

The sand continues to flow until overpressure is dissipated. At this stage the sand locks and deformation is focussed at the sand shale interfaces : commonly at the edges of the triangular sandbodies generated by sand flow to fault footwalls.

Fig. 15. Deformation model 2 illustrating the relationship between detachment faulting, sandstone diapirism and clastic intrusion.

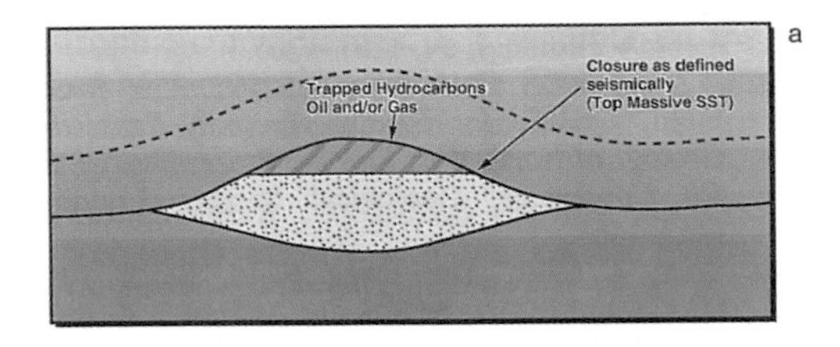

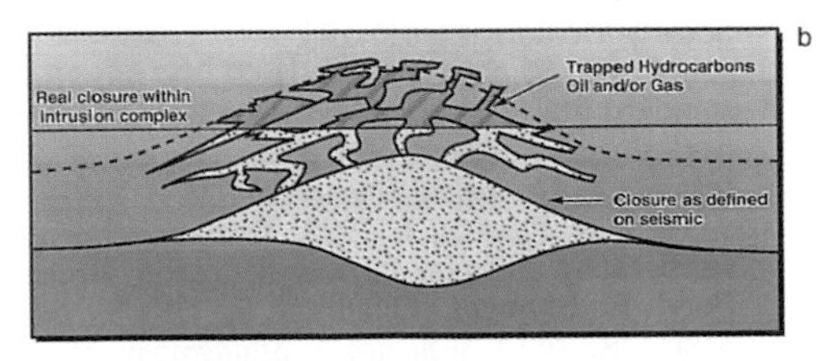

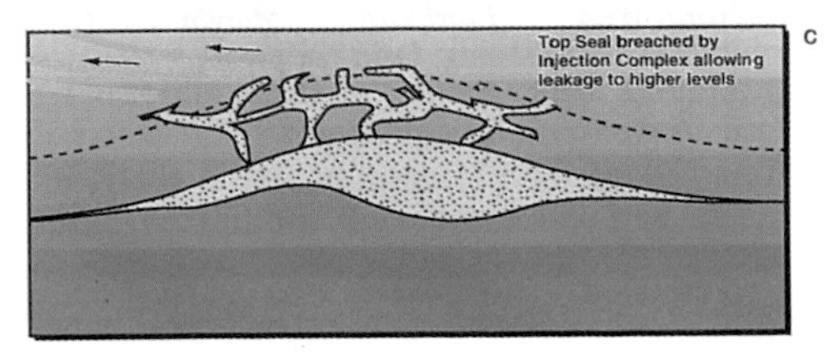

Fig. 16. Exploration significance of intrusion complexes in the Tertiary of the Bruce–Beryl Embayment.

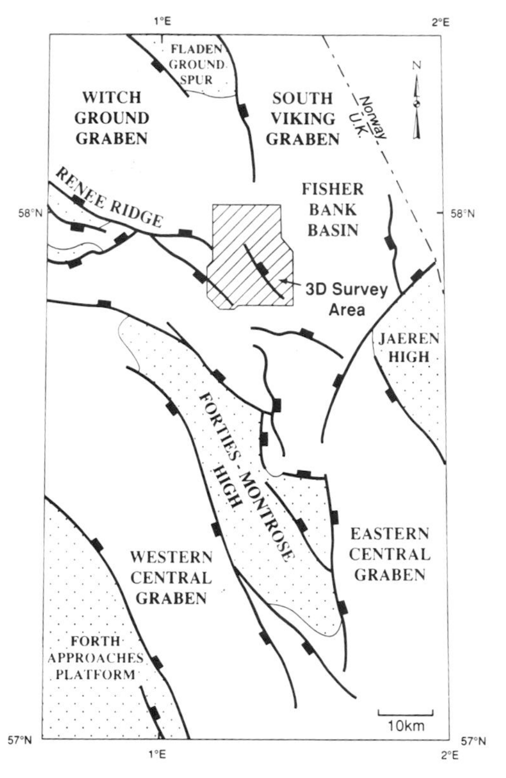

Fig. 1. Location map showing the extent of the three-dimensional seismic survey area within the Fisher Bank Basin.

conducted on a workstation permitting the generation of amplitude displays and timeslices through zones of interest. At 0.52–0.6 s two-way time (TWT) (450–550 m subsea) an apparently mounded high-amplitude horizon occurs which can be mapped regionally within the central North Sea area. The mounds lie within the Quaternary sub-littoral clays of the Aberdeen Ground Formation. They were initially interpreted as deposits of anastamosing, braided streams, with channel sand and overbank facies explaining the observed sand distribution; only thin sands and claystones are present in the 'inter-channel' area. After mapping on the original two-dimensional dataset, a tidal sand-ridge origin was considered more appropriate. However, three-dimensional data has determined that the features are circular to ovoid in plan view, and that they are therefore true mounds, with no obvious primary mechanism to account for their formation or distribution. Consequently, the original depositional environment of the sands cannot be determined. However, their geometries may provide an analogue for larger-scale mounds observed from deep water Tertiary sand environments.

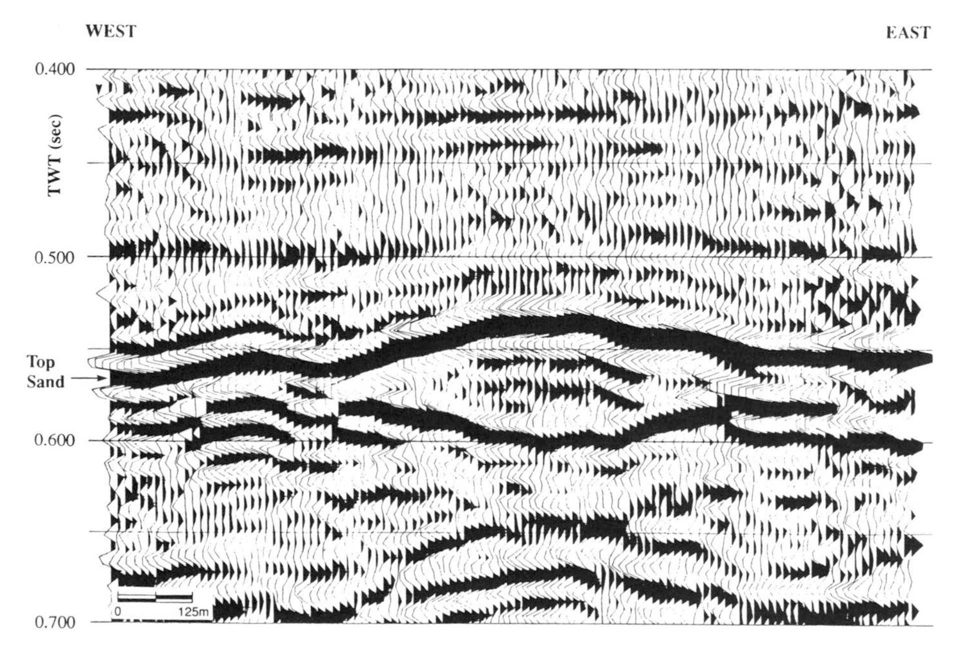

Fig. 2 Seismic line illustrating the geometry of a typical sand mound in cross section. Note conformity of reflectors immediately overlying high amplitude reflections at the top of the mound. Sand thickness is up to 45 m.

Mounded sand morphology and seismic definition

Wells drilled in the Fisher Bank Basin have been specifically located to avoid mound features owing to the high risk of encountering shallow gas; those drilled in the inter-mound area have penetrated thin, water-bearing, fine-grained sands with a velocity significantly lower than the enveloping clays. In the absence of a significant density contrast, such a reduction in velocity produces a peak on zero-phase seismic data with the polarity convention applied to the three-dimensional survey data presented here.

The detailed morphology of a typical mound can be studied in conventional time sections (Fig. 2). Amplitudes at the top of the sand

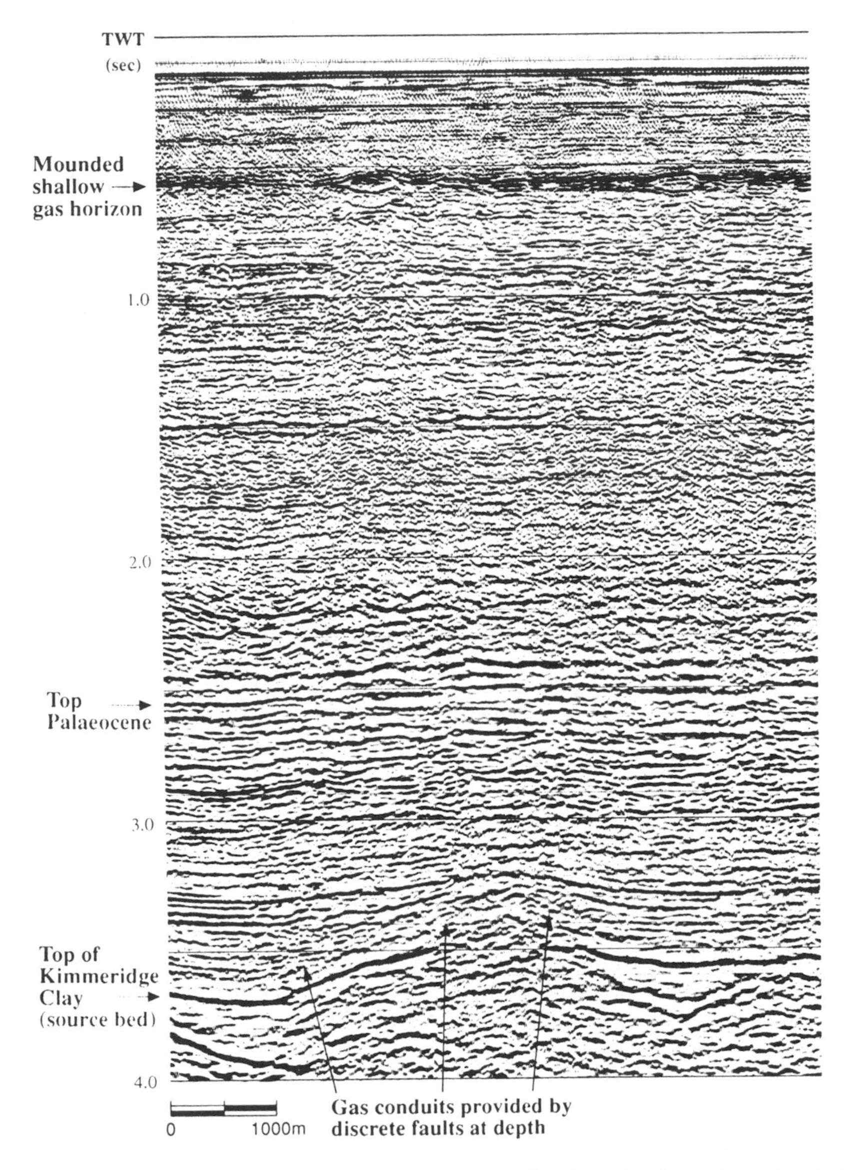

Fig. 3. Vertical time section showing possible gas conduits extending from deeply buried high at 3.5 s to the mounded shallow gas level at 0.55 s. Gas movement is believed to have taken place along discrete, small offset faults at depth, but high intensity of small-scale faulting in the post-Palaeocene section allows migration over much wider areas, giving a highly diffractive (diffuse) appearance to the younger section.

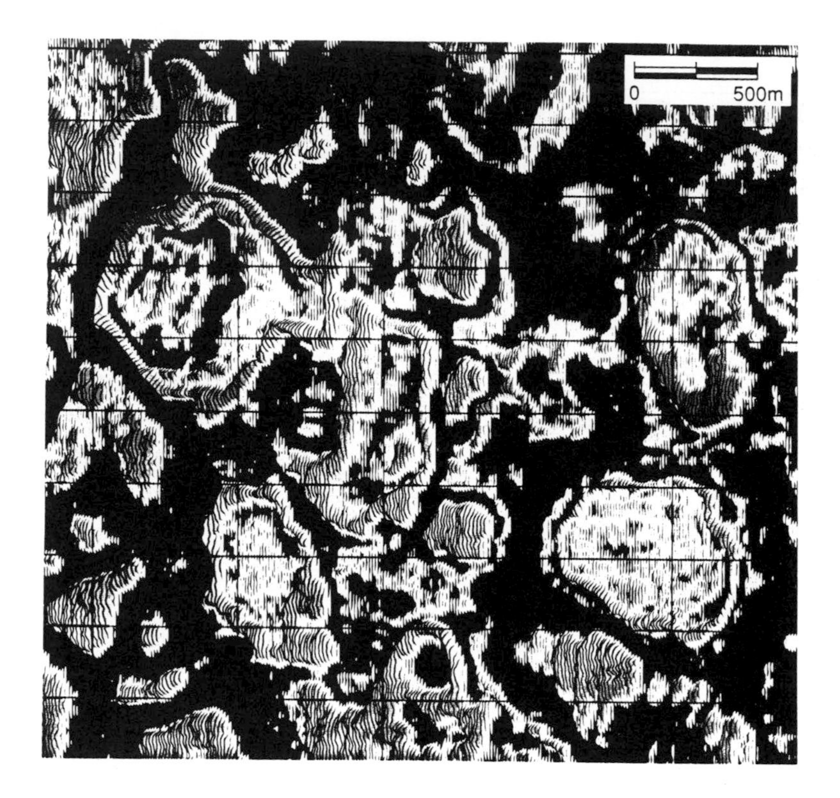

Fig. 4. Timeslice display through the crest of the sand mounds highlighting the circular to ovoid shape in plan view.

mound (0.535 s TWT) are up to twice those measured on the flanks. Thus, the acoustic impedance reduction at the crest of the mound appears greater than that either on the flanks or within the water-bearing inter-mound areas. Together with the 'push-down' of the strong underlying reflection (0.6 s TWT) this is strongly suggestive of the presence of gas. It is uncertain whether the underlying reflector represents the base of the sand or a gas–water contact, but we estimate that up to 45 m of sand could be present with a positive relief at the crest of *c.* 20 m.

Other effects compatible with the presence of gas include blanking or attenuation of both over- and underlying events and mis-stacking of events due to ray-path delays within the gas sands. As a result, the section beneath the mounds is frequently dominated by water bottom peg-leg multiples. None the less, in some areas, it is possible to identify faults which cut up from deeper levels, frequently associated with buried highs, terminating at the mounded level and are strongly suggestive of a gas feeder system (Fig. 3).

The diameter of the mounds is on average 0.5 km but can be up to 1 km (Fig. 4). Spacing is also variable but in the order of 0.5 km. Allowing for depth conversion and decompaction, the maximum observed angle at the mound periphery is 15°. Onlap and isopach changes above the mounds suggest that although they formed relatively early, at shallow depths of burial, their development was not coeval. In addition, radial, often listric, fractures occur above the mounds (Fig. 5) and are analagous to fault patterns modelled from salt-dome intrusion into sediments undergoing early extension (Withjack & Scheiner 1982). Their unusual geometry and high angles of repose leads us to the conclusion that the mounds are not primary depositional features but are the product of major post-depositional remobilization of conventionally deposited sands.

Mechanism for mounded sand formation

Shallow gas occurrences tend to be located in, or adjacent to, areas of gas generation and

Fig. 5. Chair display (half-time slice, half-vertical section) illustrating minor listric faulting above a sand mound which ties to a radial fault pattern in plan view. One fault has been illustrated as an example.

frequently occur above buried highs. The mounded sands observed in this three-dimensional dataset are in just such a setting, with an underlying intensely faulted pre-Cretaceous structure flanked by Jurassic source rocks currently mature for gas generation. The faults terminating at the shallow gas level (Fig. 3) can, in some instances, be traced down to the deeper structure via a series of 'diffuse' zones where reflections are disrupted, though often without the distinct offsets normally seen with seismic-scale faulting. Such diffuse zones have been documented feeding subsurface pockmarks (Hovland & Judd 1988). It is therefore suggested that the shallow gas anomalies are the result of thermogenic gas feeding up from a deeper structure which was the initial focus for gas migration.

Although it cannot be discounted that the mounds formed prior to gas emplacement, the close association of gas conduits with the mounded sands could indicate a causative mechanism. Gas charging of unconsolidated sediment has been recognized as providing a mechanism for liquification (*sensu* Nichols 1995) and sediment remobilization (Woolsey *et al.* 1975; Nichols 1995). We therefore tentatively suggest the following model: a sand bed is initially deposited by a conventional mechanism with minor irregularities in its upper surface (Fig. 6a). Gas is fed by fault conduits into the sand, possibly under pressure or through natural buoyancy forces, causing liquification. Migration of the gas towards subtle highs results in entrainment of sand grains. A mounded feature is thus formed by withdrawal of sand from the flank areas towards the centre (Fig. 6b). The liquification process would be facilitated by contemporaneous seismic activity or a pressure release mechanism resulting from breach of seal.

a)

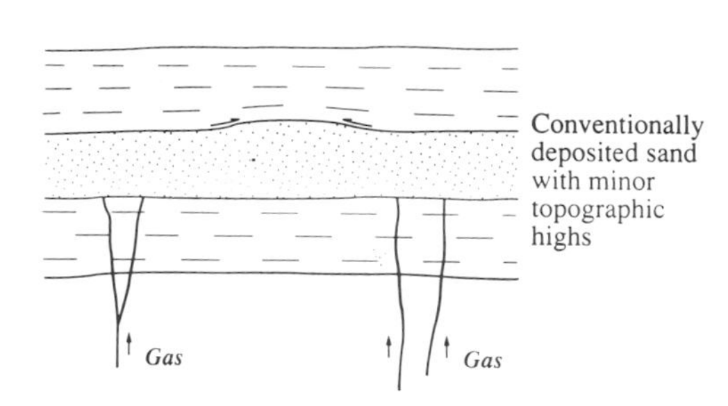

b)

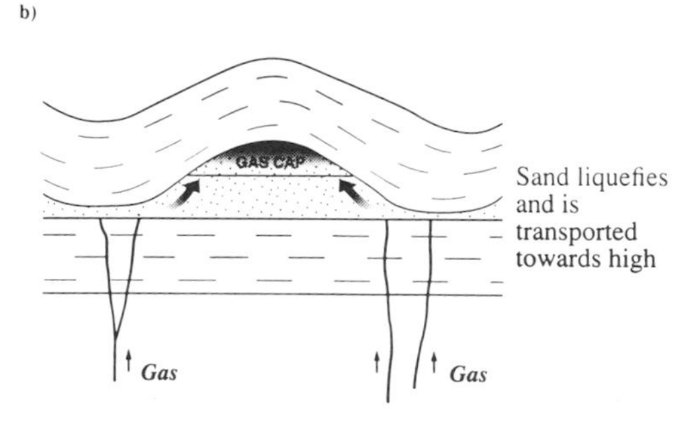

Fig. 6. Conceptual model for the formation of mounded sands by liquification. **(a)** Gas fed by fault conduits into a poorly consolidated sand. **(b)** sand liquefies and is transported towards a subtle high in the bed surface.

The seismic data indicates that the mounds formed at shallow depths, probably up to a maximum of 100 m below the sea bed. At such depths the direction of minimum principal stress (σ 3) is vertical, allowing beds to be more easily jacked apart. This would enable the remobilized sands to deform the overburden and result in the observed antiformal features.

Discussion

Recent workers have suggested that the Balder sands in the Forth (9/23b) and Gryphon (9/18b) Fields in the Beryl Embayment, South Viking Graben, northern North Sea, have been affected by large-scale remobilization and injection (Mackay & Reynolds 1992; Newman *et al.* 1993; Jaffri 1993; Dixon *et al.* 1995). This has resulted in a sandbody geometry which cross-cuts the primary stratigraphy and, particularly in the Forth Field, has given rise to a mounded feature with such high angles of repose that it could not have been stable as a body of unconsolidated sediment on the seafloor (Alexander *et al.* 1993). A mechanism has been invoked involving liquification of the originally deposited sand and is postulated to result in remobilization and doming, culminating in a breach of seal and subsequent injection. This paradigm is supported by the experimental modelling of Nichols (1995).

Several mechanisms for liquification of the Balder sands have been suggested, including seismic pumping (Jaffri 1993), migration of hydrocarbons and rapid dewatering of sediments containing high initial depositional volumes of pore water. Whatever the cause of liquification, the presence of volcanically-derived bentonite horizons within the Balder Formation claystones seems to be critical. These claystones are considered to have formed a highly competent vertical and lateral seal, preventing dewatering of the encapsulated sand. The sands at the shallow gas level are overlain by claystones of the Aberdeen Ground Formation which are overcompacted as a result of glacial ice overburden. These claystones similarly provide an excellent seal, permitting the build-up of pressure and ultimate liquification of the sands.

Hovland & Judd (1988) discuss the formation of sea-bed pockmarks and conclude that they are commonly formed by escape of thermogenic gas. However, the critical difference between pockmark development and formation of shallow subsurface mounds appears to be the competence of the overlying seal to withstand breaching to sea bed.

The proposed model for liquification differs from that suggested for the Balder sands as gas rather than water is invoked to be the driving fluid. However, we suggest that the resulting features are genetically related: in both cases a conventionally deposited sand has been deformed following large-scale remobilization as a result of liquification. In the Forth Field, injected sands above the mound testify to breach of seal and associated pressure release (Dixon *et al.* 1995). Such features are not recognized above the shallow gas charged mounds, although it is possibile that a breach of seal has occurred.

Conclusions

The shallow gas features described from the Fisher Bank Basin are true mounds in three dimensions and they are tentatively suggested to result from sand liquification by charging with thermogenic gas. Analogies can be drawn with mounded sand geometries seen in the early Tertiary reservoirs of the Forth–Gryphon area, northern North Sea. These are also considered to be the product of liquification although the process is not homologous as water rather than gas is considered to be the driving fluid. It is proposed that post-depositional liquification/ sandbody remobilization is a more widespread process than has previously been documented and the early recognition of such mechanisms provides a valuable model for the prediction of sand distribution outside the purely depositional geometries normally adhered to. Sɔch models must therefore take into account features formed by processes more akin to salt diapirism and fluid escape than conventional depositional models.

We would like to thank the directors of Clyde Expro PLC and our partners Amoco (UK) Exploration Company, and Monument Oil and Gas for permission to publish this paper. We acknowledge the contribution of colleagues both past and present, and thank Rob Nichols for discussion of the mechanism. The opinions expressed here are those of the authors and do not necessarily reflect the views of either Clyde or partners.

References

ALEXANDER, R. W. S., SCHOFIELD, K. & WILLIAMS, M. C. 1993. Understanding the Eocene reservoirs of the Forth Field, UKCS Block 9/23b. *In*: SPENCER, A. M. (ed.) *Generation, accumulation and production of Europe's hydrocarbons III.* Special Publications of the European Association of Petroleum Geoscientists, **3**.

DIXON, R. J., SCHOFIELD, K., ANDERTON, R., REYNOLDS, A. D., ALEXANDER, R. W. S., WILLIAMS, M.C. & DAVIES, K. G. 1995. Sandstone diapirism and clastic intrusion in the Tertiary submarine fans of the Bruce–Beryl Embayment, Quadrant 9, UKCS. *This volume.*

HOVLAND, M. 1993. Submarine gas seepage in the North Sea and adjacent areas. *In:*PARKER, J. R. (ed.) *Petroleum Geology of Northwest Europe: Proceedings of the 4th Conference.* Geological Society, London, 1333–1339.

—— & JUDD, A. G. 1988. *Seabed Pockmarks and Seepages: Impact on Geology, Biology and the Marine Environment.* Graham and Trotman, London.

JAFFRI, F. 1993. *Cross-cutting sandbodies of the Tertiary, Beryl Embayment, North Sea.*PhD thesis, University College, London.

MACKAY, T. A. & REYNOLDS, J. W. 1992. Postdepositional modification of deep water sands: reservoir geometry of the Gryphon Sand, Gryphon Field UK Sector 9/18b. *In*: Deep Water Massive Sands; Arthur Holmes European Research Conference Programme, 19.

NEWMAN, M. ST J., REEDER, M. L., WOODRUFF, A. H. W. & HATTON, I. R. 1993. The geology of the Gryphon Oilfield. *In:* PARKER, J. R. (ed.) *Petroleum Geology of Northwest Europe Proceedings of the 4th Conference.* Geological Society, London, 123–134.

NICHOLS, R. N. 1995. Remobilization of sandy sediments. *This volume.*

SALES, J. K. 1993. Closure vs. seal capacity – a fundamental control on the distribution of oil and gas. *In:* DORE, A. G. ET AL. (eds) *Basin Modelling: Advances and Applications.* Norsk Petroleunsforening, Special Publication, **3**. Elsevier, Amsterdam, 399–414.

WITHJACK, M. O. & SCHEINER, C. 1982. Fault patterns associated with domes – an experimental and analytical study. *American Association Petroleum Geologists Bulletin,* **66**, 302–316.

WOOLSEY, T. S., MCCALLUM, M. E. & SCHUMM, S.A. 1975. Modelling of diatreme emplacement by fluidization. *Physics and Chemistry of the Earth,* **9**, 29–42.

Sandstone megabeds from the Tertiary of the North Sea

J. C. PAULEY

PM Geos Ltd, Suite 3, 343 Union Street, Aberdeen AB1 2BS, UK

Abstract: Three cored sandstone megabeds of Tertiary (late Palaeocene) age from two quadrants in the Central Graben of the North Sea are discussed. The megabeds are upwards-fining units 3–9 m thick, underlain by gravelly mudstones (1–1.4 m thick) which are interpreted as debrites. The megabeds are divided into five units (M1–M5) based on grain size and sedimentary structures, which are organized into a type vertical sequence. The basal unit M1 comprises a clast-supported mudclast conglomerate. Unit M2 comprises very coarse- to medium-grained, occasionally gravelly, sandstone which is horizontally laminated to cross-stratified. Unit M3 is the thickest unit (2–4 m) and is a laminated, upwards-fining, medium- to fine-grained sandstone with occasional cross-lamination. Unit M4 is a fine- to very fine-grained sandstone characterized by disturbed lamination and water-escape structures. Unit M5 is an upwards-fining silty sandstone to siltstone characterized by discontinuous, irregular, and locally overfolded laminae. The megabeds are interpreted as having been deposited by turbidity currents. There are some similarities to Bouma sequences and high-density turbidite sequences, but massive sandstone is notably absent. The megabeds are associated with mudstones and debrites and only occasional thin turbidites, and as such they are not part of an organized sequence of turbidites. The beds are notably much thicker than those above and below, and lithologies are distinct from those of the surrounding sediments. These features are comparable to those of 'megaturbidites' as defined by Bouma. The debrite which immediately underlies the megabeds is significant in that this relationship is also a common feature of megaturbidites. The debrite is thought to represent slope failure which immediately preceded megabed deposition. A hypothesis is presented of earthquake-induced catastrophic slope failure, debrite deposition and associated megaturbidite generation. Such sandstone megabeds are thought to be widespread sheet-like deposits. They are exploited as a hydrocarbon reservoir in two of the wells discussed.

This paper discusses sandstone megabeds observed in three cores from two quadrants in the Central Graben of the North Sea (Fig. 1), and is believed to be the first published account of such beds of Tertiary age from the North Sea. Due to commercial constraints details of well locations and stratigraphy cannot be divulged. One core (well 'C') is from Quadrant 22. The sandstone megabeds are distinct from the associated sediments and from typical turbidites of Tertiary age from the North Sea. They show similar characteristics, a common sequence of structures and lithologies, and occur at a similar stratigraphic horizon (Fig. 2). Within the constraints of the available data it cannot be demonstrated that the sandstone megabed in the two quadrants is the product of the same event. In two wells (A and B) the megabed is assigned to the Lista Formation and is believed to be one discrete bed which is correlatable between several nearby wells (a few to several kilometres apart) on the basis of detailed biostratigraphic analysis. In well 'C', which is *c.* 50 km distant from wells A and B, a megabed occurs at the boundary between the Lista and Sele Formations.

The megabeds are compared with megaturbidites as defined by Bouma (1987), however, documented megaturbidites are commonly much thicker than the beds described here and often contain large clasts (e.g. Bouma 1987; Labaume *et al.* 1983, 1986, 1987; Séguret *et al.* 1984; Souquet *et al.* 1987). For example, the megaturbidites from the Hecho Group of Eocene age from the SW Pyrenean Basin of Spain contain slabs up to several hundred metres wide and clasts commonly several metres wide in their basal parts (Labaume *et al.* 1987). Labaume *et al.* (1987) used the term 'megaturbidite' to denote 'the deposit of an exceptionally large-volume sediment gravity flow'. Bouma (1987, 66) recommended that the term 'megaturbidite' be restricted in usage to layers which have the following characteristics (quote): (1)

From Hartley, A. J. & Prosser, D. J. (eds), 1995, *Characterization of Deep Marine Clastic Systems*, Geological Society Special Publication No. 94, pp. 103–114.

Fig. 1. Location map showing approximate location of the wells discussed.

have layer thicknesses that are normally as thick, or thicker, than the thickest layers in the host rock series; (2) are laterally extensive and thus cover a significant part of the basin in which they are deposited; (3) differ in composition to that of the host rock series; (4) do not have geometries that are typical for submarine fans, such as channels and depositional lobes; (5) have internal properties that suggest they are the deposit of one single transport event, even if that event covers several depositional processes. Consequently, megaturbidites should be excellent synchronous marker horizons; (6) have all the characteristics to make them good markers for stratigraphic and seismic correlations, and thus can be extremely useful in basin analysis, mapping and structural studies (unquote).

The sandstone megabeds described here are only associated with rare, thin turbidites, being isolated within hemipelagic mudstones and debrites. In this respect they are different from the majority of turbidite deposits of Tertiary age from the North Sea. The megabeds are mixed siliciclastic and calciclastic deposits intimately associated with an underlying debrite which is of importance with regard to the interpretation of depositional mechanisms. The sequences of sedimentary structures within the megabeds can be compared with turbidite Bouma sequences, although significant differences exist which are discussed later, and a type sequence for these megabeds is proposed.

Facies analysis of the cored intervals

Well A

An entire section through the sandstone megabed was cored in this well (Fig. 3). Mudstones with occasional thin sandstone beds occur in the

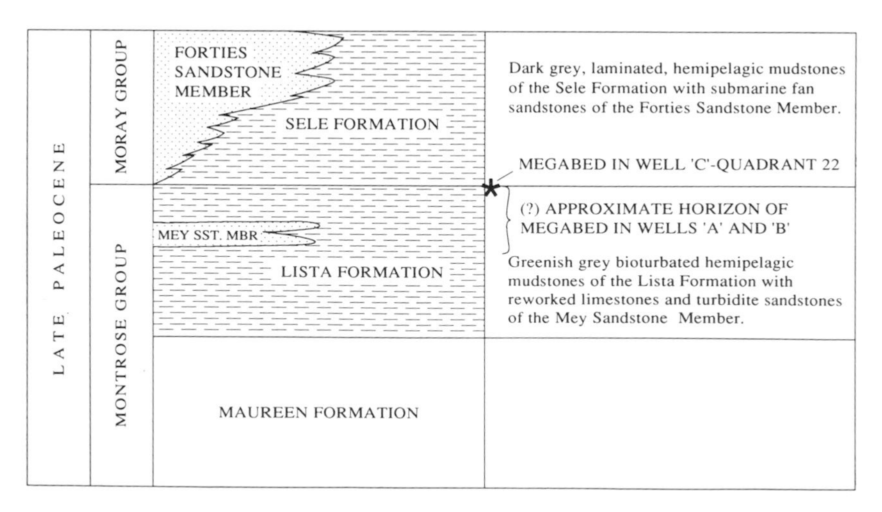

Fig. 2. Lithostratigraphic context of the sandstone megabeds (based on the lithostratigraphic nomenclature of Knox & Holloway 1992).

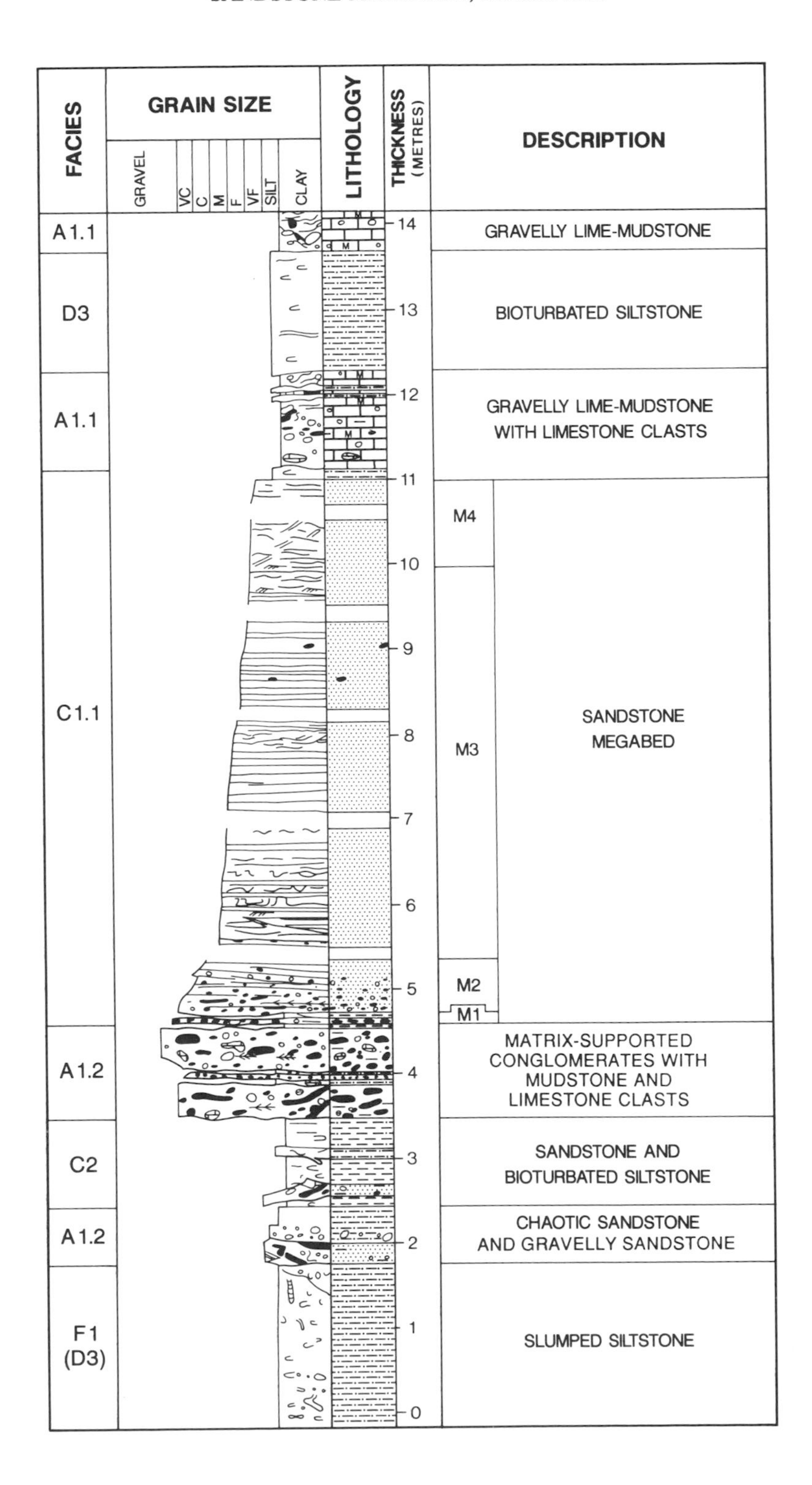

Fig. 3. Core log for well 'A' (see p. 106 for key to logs).

Key to Logs A, B and C.

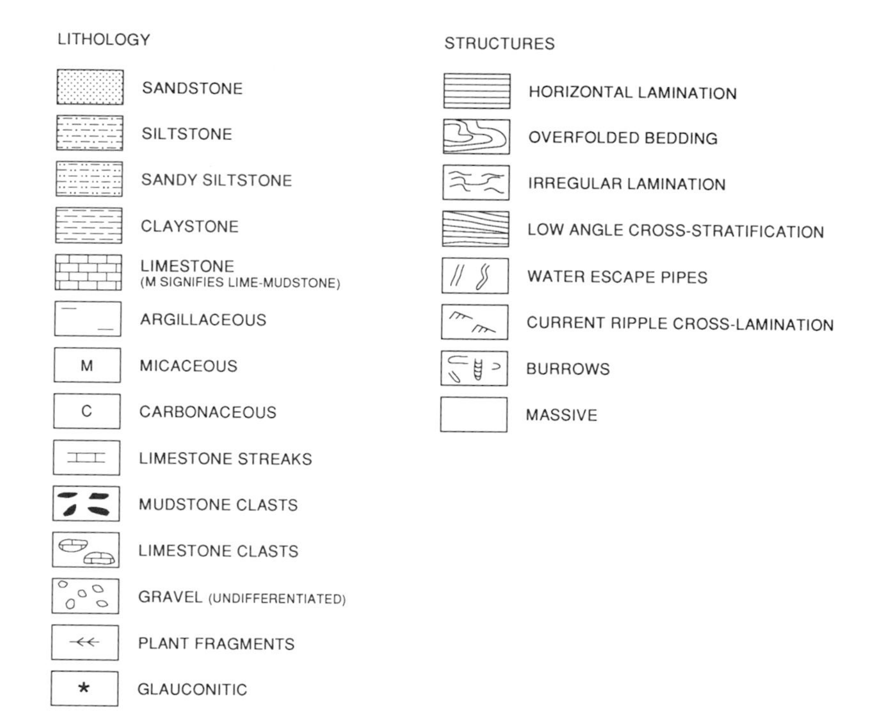

lower part of the section. A matrix-supported conglomerate, 1 m thick, immediately underlies the sandstone megabed. The megabed itself is 6.4 m thick and fines upwards from a granule conglomerate to a very fine-grained sandstone. Above this are bioturbated mudstones with beds of gravelly lime–mudstone within which are numerous lime–mudstone clasts.

The facies have been classified according to the facies scheme detailed in Table 1. This classification is a modified version of the turbidite facies classification scheme of Pickering *et al.* (1989).

Chaotic facies (Facies F1) This is composed of bioturbated greenish grey siltstone with steeply dipping *Zoophycos* burrows which indicate that the siltstone has been affected by slumping and/or sliding. The siltstone occurs at the base of the section in a bed at least 1.8 m thick. On the core

Table 1. *Facies classification*

Facies group	Facies	Subfacies
A Gravel-rich sediments	A1 massive muddy gravels	A1.1 gravelly lime-mudstones A1.2 gravelly siliciclastic mudstones
C Interbedded sand and mud	C1 thick-bedded sandstones C2 medium-bedded sandstones	C1.1 sandstone megabeds
D Siltstones	D2 thinly, regularly laminated D3 massive, bioturbated	
E Claystones	E1 massive, bioturbated E2 laminated	
F Chaotic deposits	F1 coherent folded and contorted strata	

log this facies is designated F1(D3) which signifies a chaotic facies (F1) composed of siltstone (facies D3).

Bioturbated Siltstone (Facies D3). This siltstone is similar to that of facies F1 except that *Zoophycos* burrows are undisturbed and horizontal, and the upper and lower bedding contacts are straight and horizontal indicating that this facies is *in situ.* The siltstone occurs in a bed 1.2 m thick in the upper part of the section above the sandstone megabed.

Medium-bedded sandstones with interbedded mudstones (Facies C2). This facies comprises disturbed beds of very fine-grained, mainly massive, sandstone which is in abrupt, sometimes interpenetrative, contact with bioturbated to poorly laminated silty claystone to siltstone. The disturbed bedding suggests that there has been some degree of sliding and/or slumping within the facies. The facies occurs in a unit *c.* 0.9 m thick in the lower part of the section.

Muddy Gravel (Subfacies A1.2). This occurs in an interval, 1.1 m thick, directly beneath the sandstone megabed. Thin mudstone bands are present which appear to separate the interval into beds up to 0.6 m thick. However, these mudstones could represent rafts within a single gravel bed. The interval comprises a clay to clayey-sand grade matrix which supports a variable amount of dispersed gravel. Clasts reach 6 cm in diameter and comprise greenish grey mudstone, small lime-mudstone clasts, and occasional coaly fragments. Although the interval is mainly massive, overfolded mudstone streaks are evident. The gravel shows inverse clast grading where larger clasts are concentrated towards the top of the bed. A similar lithology occurs in a thin bed lower in the sequence (Fig. 3) where it appears to be associated with a thin fine-grained sandstone containing overfolded mudstone streaks and a minor amount of gravel.

Sandstone megabed (Subfacies C1.1). This bed is 6.4 m thick and is gradually normally graded from a mud-clast rich, granule to small pebble conglomerate at the base to a very fine-grained sandstone at the top. The bed has been divided into four units designated M1–M4 (Fig. 3). The basal unit (M1) is 0.2 m thick and consists of clast-supported conglomerate. The clasts comprise greenish grey mudstone, lime–mudstone, and occasional coal and plant debris. The overlying unit M2 (0.6 m thick) is a laminated coarse to very coarse-grained sandstone with granule grade clasts similar in composition to those of the underlying conglomerate. There is some poorly developed cross-stratification within the coarse-grained sandstone. The following unit M3 comprises 4.6 m of upwards-fining medium to fine-grained sandstone. The majority of the unit is horizontally laminated with occasional convolute lamination and ripple cross-lamination. The latter is particularly evident at the top of the unit. Unit M4 at the top of the megabed is 1.1 m thick and is composed of very fine-grained sandstone with discontinuous, horizontal, wavy, non-parallel lamination cut by water-escape pipes, which are inclined and parallel to each other. A thin bed of bioturbated siltstone (facies D3) rests abruptly on top of unit M4, and unit M5 is absent.

In thin-section, grains of lime-mudstone form a significant component of the sandstones, but quartz grains are dominant. Dispersed lime-mud forms a detrital matrix in the upper fine-grained portions of the megabed and is interpreted to represent disaggregated chalk clasts.

Gravelly lime-mudstones (Subfacies A1.1). Gravelly lime-mudstones are found in the upper part of the section in beds up to *c.* 0.9 m thick. They are occasionally separated by thin beds of bioturbated siltstone similar to that of facies D3, with which they are in abrupt contact. The facies comprises matrix-supported intraclastic chalks with fragments of lime-mudstone and occasional mudstone up to several centimetres in diameter. The gravelly lime-mudstones are variably argillaceous and are either massive or contain argillaceous streaks and laminae which are commonly irregular, and sometimes overfolded.

Well B

The cored interval in this well (Fig. 4) does not contain the base of the sandstone megabed. The megabed (subfacies C1.1) is overlain by mudstone (facies D3), and gravelly lime–mudstones (subfacies A1.1) with interbedded bioturbated mudstones (facies D3). These facies are similar to those in well A.

Wireline logs indicate that the base of the megabed is *c.* 1.8 m below the cored interval and that the megabed is 9.1 m thick. An increase in gamma-ray values towards the base of the megabed in well B is also observed in the case of well A, where this is due to the development of mud-clast rich conglomerates. Therefore, a basal unit M1 is believed to be present in

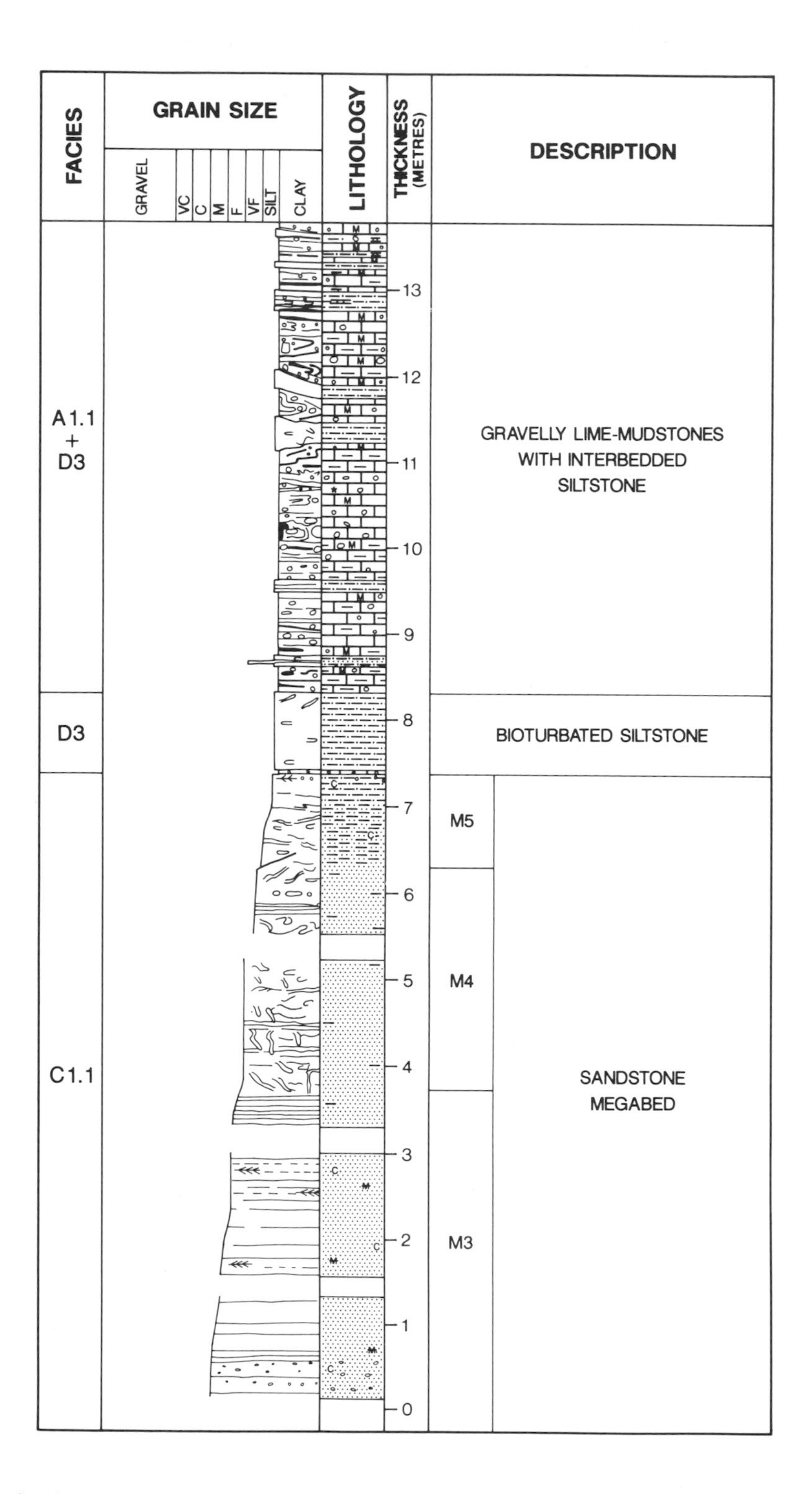

Fig. 4. Core log for well 'B' (see page 106 for key to logs).

well B, and also possibly an underlying muddy gravel (subfacies A1.2), though this is less clearly differentiated from the siltstone of facies D3 on the basis of log character alone. Gravelly coarse-grained sandstones comparable to unit M2 are not present in the core but are possibly present in the uncored section below.

The cored section of the megabed consists of units comparable to those observed in well A. Unit M3 (at least 3.5 m thick) comprises horizontally laminated, upwards-fining, medium- to fine-grained sandstone with concentrations of fine carbonaceous detritus and mica along many of the laminae, and with small greenish grey mudstone clasts towards the base. Horizontal lamination becomes more clearly defined and more closely spaced at the top of the unit. Ripple cross-lamination is not as well developed as in well A. Unit M4 (2.6 m thick) is very fine to fine-grained, broadly fines upwards, and is characterized by water-escape structures (mainly water-escape pipes) which cut horizontal to irregular lamination. The water-escape pipes tend to be inclined in the same direction and dip at *c.* 70°. At the top of the megabed in well B there is an upwards-fining sandy siltstone unit, 1.1 m thick, which is characterized by discontinuous, irregular, and locally overfolded laminae. At the very top of this siltstone is a thin layer of scattered, small, lime–mudstone clasts and carbonaceous detritus. An equivalent sandy siltstone is not evident in well A and this is assigned to an additional unit, M5.

Gravelly lime-mudstones (subfacies A1.1) are predominant in the upper part of the cored interval. The centimetre-scale clasts are matrix-supported and are composed of lime–mudstone, greenish grey mudstone and glauconitic marl. The beds are massive to laminated with occasional overfolded and disturbed lamination defined by changes in clay content. Thin beds of siltstone are present within the upper part of the core. These have abrupt, horizontal to disturbed contacts with the gravelly lime-mudstones and are either greenish grey and bioturbated, or dark grey, calcareous, and horizontally laminated.

Well C

This well is located in Quadrant 22. The illustrated cored section from this well (Fig. 5) contains a sandstone megabed which is 3 m thick. This megabed is similar to the megabed in wells A and B. Underlying the megabed is gravelly mudstone (subfacies A1.2) and bioturbated claystone (facies E1) of the Lista Formation. The overlying Sele Formation contains laminated silty claystone (facies E2) and laminated siltstone (facies D2).

The bioturbated claystone (facies E1) of the Lista Formation occurs in a bed 2.3 m thick beneath the sandstone megabed. It is greenish grey to medium grey in colour with reddish brown mottling. *Chondrites* burrows are evident, and there are occasional thin beds containing small mudstone fragments, thin beds of glauconitic limestone and rare glauconitic horizons.

The gravelly mudstone subfacies (A1.2) underlies the sandstone megabed in an interval 1.4 m thick. It has abrupt contacts with the bioturbated claystone below, and with a thin glauconitic claystone which underlies the sandstone megabed. The gravelly mudstone is dark grey in colour with floating clasts of lime–mudstone up to small pebble grade and with rafts and clasts of mudstone. The gravelly mudstone is massive to poorly laminated and the laminae are often irregular to overfolded. Beneath the part of the core shown on Fig. 5, sandy beds are commonly developed which contain clasts and streaks of lime-mudstone. These are interbedded with bioturbated claystone and grade in places to gravelly mudstones similar to those described above.

The sandstone megabed can be divided into units comparable with those observed in wells A and B. At the base is a 0.3 m thick interval of clast-supported, mud-clast conglomerate which has a silt matrix. This interval is assigned to unit M1. The clasts are of dark grey to greenish grey mudstone which are commonly of small to medium pebble grade. This unit is overlain by 0.6 m of horizontally laminated, clast-rich sandstone assigned to unit M2, which fines upwards from coarse to medium-grained. Clasts in unit M2 comprise grey mudstone and white lime-mudstone. The unit grades upwards into a 2 m thick, fine-grained, horizontally-laminated sandstone, assigned to unit M3. Current ripple cross-lamination is occasionally evident in unit M3 and there are local concentrations of small plant and mudstone fragments. Horizontal lamination becomes more evident towards the top of the unit. At the top of the megabed is a thin, horizontally-laminated unit (0.2 m thick) which fines upwards from very fine-grained sand to silt. This is assigned to unit M5. Unit M4 is not evident in the megabed of well C.

Dark grey, carbonaceous, horizontally-laminated silty claystone (facies E2) of the Sele Formation overlies the sandstone megabed. This passes upwards into dark grey laminated siltstone of facies D2, which in turn is overlain by thin-bedded turbidites (not shown on Fig. 5)

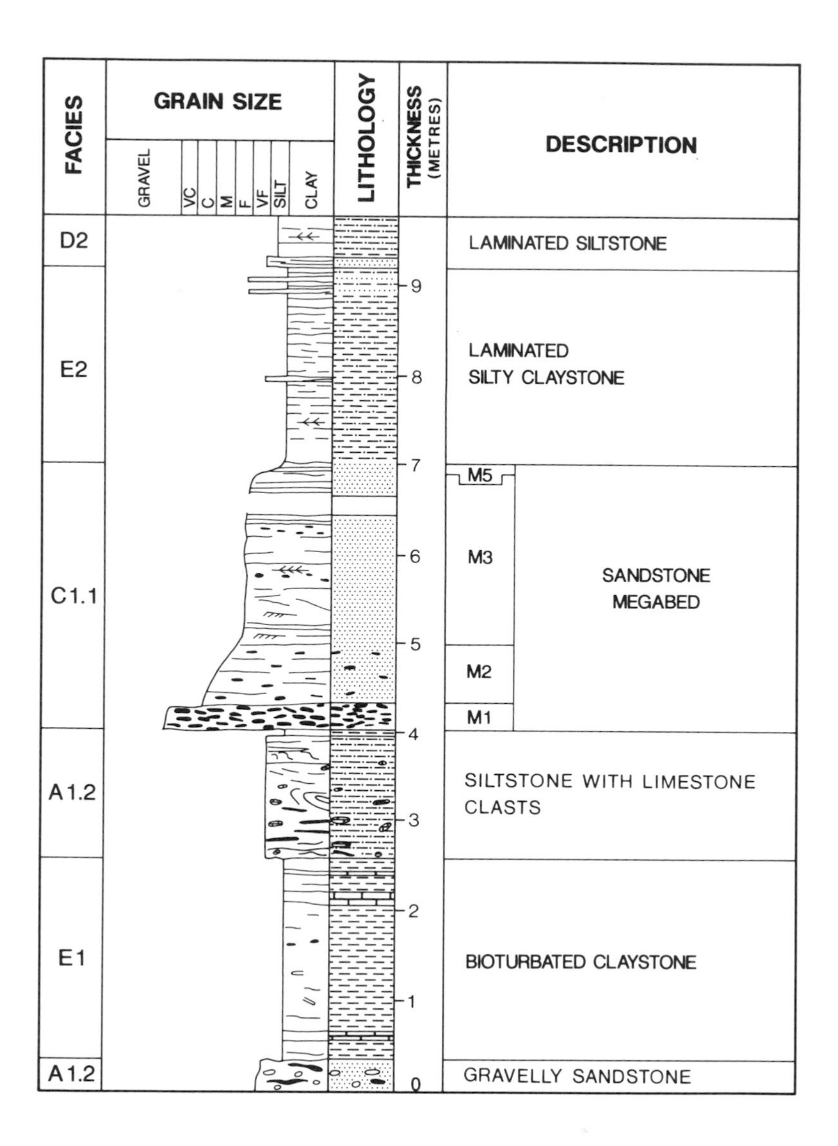

Fig. 5. Core log for well 'C' (see page 106 for key to logs).

which are assigned to the Forties Sandstone Member.

Interpretation

The three described cores show common characteristics: (1) matrix-supported gravelly mudstones and mudstones occur beneath the megabeds; (2) a matrix-supported gravelly mudstone immediately underlies the megabed; (3) the megabeds are predominantly siliciclastic with a subordinate component of lime–mudstone clasts and lime–mudstone matrix; (4) the megabeds fine upwards from a gravelly base to a very fine-grained sandstone or siltstone; (5) they are characteristically parallel laminated; (6) they are unusually thick in comparison to the surrounding beds; and (7) the sediments above the megabeds are different from those below.

Beneath the sandstone megabeds the Lista Formation is characterized by hemipelagic, bioturbated, greenish grey mudstones (facies E1 and D3). These are interbedded with occasional gravelly deposits (subfacies A1.2) which are interpreted as debrites on the basis of their poor sorting, muddy, matrix-supported character, occasional contorted to overfolded lamination, and non-graded to inversely graded character. Limestone clasts are present in the debrites, but mudstone clasts predominate, indicating a different source-rock composition than for subfacies A1.1 above the sandstone megabed. Associated with the debrites are occasional sandy beds (facies C2) and slumped siltstone (facies F1). The former are turbidite-like, being sandy and massive with a few 'floating' clasts.

The sandstone megabed in each instance is an integral upwards-fining unit, which is interpreted as the product of a single event. A idealized sequence of sedimentary structures and grain-size changes is present (units M1–M5) but some units are absent within individual instances. The sedimentary structures can be compared with those of turbidites, but the unusual characteristic of the sandstone megabeds, in comparison with typical turbidites of Tertiary age from the North Sea, is the absence of massive sandstone. The sequences for high-density turbidites described by Lowe (1982) can contain clast-supported conglomerates at the base (divisions R2 and R3) and partly laminated to cross-stratified gravelly sandstones (divisions S1 and S2). The sandstones of divisions S1 and S2 were interpreted, respectively, as traction carpet and traction deposits (Lowe 1982). These conglomerates and sandstones are similar to the units M1 and M2 described here. However, high-density turbidite deposits commonly have a thick interval of massive sandstone (division S3 of Lowe 1982) which is not evident in the sandstone megabeds. The horizontally laminated unit M3 could be compared with Bouma division T_b and could be the result of upper flow regime traction sedimentation, but it is much thicker than is usually observed in North Sea turbidites of Tertiary age. An alternative explanation for the horizontal lamination of unit M3 is that it is the result of shear at the base of a high-concentration turbulent suspension (Postma 1986; Hiscott 1994). Spaced stratification 5–10 cm thick in some turbidites was interpreted by Hiscott (1994) to be the result of 'vigorous burst/sweep cycles' caused by the impingement of large turbulent eddies on the bed, resulting in inversely graded layers 1 cm thick. Although this mechanism is possible for some of the lamination in unit M3, distinct inversely graded laminae are not apparent. Rare current ripple cross-lamination associated with the horizontal lamination of unit M3 indicates that there was at least some traction sedimentation. The more closely-spaced horizontal lamination, and in some places well-developed current ripple cross-lamination, at the top of unit M3, is interpreted to be due to traction sedimentation. The dewatered unit M4 indicates the rapidity of megabed deposition. The inclination of the water-escape structures could be explained by shear caused by the action of the current on the underlying sediment closely following deposition. Unit M4 is possibly an equivalent of Bouma division T_c and the laminated to dewatered argillaceous unit M5 is possibly an equivalent of Bouma division T_d.

The occurrence of a debrite beneath the sandstone megabed in wells A and C is considered to be significant rather than fortuitous. The megabed unit M1 and the underlying debrites differ only in that the former are clast-supported, and the latter are matrix-supported. Clast types in the two deposits are similar; this suggests a genetic link which is discussed in a following section.

In all the wells there is a significant change in sedimentation following deposition of the megabed but this is expressed differently in wells A and B compared to well C. In wells A and B lime-mudstone debrites (subfacies A1.1) are common and interrupt the relatively oxic, background siltstone of facies D3, which is similar to that of the underlying Lista Formation. In well C, lime-mudstone debrites are absent and deposition is marked by hemipelagic anoxic mudstones (facies E2 and D2) of the Sele Formation closely followed by a prograding upwards-coarsening turbidite unit of the Forties Sandstone Member.

Comparison with megaturbidites

The characteristics shared between the megabeds and 'megaturbidites' as defined by Bouma (1987) are: (1) they are unusually thick in comparison to the surrounding beds; (2) they are different in composition (this is most apparent in wells A and B where the megabed is mainly siliciclastic as opposed to the mainly calciclastic nature of the associated gravity flow deposits); (3) they do not appear to be part of turbidite lobe or channel deposits (i.e. they are isolated units); and (4) they were deposited following a single transport event. The extent of the bed is uncertain, however, it is found within an area of several tens of square kilometres within one quadrant. If the megabed in the studied wells is the same bed (which is not known at present) then it has a lateral extent of at least 50 km. The areal extent of the bed, which is more debatable, would be several hundreds of square kilometres.

Bouma (1987) noted that the term 'megaturbidite' normally refers to resedimented carbonate layers that are intercalated within siliciclastic turbidites, and that many are matrix-supported conglomerates which could be interpreted as debrites. However, not all megaturbidites are predominantly calciclastic. Examples of siliciclastic megaturbidites include the 'Black Shell turbidite' from the Hatteras Abyssal Plain, which is < 4 m thick and comprises sand and mud with a minor amount of skeletal material, and the 'Contessa' bed, from the Miocene Marnoso-Arenacea Formation in Italy, which is a 16–20 m thick, fossiliferous quartz-lithic sandstone (Mutti *et al.* 1984). Examples of carbonate deposits of Eocene age from the South Pyrenean Basin of Spain are described by Séguret *et al.* (1984) and Labaume *et al.* (1987). These authors recognized several megaturbidites 10–200 m thick which were thought to have been mobilized by large earthquakes. They recognized internal organization of the megaturbidites into several divisions which include a basal 'mega-breccia' and an overlying thick, graded calcarenite which passes upwards from a breccia via a graded calcarenite to a calcareous mudstone. The basal breccia contains large carbonate slabs and clasts, and was noted to be missing in some cases. Rupke (1976) described the same megaturbidites discussed by Séguret *et al.* (1984) and Labaume *et al.* (1987). The graded calcarenite and mudstone divisions of the beds were attributed to turbidite deposition and were noted to have a combined thickness of up to 41 m. Rupke (1976) recognized a repetitive, though systematically changing, sequence of Bouma divisions which created the impression of internal beds in some instances. This was attributed to oscillatory flow, caused in part by ponding and reflected flows, but possibly also by tributary turbidity currents which were thought to have contributed to the megaturbidite. Johns *et al.* (1981) described one of the thick carbonate beds, the Roncal Unit from the Hecho Group of Eocene age from the south-central Pyrenees. They reinterpreted the basal breccia unit in this case to be the result of transportation in a highly concentrated suspension at the base of a 'giant turbidity current'. Labaume *et al.* (1987) thought that the interpretation of such 'mega-breccias' was problematic but referred to probable debris flow and sliding mechanisms.

Souquet *et al.* (1987) described facies sequences in large-volume debris and turbidity-flow deposits from the Pyrenees of Cretaceous age. They recognized debrites, debrite–turbidite couplets, and turbidites of mixed siliciclastic and carbonate composition. They proposed that there was progressive incorporation of water in a subaqueous debris flow such that it evolved downslope into a turbidity current resulting in a variety of gradational vertical and downcurrent facies sequences and debrite–turbidite couplets. During sea-level lowstands megaturbidites were deposited on the basin plain. These comprise a basal debrite overlain by a normally graded bed showing Bouma sequences.

Bouma (1987) implied that often the term megaturbidite has been applied to debris flow deposits although some megaturbidites were thought to have been deposited by turbidity currents. Mutti *et al.* (1984) illustrated fine- to very fine-grained megaturbidites of Marnoso–Arenacea in Umbria which were described in terms of the Bouma sequence as Tc-e and Tb-e turbidites.

Debrites commonly appear to constitute an important part of megaturbidites, and debrite-turbidite couplets and carbonate or mixed carbonate siliciclastic lithologies appear to be common. The sandstone megabeds described here are also of mixed carbonate/siliciclastic composition and are found above debrites.

Discussion

In well C the megabed occurs at the boundary between the green mudstones of the Lista Formation and the dark grey mudstones of the Sele Formation which are immediately overlain by prograding submarine fan deposits of the Forties Sandstone Member. Stewart (1987) and Hartog Jager *et al.* (1993) interpreted the Forties fan as having been deposited during a relative sea-level lowstand, whereas the upper part of the Lista Formation was deposited during a relative highstand. It therefore follows from this interpretation that the megabed in well C was deposited during, or immediately following, a fall in relative sea level. Although there appears to be a link between relative sea level and megabed deposition in this case such a link has not been noted for other megaturbidites. For example, Labaume *et al.* (1987) noted that megaturbidites in the southwest Pyrenees were randomly scattered throughout the stratigraphic succession, and in particular are not associated with, or mark, sequence boundaries.

A link between megaturbidite deposition and major earthquakes has been proposed by Mutti *et al.* (1984), Séguret *et al.* (1984) and Labaume *et al.* (1987). This link led Mutti *et al.* (1984) to define some megaturbidites within highly tectonically mobile basins as 'seismoturbidites'. The megabeds described here could have been triggered by a major earthquake, but there is no direct evidence for this. The debrite which underlies the megabeds is interpreted to represent the failure of slope sediments by sliding and slumping. This would have resulted in the instability of sand deposits at the shelf edge, and gullies and slump scars resulting from slope failure would have acted as conduits for turbidity currents which carried sand into the basin. Souquet *et al.* (1987) explained megaturbidite–debrite couplets in sediments of Cretaceous age from the Pyrenees as the result of the downslope transition of debris flows into turbidity currents. However, such a mechanism does not adequately explain the spatial and temporal relationships. If a viscous debris flow were to evolve into a turbidity current in the manner proposed then why should turbidity current deposits overlie the deposits of the debris flow at the same point in space? In one scenario this would require the debris flow to partially evolve into a turbidity current which then lagged behind the debris flow, which does not appear likely. Alternatively, the relationships could be explained if the turbidity current was ponded and reflected. Rupke (1976) did note some evidence for flow oscillation in the megaturbidites of Tertiary age from the Pyrenees and proposed such a ponding mechanism. In the case of the megabeds described here, evidence of flow reversal and ponding, in the form of repetitive grain-size and structure changes, is not discernible. No palaeocurrent data from the ripple cross-lamination in the megabeds that might indicate flow reversal are available. The sandy nature of the turbidite argues against derivation from the mud-dominated debrite. Therefore, it is proposed that there were two separate but closely linked events, a debris flow resulting from slope failure, followed closely by a turbidity current. The exact mechanisms which link the two are debatable.

The mainly siliciclastic composition of the megabed indicates derivation from shelf/deltaic sands bordering the Shetlands Platform or Scottish Highlands regions. There are chalk fragments and some lime-mud matrix in the sandstone megabed of wells A and B which represent whole and disaggregated rip-up clasts, respectively. In wells A and B there is a significant increase in carbonate debris-flow activity following deposition of the sandstone megabed. This is interpreted to be due to the uplift and subsequent instability of local chalk highs which acted as a source for these debrites, and this could support a potential earthquake trigger for the megabed. In this case the trigger for megaturbidite deposition was basinwide and not only affected local carbonate highs, but also initiated slope failure at the shelf margin. Alternatively, increased carbonate debris-flow activity may reflect sea-level fall where the highs were brought above storm wave base, and both mechanisms might have operated at the same time.

Conclusions

Three sandstone megabeds of late Palaeocene age from the Central Graben of the North Sea are interpreted as the deposits of turbidity currents. These beds share unusual characteristics in comparison with typical turbidite fan deposits of similar age. They are interpreted as megaturbidites, as defined by Bouma (1987). Deposition was from a single turbidity current which waned in energy, resulting in the deposition of a type vertical sequence of grain size and sedimentary structures defined as units M1–M5. A debrite is found directly beneath the megabeds and this is interpreted to be genetically linked to the megabeds. The debrites are interpreted to represent shelf-margin slope failure, possibly triggered by earthquake activity. Megabed deposition closely followed debrite deposition and is interpreted as a separate depositional event. The exact mechanisms which linked debris flow and turbidity current generation are conjectural without further data.

There is strong evidence from well C for a link between a fall in relative sea level and megabed generation. This is not demonstrable at present for wells A and B. Further data that can be released in the public domain are necessary to explore the possible link between megabed generation and allocyclic mechanisms such as sea-level fall and tectonic activity.

I thank Shell UK Exploration and Production Limited, Esso Exploration and Production UK Limited, Phillips Petroleum Company UK Limited, AGIP (UK) Limited, and British Gas Exploration and Production Limited for permission to utilize core data. All interpretations and opinions are those of the author and do not necessarily reflect those of the companies concerned.

References

Bouma, A. H. 1987. Megaturbidite: an acceptable term? *Geo-Marine Letters*, **7**, 63–67.

Hartog Jager, D. Den, Giles, M. R. & Griffiths, G. R. 1993. Evolution of Palaeogene submarine fans of the North Sea in space and time. *In:* Parker, J. R. (ed.) *Petroleum Geology of Northwest Europe: Proceedings of the 4th Conference,* Geological Society, London, 59–71.

Hiscott, R. N. 1994. Traction-carpet stratification in turbidites – fact or fiction? *Journal of Sedimentary Research,* **A64,** 204–208.

Johns, D. R., Mutti, E., Rosell, J. & Séguret, M. 1981. Origin of a thick, redeposited carbonate bed in Eocene turbidites of the Hecho Group, south-central Pyrenees, Spain, *Geology,* **9,** 161–164.

Knox, R. W. O'B. & Holloway, S. 1992. 1. Palaeogene of the Central and Northern North Sea. *In:* Knox, R. W. O'B. & Cordey, W. G. (eds.) *Lithostratigraphic Nomenclature of the UK North Sea.* British Geological Survey, Nottingham.

Labaume, P., Mutti, E. & Séguret, M. 1987. Megaturbidites: a depositional model from the Eocene of the SW-Pyrenean Foreland basin, Spain, *Geo-Marine Letters,* **7,** 91–101.

—— Séguret, M. & Mutti, E. 1986. Calcareous megaturbidites of Eocene South Pyrenean foreland basin. *American Association of Petroleum Geologists Bulletin,* **70,** 609.

—— Mutti, E., Séguret, M. & Rosell, J. 1983. Mégaturbidites carbonatés du bassin turbiditique de l'Éocène inférieur et moyen sud-Pyrénéen, *Bulletin Societé Geologie, France,* **XXV (6),** 927–941.

Lowe, D. R. 1982. Sediment gravity flows: II. Depositional models with special reference to the deposits of high-density turbidity currents. *Journal of Sedimentary Petrology,* **52,** 279–297.

Mutti, E., Ricci Lucchi F., Séguret, M. & Zanzucchi, G. 1984. Seismoturbidites: a new group of resedimented deposits. *Marine Geology,* **55,** 103–116.

Pickering, K. T., Hiscott, R. N. & Hein, F. J. 1989. *Deep Marine Environments, Clastic Sedimentation and Tectonics.* Unwin Hyman, London.

Postma, G. 1986. Classification for sediment gravity-flow deposits based on flow conditions during sedimentation. *Geology,* **14,** 291–294.

Rupke, N. A. 1976. Sedimentology of very thick calcarenite-marlstone beds in a flysch succession, southwestern Pyrenees. *Sedimentology,* **23,** 43–65.

Séguret, M., Labaume, P. & Madariaga, R. 1984. Eocene seismicity in the Pyrenees from megaturbidites of the South Pyrenean basin (Spain). *Marine Geology,* **55,** 117–131.

Souquet, P., Eschard, R. & Lods, H. 1987. Facies sequences in large-volume debris- and turbidity-flow deposits from the Pyrenees (Cretaceous; France, Spain). *Geo-Marine Letters,* **7,** 83–90.

Stewart, I. J. 1987. A revised stratigraphic interpretation of the Early Palaeogene of the central North Sea. *In:* Brooks, J. & Glennie, K. (eds.) *Petroleum Geology of North-West Europe.* Graham and Trotman, London, 557–576.

Structurally-controlled deep sea channel courses: examples from the Miocene of southeast Spain and the Alboran Sea, southwest Mediterranean

BRYAN T. CRONIN

The Marine Geoscience Research Group, Department of Earth Sciences, Cardiff University, PO Box 914, Cardiff CF1 3YE, UK

Abstract: Two deep water channels, the Tortonian (Upper Miocene) Solitary Channel in the Tabernas Basin in southeast Spain, and Almeria Canyon (Recent) off southeast Spain, are compared and contrasted. Both channels are developed at a similar scale and allow comparison between modern and ancient. Both are sinuous, trunk bypass feeders located in tectonically active, high-gradient turbidite settings, where strike-slip fault patterns appear to deflect channel courses by up to 90°. Sandy turbidites and debris flows dominate the channel-fills, with thick sequences of mudstone deposits outside the channels.

The exhumed Tabernas Solitary Channel is associated with the margin of a Tortonian strike-slip basin in the Betic Cordillera of southeast Spain. It is up to 200 m wide and 40 m thick, and can be mapped almost continuously for 11 km in the field. Half-way along its course the north-south oriented channel is deflected through almost 60° by a strike-slip fault, and from here it meanders southwestwards to terminate abruptly in basin-floor mudstones. The Solitary Channel fill is characterized by four separate fill units which are identified along the length of the channel course, and these are: (1) clinoform surface unit; (2) thin-bedded turbidite unit; (3) massive thick tabular sandstone and debris flow unit; and (4) thin-bedded turbidite unit. The units are inferred to be related to fluctuations in transport processes within the canyon, from large bedform migration to low-density turbidity currents. A conglomeratic thalweg plug is found in the lower channel reaches.

The Almeria Canyon, in the southwest Mediterranean, has a similar planform geometry. The channel is > 24 km long, varies in width from 250–500 m, and varies in depth from 40 to 10 m. The channel is deflected 90° half-way along its length by a fault escarpment. The channel has a variable sinuosity. One meander has a sinuosity value of 3.8, three meanders have sinuosities within the region of 2.0, and elsewhere sinuosity is more commonly in the order of 1.2. The channel is largely, unfilled, though upper reaches contain material which has either slumped into the channel from the walls or been transported down the channel by debris flow. The channel also has a thalweg channel in its lower reaches, which appears braided.

The observations have a number of implications. (1) The planform geometry of the Solitary Channel is very similar to that of the Almeria Canyon, indicating predictable effects of structural control on conduits in tectonically-active settings. (2) The Solitary Channel extends 2 km further into the basin than two other systems previously interpreted from one palaeogeographic timeslice, and this could be explained by structural confinement of the channel course. (3) The Solitary Channel fill is characterized at lower levels by large sandy bedforms indicating strong current conditions. (4) The distinction can be made between relatively long-lived sand-conduits in otherwise mud-prone sectors of basins, such as the two examples in this paper, and short-lived, sand-prone and mobile channels in more sandy areas.

The comparison of modern and ancient deep water channels has traditionally been frustrated by the nature of modern and ancient datasets. Even exceptional ancient exposures of channels usually show two-dimensional cross-sections of filled channels, and most published work on modern channels has focused on processes operating within the channels at the present day, and the planform characteristics of unfilled channels. The scale bias of modern and ancient datasets has led to the development of deep marine clastic models which differ markedly

From Hartley, A. J. & Prosser, D. J. (eds), 1995, *Characterization of Deep Marine Clastic Systems*, Geological Society Special Publication No. 94, pp. 115–135.

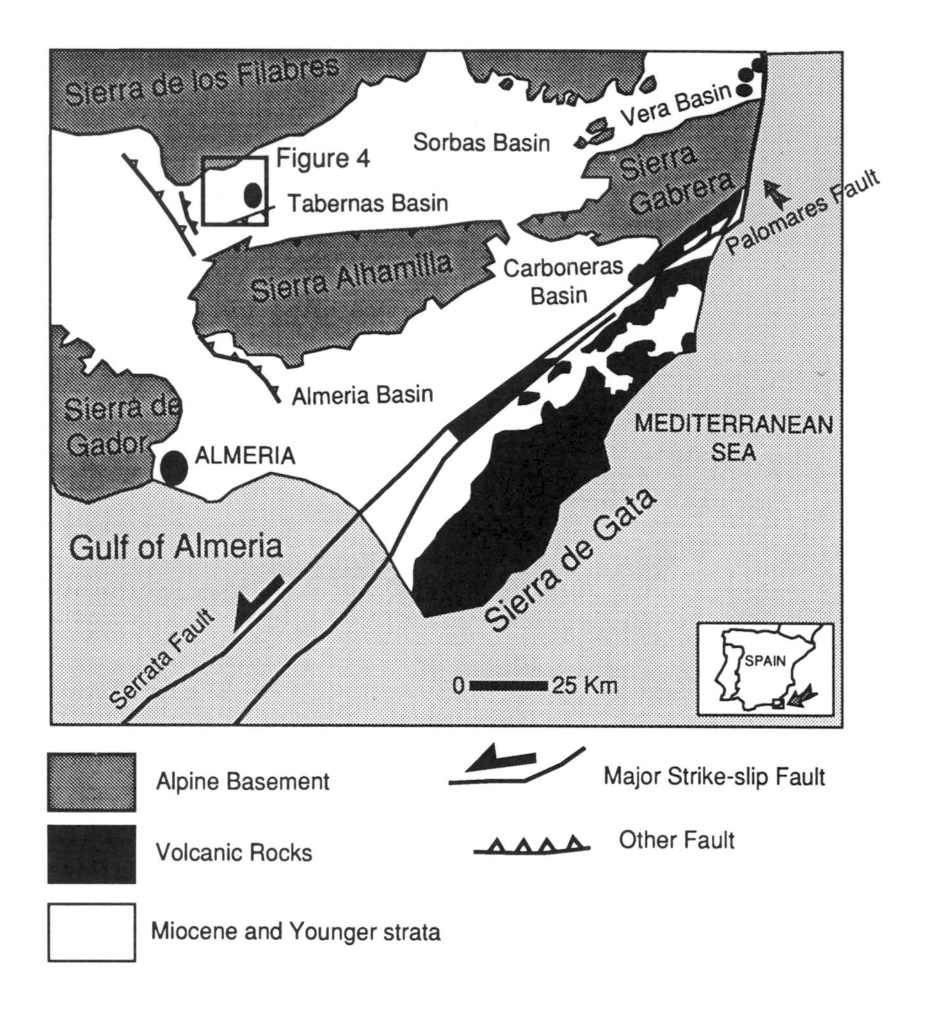

Fig. 1. Location map of the Tortonian (Miocene) Tabernas Basin, SE Spain.

from each other, and this has led to similar treatment of channels. The recent developments in sidescan sonar technology now allow field-scale levels of planform and sometimes cross-sectional analysis of channels.

The objectives of this paper are to compare two deep sea channels from similar tectonic settings, one from the Recent in relatively deep water on a slope dominated by gravity-flow processes, the other from a small Miocene basin infilled with debris flows and turbidites. These modern and ancient datsets can be integrated to produce models which improve our understanding of deep water systems, by using course deflection of modern and ancient deep water channels as an example; and to discuss the nature of long, sand-filled conduits in otherwise muddy sequences by studying the architecture of their fills. The abundance of linear bodies of coarse material, propagating long distances from the basin margin in comparison with standard fans, has implications for deep marine clastic hydrocarbon reservoirs as a largely unexplored stratigraphic hydrocarbon trap.

Two long (> 11 and > 30 km) deep water channels of similar dimensions and geometries from the same orogenic front are compared: the Tabernas Solitary Channel, Tabernas Basin, SE Spain (Kleverlaan 1989; Cronin 1994) and the Almeria Canyon in the SW Mediterranean. The

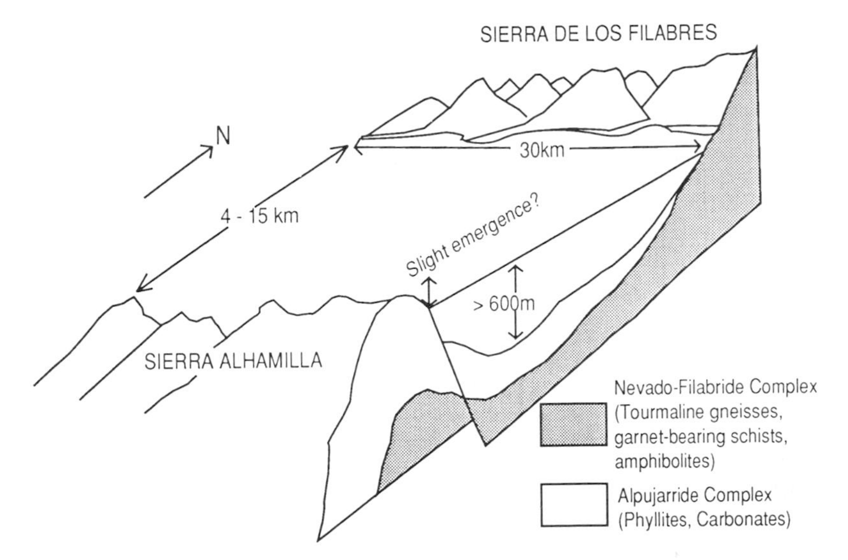

Fig. 2. Diagram showing the relative locations of the primary sediment source areas for the Tortonian–Messinian Tabernas Basin.

paper focuses on the Solitary Channel dimensions, fill architecture and planform nature, the similarities with the Almeria Canyon planform and scale are then discussed.

Example 1: the Solitary Channel, Tabernas Basin, SE Spain

The Tabernas Basin is a small, elongate, Neogene basin in the Betic Cordillera of southeast Spain (Fig. 1) (02°23′37°03′). It is situated between two mountain ranges which form the basin margins, and which comprise part of a crustal shear zone (Montenat *et al.* 1987), the Sierra de los Filabres to the north and the Sierra Alhamilla to the south (Fig. 2). The eastern foothills of the Sierra Nevada rise up further to the west. Two other Neogene basins, the Sorbas and Vera Basins, are situated to the east. The Tabernas and Sorbas Basins were contiguous in the Tortonian but had different basin histories in the Messinian (Weijermaars *et al.* 1985). The Tabernas Basin stretches for over 30 km east–west and has a maximum north–south width of 15 km. It consists of up to 1500 m of continental, shallow and deep marine sediments of Serravallian to early Messinian age. Almost continuous exposure of the sedimentary sequence is afforded by the barren nature of the area and by a complicated network of wadis (called Ramblas), which improve the three-dimensional aspect of the outcrop.

Regional palaeogeographic setting

The Betic Cordillera is the westernmost of the orogenic belts created by the relative compressive motion between the European and North African plates (Ziegler 1982). This tectonic setting was dominated by compressive stresses which began in the Eocene and continued through various stages into the Pliocene. As part of an active fold belt, many mountain ranges formed during this period, and these were separated by small but often quite deep intervening basins. These Neogene basins are therefore located on the sites of Late Cretaceous to Palaeogene orogens generated by collisional stacking. For most of the Late Miocene in particular, much of southeast Spain consisted of small interconnected intramontane basins which were separated by 'islands' of the metamorphic basement. This basin network has been referred to as a series of Miocene 'corridors' which connected the Atlantic to the Mediterranean while the Gibraltar Straits were closed (Weijermars 1991). The emergent areas would have been sediment source areas for the intervening basins. The present-day situation is different, with the Alboran Sea being the oceanographic gateway between the Atlantic and the Mediterranean. It is thought that the Alboran Sea was not the direct gateway until after Messinian isolation of the Mediterranean (Adams *et al.* 1977).

The Tabernas Basin is an ancient 'gateway' basin. The marine fill of the Basin is *c.* 900 m

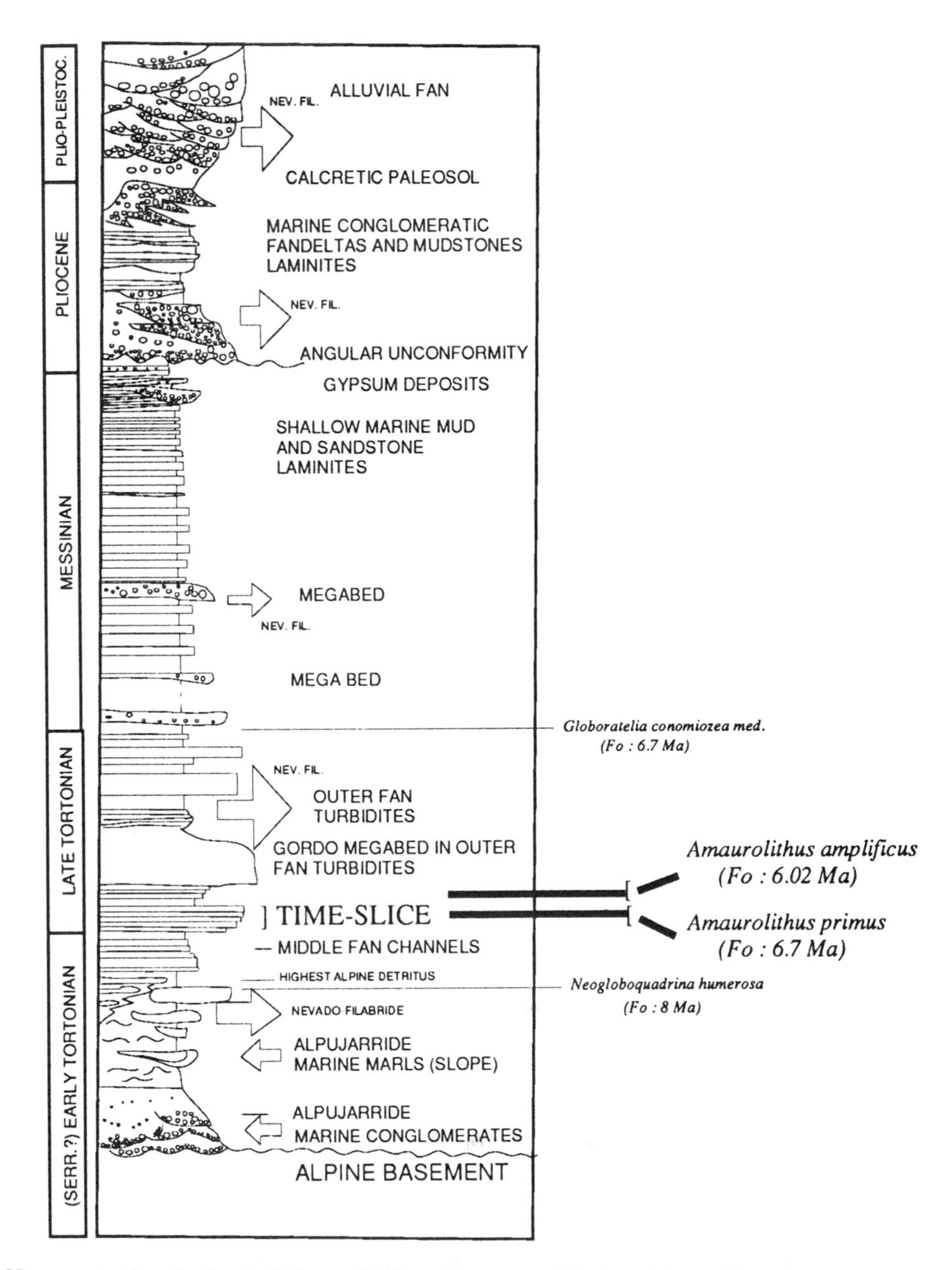

Fig. 3. Neogene stratigraphy for the Tabernas Basin, with new coccolith dates (*A. amplificus* and *A. primus*). Foraminifer dates (*G. conomiozea* and *N. humerosa*) from Kleverlaan (1989). After Kleverlaan (1989).

thick (Kleverlaan 1989), and is mainly composed of a deep water fan complex overlain by shallow marine deposits (Fig. 3). The fill extends from the Late Serravallian to the Early Messinian, overlying the Alpine Basement (Triassic and older) and overlain by Pliocene transgressive deposits. Of this 900 m, only the upper and lower 50 m are not 'submarine fan' deposits. The lowest *c.* 50 m are shallow marine marls and conglomerates and the uppermost *c.* 50 m are Messinian shallow marine muds and sandstone laminites.

Previous work

The principal features of the Tabernas fan complex and the controlling Tortonian basin relief have been described by Kleverlaan (1982, 1987, 1989). He estimated that the northernmost coastal sequence, which includes limestone reefs and oyster beds, was situated at least 10 km from the most distal fan lobes (Fig. 4). The Sierra de los Filabres were at this time emergent and the northern margin of the basin had a steep cliff coast fringed by coral reefs. The basin slope faced south and southwest, and the basin was unconfined or semi-confined to the south. The main sediment source area was to the north and the basin deepened to the south and abutted against a non- to slightly-emergent ridge which later went through its major phase of growth. The structural evolution of the Betic Cordillera is the product of a complicated pattern of relative motions between the African and European plates (Dewey *et al.* 1973). The relationship between the Betic orogen and the evolution of the Alboran Sea is summarized elsewhere (Weijermars *et al.* 1985; Channel & Mareschal 1989; Platt & Vissers 1989).

The Tortonian–Messinian submarine fan deposits within the Tabernas Basin are composed of thick sequences of thickly-bedded coarse-grained deposits. Megabeds have been found in the fill (cf. Mutti *et al.* 1984), which suggests that the basin was strongly influenced by tectonic activity (e.g. Kastens 1984; Bourrouilh 1987; Kleverlaan 1987; Labaume *et al.* 1987). One such megabed (Fig. 3), the Gordo Megabed, was described by Kleverlaan (1987). The abundance of such thick beds implies a steep and unstable slope margin to the north. Three distinctive deep water fan systems have been described from the Tabernas Basin by Kleverlaan (1982, 1989). Two of the three systems have recognizable radial geometries, both with feeder channels and related channel mouth deposits. They are distinguished from each other by different sandstone-to-shale ratios and channel-fill characteristics. The first system is sand-rich (Cossey & Kleverlaan 1994) and the second is a mixed sand-mud system. The third is a sand-rich elongate system which principally consists of one long channel completely isolated within basinal (sand-starved) mudstones. The stratigraphic relationship between the third (solitary) system and the other two is not clear. All three occur between two known biostratigraphic markers, the planktonic foraminiferal datums *Globoratalia conomiozea med.* and *Neogloboquadrina humerosa* (Kleverlaan 1989) (Fig. 3). This biostratigraphic control, combined with the use of common basinwide and local lithostratigraphic markers such as the Gordo megabed, make possible the reconstruction of a summary palaeogeography for the Late Tortonian (7–8 Ma), covering *c.* 6000–60 000 years (Kleverlaan 1982, 1989).

The palaeogeographic reconstruction of Kleverlaan (1989) was used as a basis for this field study. His System Three (Solitary Channel) was analysed in detail along its length using photomosaics and closely-spaced sedimentary logs for semi-quantitative analysis of the channel-fill architecture. Kleverlaan (1989) described the channel as being linear, slightly sinuous and usually sand filled, 10–30 m wide and 10–25 m thick, extending at least 2 km beyond Systems 1 and 2, and ending in a small lobe. He interpreted the channel to have been filled by sandy turbidites and to have persisted for some time (recording 66 major cut-and-fill sequences). The fill was considered to have plugged the channel with the subsequent subdued channel relief resulting in flows spilling on to the unconfined channel margins to form overbank deposits. There are no extensive channel mouth deposits, which Kleverlaan (1989) interpreted as being the result of flow-stripping. (cf. Piper and Normark 1983). Kleverlaan suggested that the channel form was confined within the basin, perhaps following an intrabasinal faultline scarp.

The Solitary Channel provides an excellent example of a long-lived channel, isolated in a mud-prone setting. It has a complicated multistorey fill architecture, and comprises many planform and cross-sectional features which can be compared with the other, modern, case study in this paper.

Mapped localities

A re-interpreted planform of the Solitary Channel is shown in Fig. 4. It was mapped over early summer when vegetation was minimal, using aerial photographs, walking out ridges, and tracing sandstone bodies across the many strike-slip faults that criss-cross the area. Sites 1 and 2 were chosen for detailed analysis (Figs 4 & 5). Site 1 (Fig. 6) was chosen to illustrate the cross-section near where the channel course is deflected half-way along its course. Most of the fill architecture analysis was conducted at Site 2, due to the lateral continuity of the channel through a series of hills, cliffs and Rambla cuts. A series of detailed photomosaics and line-drawings were constructed and detailed sedimentary logs were drawn through the deposits

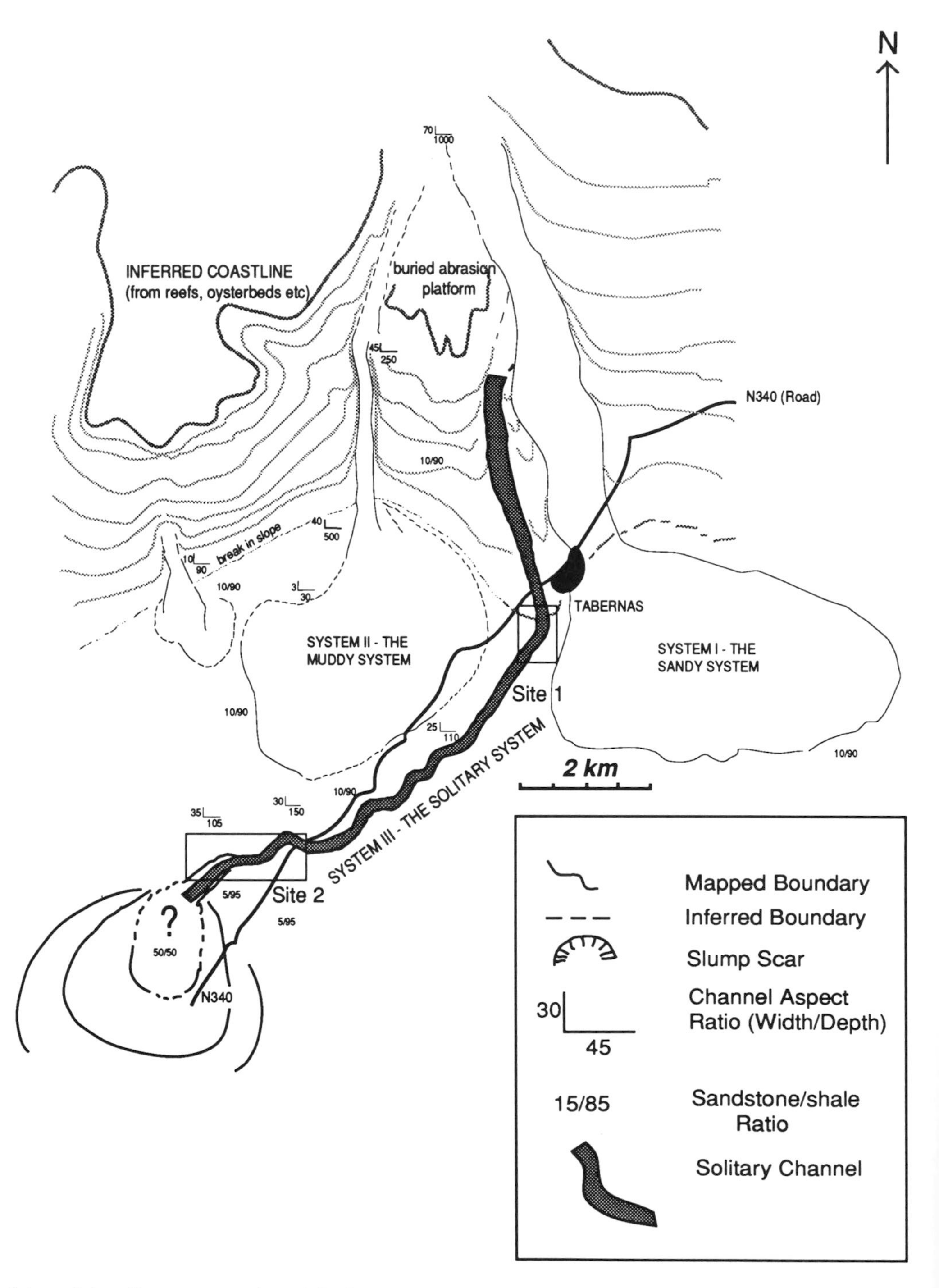

Fig. 4. Adapted timeslice reconstruction of the Tabernas Basin, showing the reinterpretation of the Solitary Channel (after Kleverlaan 1989).

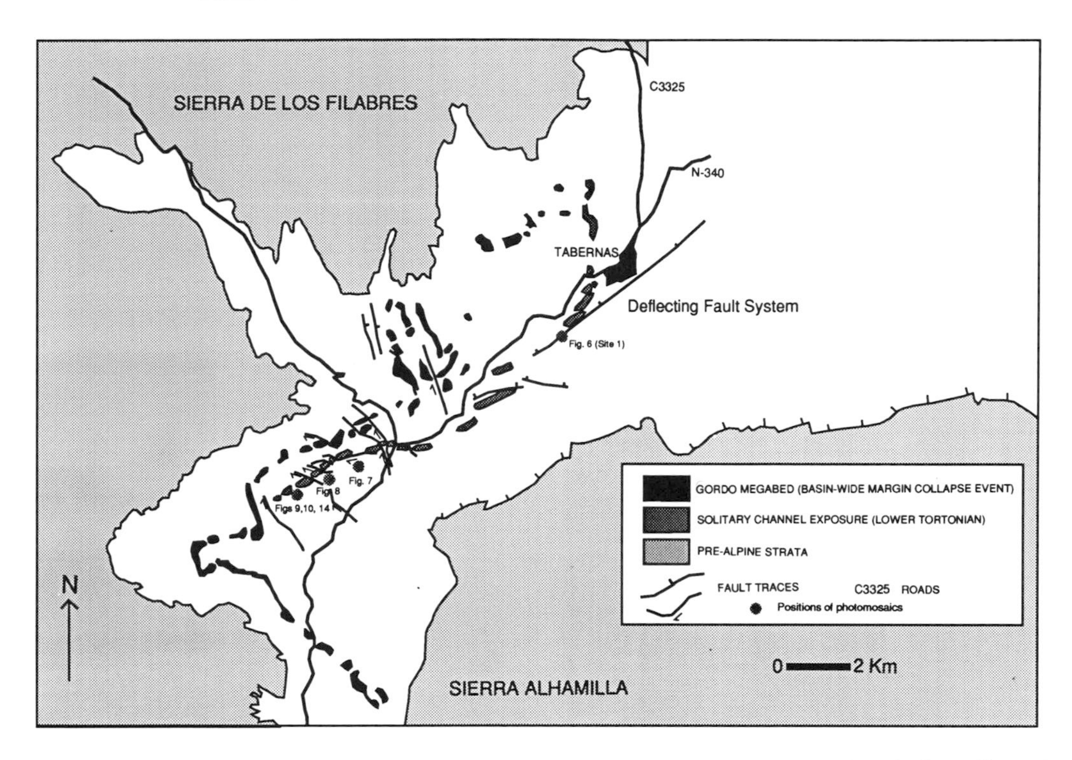

Fig. 5. Sketch geological map of: 1, outcrops of the Gordo Megabed; and 2, outcrops of the Solitary Channel, in the Tabernas Basin (after Cronin 1994).

Fig. 6. Site 1. View towards the north of the Solitary Channel, immediately south of Tabernas village. The right-hand (northern) margin is seen. Cliff section above muds is almost 35 m high.

below the channel, across the channel base and through the fill sequence (Figs. 7–13). Photography was used to connect surface areas not transected by logs.

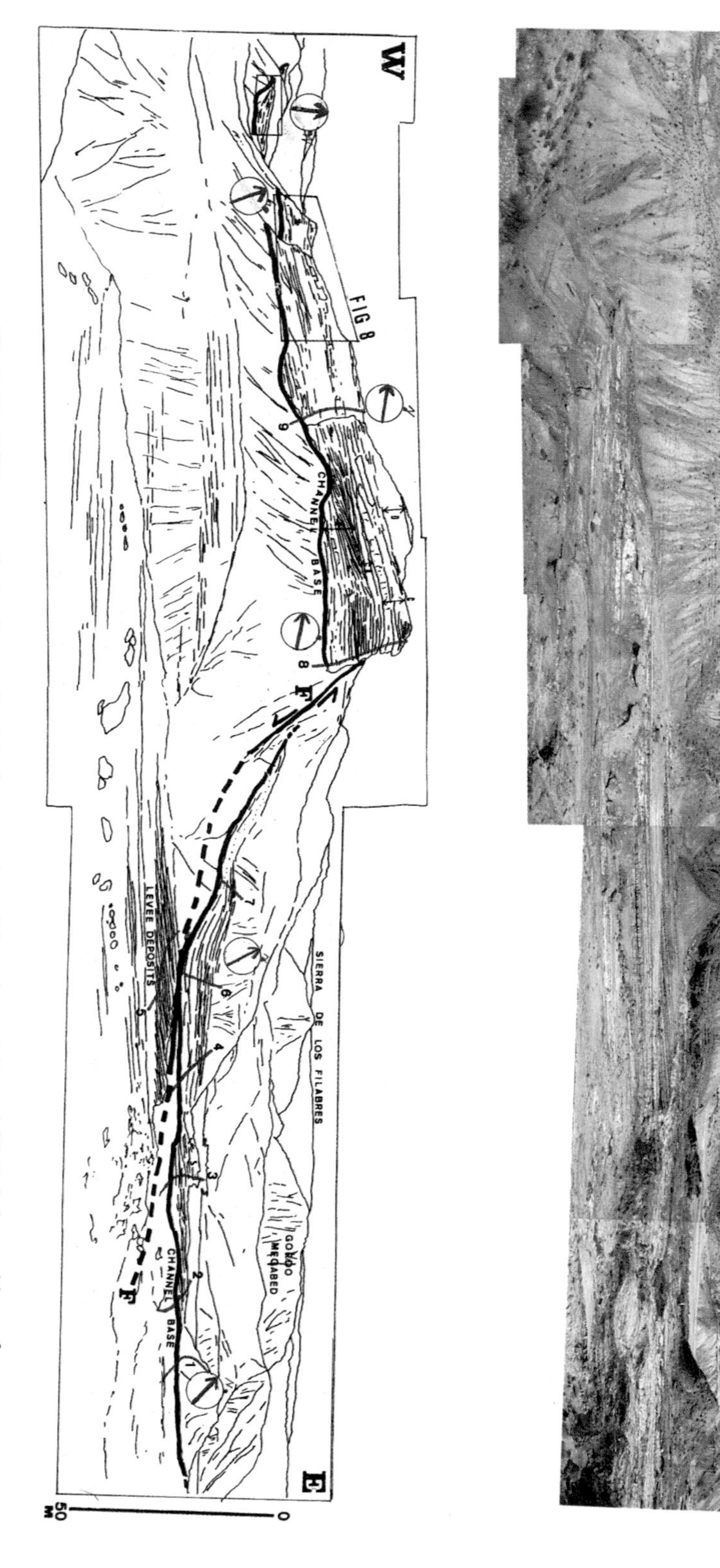

Fig. 7. Site 2. Photomosaic showing much of the lower reaches of the exhumed channel fill. Relative positions of sample sedimentary logs (Fig. 11), mean palaeocurrent direction, and fill units are indicated. Note the position of the Gordo Megabed and the northern source area (Sierra de los Filabres).

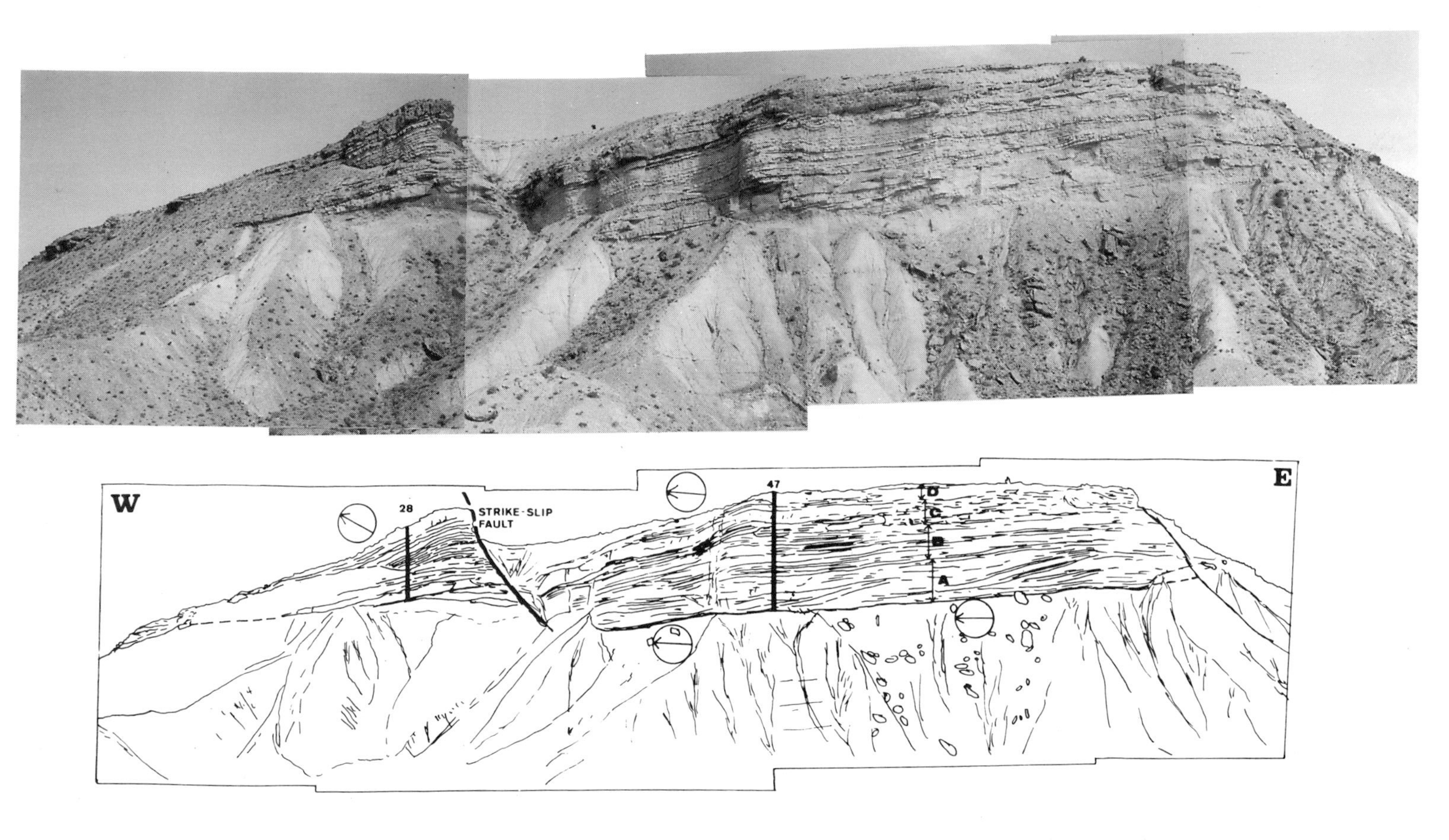

Fig. 8. Site 2. Photomosaic of section of the Solitary Channel mostly cut parallel to palaeocurrent direction, showing prominent downstream accretion surfaces (fill sequence A in figure). The channel meanders from right to left and bears into the cliff face. Four fill units have been identified: A, clinoform surface unit; B, thin-bedded turbidite unit; C, massive thick tabular sandstone and debris flow unit; and D, thin-bedded turbidites. The channel overlies sand-starved slope/basin plain deposits. The sections with numbers 28 and 47 are number of counted sandstone beds. The thickest fill section is 37 m. See fig. 7 for location.

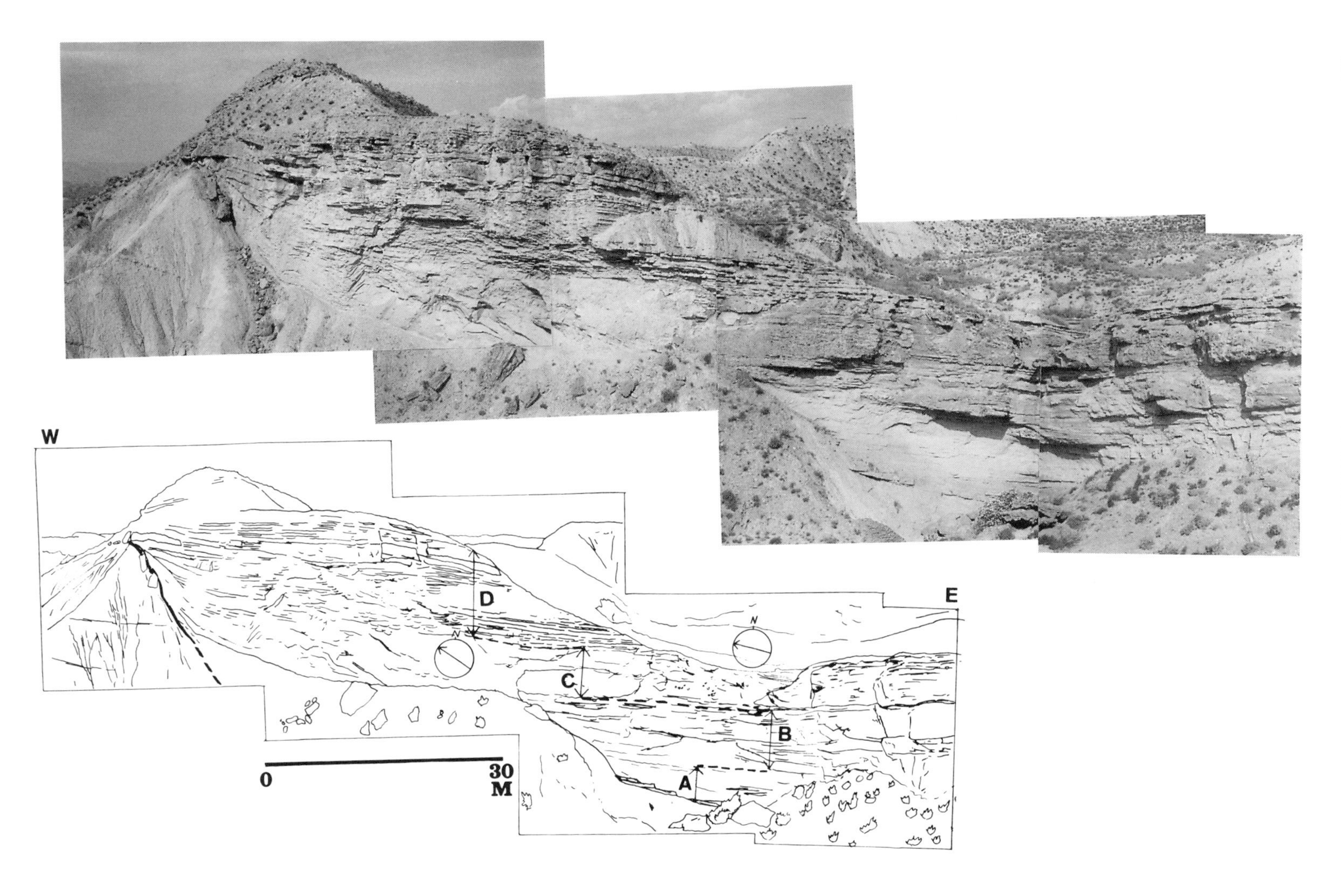

Fig. 9. Site 2. Photomosaic of distal exposure of the channel. Palaeocurrent direction is obliquely (20–25°) away to the left. Note the steep left-hand margin, where beds onlap and overtop during deposition of sequence D.

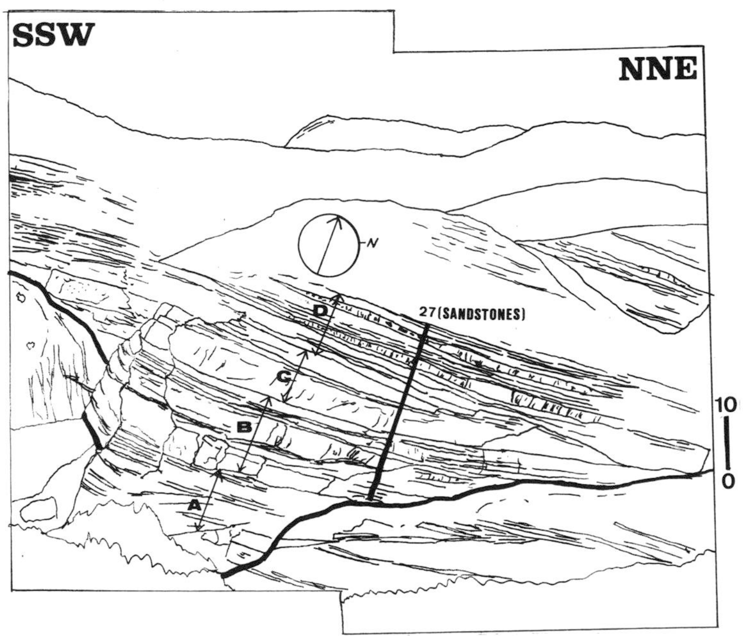

Fig. 10. East-facing outcrop showing the right-hand margin of the channel perpendicular to palaeocurrent direction. Figure 9 shows the continuation of this section to the left. Note repetition of the sequences A–D. Also note the complex cut-and-fill relationships and onlap surfaces on the right-hand margin. Outline shows position of Fig. 12.

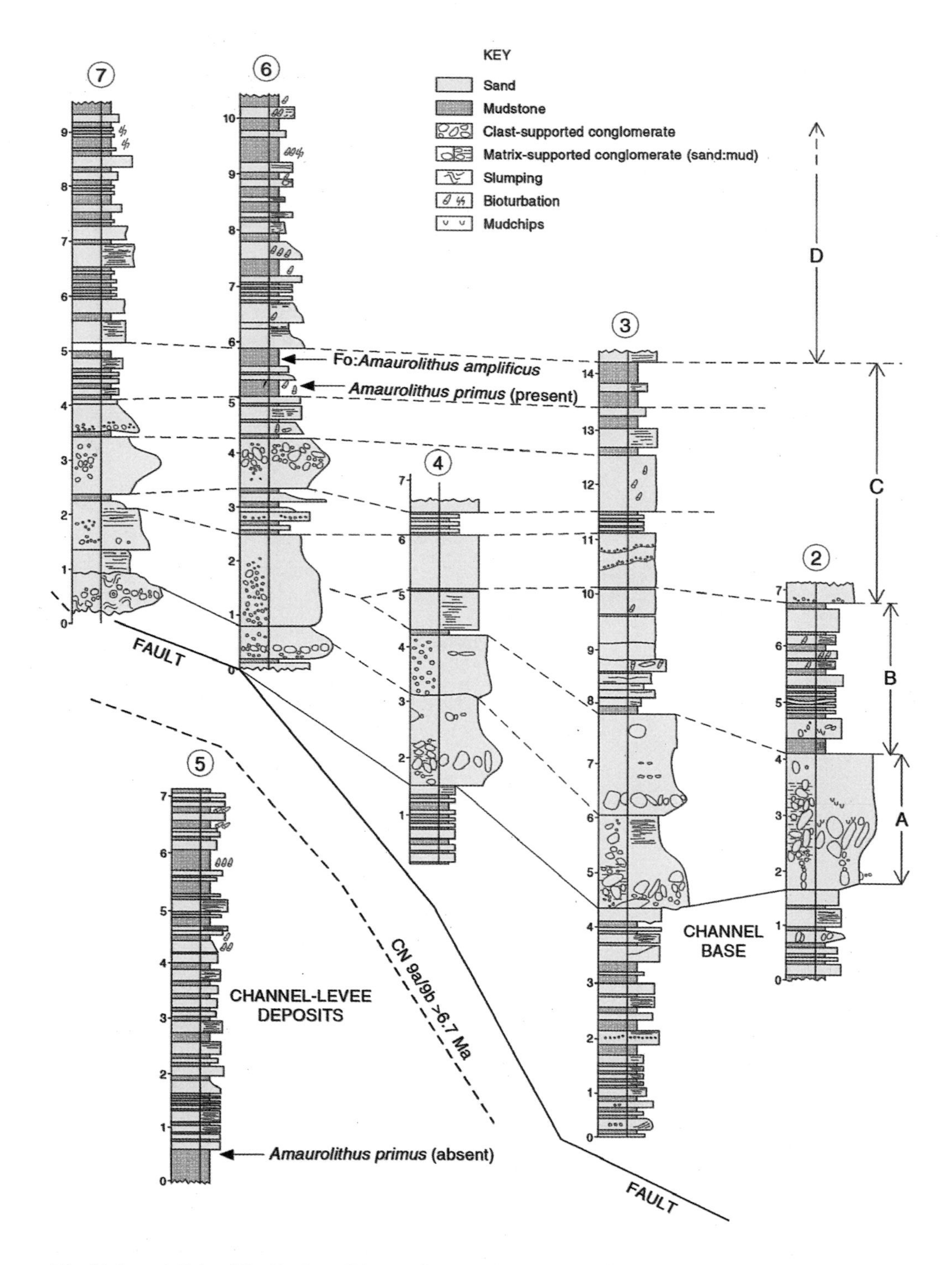

Fig. 11. Logs 1–7 (see Fig. 7). Coccolith sampling positions are indicated. Channel levee deposits in log 5 pinch out over 50 m to the west. See text for discussion. Fo, first occurrence.

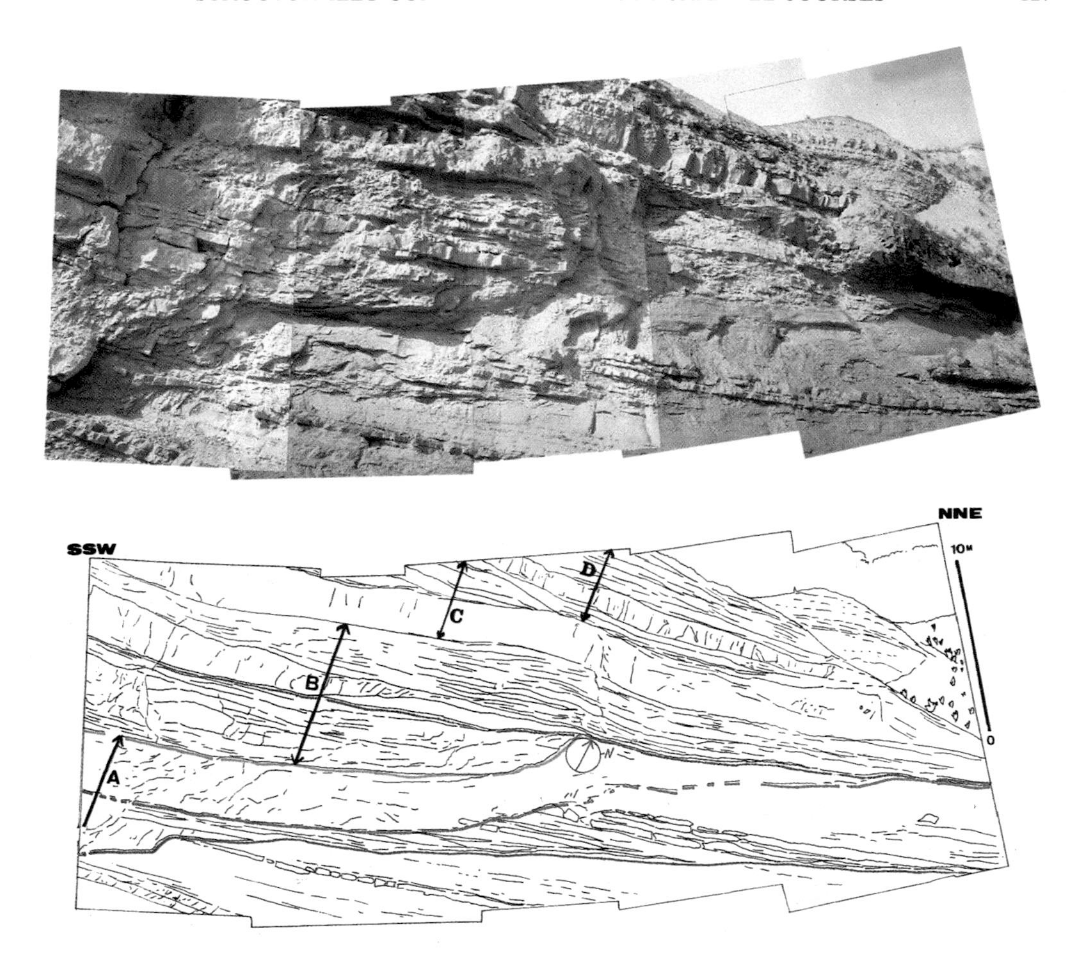

Fig. 12. Detail (from Fig. 10) showing successive overprinting of cut-and-fill packages at the right-hand margin of the Solitary Channel.

Solitary Channel dimensions and extent

The Solitary Channel can be mapped from north of Tabernas village, to the west of System 1 (Fig. 5). Immediately south of Tabernas village, a complete two-dimensional cross-section of the channel is exposed in the Rambla de Tabernas (Fig. 6), where the channel fill sequence is 30 m thick. Although not logged in detail, upwards of 30 sandstone and conglomerate beds can be seen on the photomosaics. At this locality the channel has reached a prominent strike-slip zone (Platt *et al.* 1983), with > 40 m of dip-slip component, which has deformed much of the sedimentary sequence immediately south of the locality. The channel 'dog-legs' at this locality, and its course is mapped towards the southwest.

Beyond Site 1 and further to the southwest (Fig. 4), the channel follows the northeast–southwest fault trend for a further 8 km (Fig. 5). It is exposed mainly along escarpments, and is very prominent as it is nested in sand-starved basin muds (Figs 7–10, 12 & 13). Occasionally it is dislocated by N/NW-trending Quaternary strike-slip faults which rarely display > 10–20 m displacement. The channel-deflecting fault is a reactivated normal fault/strike-slip fault active from Tortonian-Pliocene time (Bousquet 1979). The cross-sectional profile of the channel is often asymmetric. The southern bank is very steep (> 80°) and the northern margin is typically 45°. At the channel mouth to the extreme southwest, no evidence was found for the presence of channel mouth sandstones or a more well-developed lobe as is the case for the other two systems.

Fig. 13. Bedforms in section parallel to mean palaeocurrent direction in Unit A of the Solitary Channel. Note that surfaces, dipping to the left (west), are truncated at their tops. The cliff is 30 m high.

Solitary Channel fill architecture

The fill architecture of the Solitary Channel has been studied in detail at two sites (Fig. 4) and is illustrated in detail in Figs 6–13. North of Site 1 the channel is not well exposed, and the fill comprises thick conglomeratic units (< 4 m) which are at least 50 m wide. These units are normally capped by thin (0.05–0.35 m) yellow pebbly sandstones. The conglomerates have non-erosive bases. The channel is overlain and underlain by slope mudstones. It has a north-south orientation and is exposed parallel to the larger sandy system's feeder channel 0.5–0.8 km to the east (Kleverlaan 1989) but cannot be traced to the inferred source area, which is considered to be a valley cut into metamorphic basement, north of the sandy system feeder channel (Fig. 4). The distance between the two sites south of this (7 km) allows comparison to be made between channel-fill architecture in a downcurrent direction. Four distinct sedimentary packages (termed Units A–D from the base upwards) can be identified at each site and correlated along the channel course. The exception to this arrangement of fill packages is found at the most distal parts of Site 2, where a conglomeratic plug is found beneath the main channel. The plug is 2 m thick and < 10 m wide. It comprises subrounded to subangular blocks of metamorphic basement up to 0.7 m in diameter, quartz pebbles and pecten shells and other bioclastic material. The conglomerate is clast-supported, and interpreted to represent channel thalweg deposits adjacent to the southern bank.

Unit A. At Site 1 this unit comprises a sequence of thick (> 1.5 m) erosively or sharply based, massive, medium-coarse grained tabular sandstones. The unit has an overall thickness of < 5 m (Fig. 6). At Site 2, Unit A also comprises thick erosive and sharply based, coarse-grained tabular sandstones, with local lenses of pebbles and conglomerate (0.3–0.7 m thick). The unit is 5–5.5 m thick in these Site 2 sections (Figs 7–11). In sections in the southern part of Site 2, Unit A also displays evidence for the development of large and small-scale, downstream-dipping accretion surfaces (Figs 7, 8 & 13). The bedforms extend from the top of the bottom of massive 1.5–2 m thick sandstone beds. Each individual clinoform extends 4–7 m downstream. They comprise alternating beds of medium-coarse sandstones with 'floating' quartz pebbles and 0.1–0.25 m shale beds. They can be traced throughout the lower parts of the channel system, on sections which are parallel to palaeocurrent direction. At more distal parts of the channel, smaller-scale cross-bedding is seen in Unit A, dipping towards the channel axis from the southern margin.

The tabular sandstone beds are interpreted to represent deposition from high-density turbidity currents, whereas the large-scale accretion surfaces indicate periodic bedform migration downstream. The smaller-scale channel-axis dipping surfaces represent lateral accretion of material from the channel margin.

Unit B. At Site 1 this unit comprises a sequence of interbedded muds and sandstones (Fig. 6) with localized lenses of conglomerate. The unit is locally erosive into the tabular sandstones of Unit A below it and has a maximum thickness of 11 m. Sandstone beds are normally medium-to-fine grained with sharp or flamed bases, with thicknesses ranging from 0.2–0.5 m. At Site 2, the unit comprises interbedded mudstones and fine-medium grained sandstones. It has a maximum thickness of 10 m (Figs 7 & 8). Palaeocurrent indicators are common with transport directions to the west (Site 1) and southwest (Site 2). This unit is interpreted to represent lower energy (relative to Unit A) deposition from turbidity currents.

Unit C. At Site 1 this unit comprises thick (1–1.5 m) massive sandstones and conglomerates with some interbedded shales (Fig. 6). The conglomeratic beds are stacked, 0.5–1.5 m thick, and grade laterally into sandstones with 'floating' pebbles and cobbles derived from the Sierra de los Filabres (Fig. 5) to the north. The unit has a maximum thickness of 8 m. At Site 2 the unit is comprised of thick (< 2.5 m), massive, amalgamated sandstones, thin-bedded graded sandstones and shales. Many beds are cross-bedded, or make up parts of thicker (< 1.5 m high) inclined units similar to the larger features identified in Unit A. The base of the unit is a channel-wide erosional surface (Fig. 10).

The thick, massive sandstones at Site 2 and the massive sandstones and conglomerates at Site 1 are interpreted to represent renewed deposition by high-energy, high-density turbidity currents. The cross-bedded sandstones and small-scale accretion surfaces at Site 2 indicate migration of bedforms in the lower, western reaches of the Solitary Channel.

Unit D. At Site 1 this unit comprises thin-bedded turbidites at the top of the channel-fill, and the unit is capped by mudstones. The unit is < 5 m thick. At Site 2 the unit also comprises thin-bedded turbidites, and is 6–7 m thick. It is interpreted to represent the final stages of filling of the Solitary Channel by low density turbidity currents.

Biostratigraphy

A detailed biostratigraphy has not been established for the Tabernas Basin. Planktonic–benthic foraminiferal assemblages from mudstones sampled below the channel in the Site 2 area by Kleverlaan (1989) indicate minimum water depths of 600 m. The Kleverlaan timeslice was defined by two biostratigraphic markers. Figure 3 is a chart showing the stratigraphy of the Tabernas Basin (Kleverlaan 1989), which describes the basin lithostratigraphy in detail. The timeslice is located in the lower Late Tortonian, below the Gordo Megabed. Two foraminiferal datums were used to date the sediments and estimate the rates of deposition. The first appearance of *Neogloboquadrina humerosa* at 8 Ma defines the base of the Late Tortonian. The first appearance of *Globoratalia conomiozea* med. occurs at 6.7 Ma, and was located at over 100 m above the Gordo Megabed according to Kleverlaan (1989).

A number of samples were taken of clean, non-indurated shales from between the sandstones below and within the fill at Site 2, in order to further refine the biostratigraphic study of Kleverlaan (1989), at spacings of 5 m (Fig. 11).

(1) Datums. Three coccolith datums have been used to date the sediments (following the biostratigraphic scheme of Rio *et al.* 1990):

Fo (first appearance)	*Amaurolithus amplificus*	*6.02 Ma*
Lo (last appearance)	*Amaurolithus amplificus*	*5.33 Ma*
Fo	*Amaurolithus primus*	6.7 Ma

(2) Zones. The CN 9a/9b boundary is defined by the first appearance of *Amaurolithus primus* and indicates an age of 6.7 Ma (Okada & Bukry 1980).

The Solitary Channel, within the previously interpreted timeslice, is exposed at a number of localities, often continuously, over 150 m below the Gordo Megabed. Dating of the non-indurated muds extracted from a number of sites below and above the base of the channel, at least 0.2–0.3 m behind the face of the exposure, has identified a coccolith zone, CN 9a/9b, across the base of the channel, dated at 6.7 Ma. This was the age assigned to a foram first appearance at least 300 m up stratigraphy (Kleverlaan 1989). A first appearance coccolith date of 6.02 Ma was also identified from the channel fill, again *c.* 300 m below the previous date of 6.7 Ma. These new dates indicate that the sedimentation rates published in Kleverlaan (1989), which relied on foram dating and the application of dates and the time chart used in Fig. 3, in relation to lithostratigraphy, should be revised.

Interpretation

The Solitary Channel was previously described as a channel complex with a minimum length of 8 km, a width of 25–130 m and a depth of 10–30 m. It was thought that flows followed some intrabasinal confinement, perhaps a fault escarpment trending obliquely to slope. It was suggested that the tectonic tilt may explain the basinward extent of the channel (Kleverlaan 1989).

The Solitary Channel is more extensive than previously thought; it can be mapped for over 11 km. The north–south oriented channel was deflected by 60° half-way along its length by a northeast–southwest oriented strike-slip fault, from where the channel meandered southwestwards. The fill of the channel is consistent, with four main packages of turbidite sandstones, intraformational conglomerates and thin mudstones identified at every exposed cross-section of the channel. Debris flow deposits increase in volume and percentage of the fill towards the source area. The four packages are defined on the basis of their sedimentology and geometry within the fill architecture. The base of each package is erosive into underlying packages. This sand-prone channel has a fill composed of in excess of 70 turbidite sandstones, with thin, non-extensive levee deposits, suggesting that significant volumes of sand were transported through the channel and that it persisted for some time. This is supported by the biostratigraphic data, which show that sediments in the highest parts of the fill have an age of 6.02 Ma, whereas sediments in the highest part of the channel wall (where incision is minimal) have an age of 6.7 Ma (Fig. 11).

In the lower parts of the fill, large isoclinal surfaces are seen, dipping downstream and downlapping on to the channel base. The dimensions suggest that the bedforms, which were probably sandwaves, extended across the floor of the channel. The dip of accretion surfaces indicates that the bedforms migrated down the channel. Bedforms of similar dimensions (2–5 m high, 5–10 m across), which may be analagous, have been identified in a small number of modern submarine fan/channel settings and have been attributed to tidal currents (Belderson *et al.* 1972; Kenyon & Belderson 1973; Prior *et al.* 1986, 1987; Kostaschuk *et al.* 1992).

A channel thalweg plug is interpreted to be present at a number of localities, it is steep walled and comprises clast-supported conglomerates, with large boulders even within the interpreted lower channel reaches. The plug is thought to correspond to the initial eroding flow to follow the faultline scarp along the basin floor. The angularity of the clasts suggests that the coarser material was transported beneath wavebase by debris flows. This hypothesis is supported by the frequency of megabeds in the basin stratigraphy (see Fig. 3). Modern examples of incised thalwegs are common in channels, and are normally considered to represent a change in channel axis gradient and a downsizing of flows within the larger channel (e.g. Le Pichon & Renard 1982; Droz & Bellaiche 1985; Kostaschuk *et al.* 1992).

Example 2: The Almeria Canyon, Alboran Sea, SW Mediterranean

The Almeria Canyon, in the southwest Mediterranean (Fig. 14), was analysed using deep-towed sidescan sonar, a 5.5 kHz sub-bottom profiler and gravity cores from the R/V Gelendzhik in 1992. The position of the canyon on the continental slope was known from a previous long-range GLORIA sidescan survey of the Alboran Sea. The canyon was known to be incised further out into the Alboran Sea than the shorter, more closely-spaced canyons and gullies that typify this tectonically active margin. The canyon has been described in more detail elsewhere (Cronin 1994; Cronin *et al.*, 1994). This section of the paper summarizes the geological setting and comparative planform and cross-sectional features of the Almeria Canyon.

Two sidescan sonar systems, the OKEAN medium-to-long-range system, and the deep-towed MAK-1 system, with sub-bottom profiler, were used. Seismic profiling and gravity cores supplemented the sidescan dataset.

Geological setting and previous work

The Andarax System is situated off the southeast Iberian Continental margin, and comprises the Almeria Canyon and the Andarax Fan. The system is fed by the Andarax River, which drains the southeast Iberian Peninsula (Fig. 14), primarily from the eastern Sierra Gador, the Sierra de los Filabres and the Sierra Alhamilla,

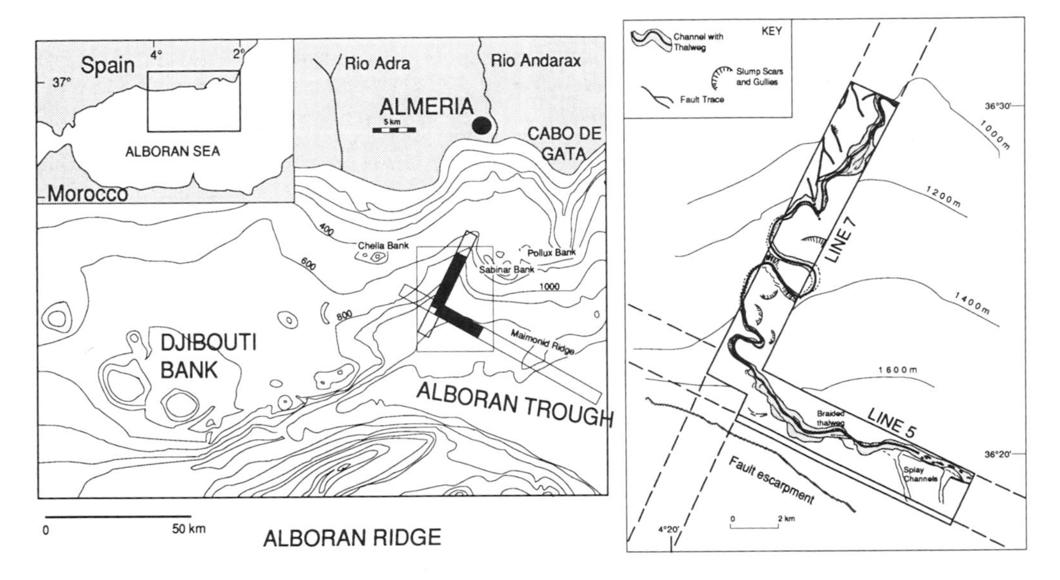

Fig. 14. Location map of the details from two MAK-1 deep-towed sidescan sonographs which view the main parts of the Almeria Canyon from the upper slope to the Alboran Trough. After Cronin *et al.*, 1994.

which surround the Tabernas Badlands. The Andarax River is normally dry, and sediment is discharged into Almeria Bay during seasonal floods. A sand and gravel fan extends into Almeria Bay with its apex at the mouth of the Andarax River, and the active channel is along the axis of the fan-delta plain (Maldonado & Zamarreno 1983). A wedge of finer clastic material extends basinwards from the fan-delta across the inner continental shelf.

This continental margin is typically narrow, 4–6 km, west of Almeria towards Malaga, and incised by closely-spaced, parallel canyons and gullies (Alonso & Maldonado 1992). The exception to this margin physiography is the Almeria Bay shelf area, which is 20 km wide and flat, comprising a carbonate shelf. The carbonate and mixed carbonate-terrigenous character reflects the temperate climate and the low terrigenous input. The shelf break is at 100 ± 30 m (Woodside & Maldonado 1992), and the slope is smooth in the Almeria Bay area, but abrupt elsewhere. The margin in this area is bound to the south by the Alboran Trough, a flat abyssal plain (1800 m) restricted to the south by the eastern Alboran ridge and by part of the African margin (Fig. 14). The trough opens eastwards into the Algerian Basin. Most of the continental margin is disrupted by seamounts and other ridges and escarpments, some of which have been interpreted as faults with morphological expression (Dillon *et al.* 1980; Woodside & Maldonado 1992; Cronin *et al.* 1994).

The northeastern Alboran Sea is characterized by active tectonics (see Maldonado 1992 for a review). The principal neotectonic elements in the area are considered to have resulted from north–south compression between Africa and Iberia (Woodside & Maldonado 1992). The tectonic elements include northeast–southwest oriented wrench faults separated by intervening north–south and north-northwest–south-southeast wrench faults. It is these latter trends that can be identified on sonograph images. Variations in the direction of the principal stresses have resulted in switching of activity between the two fault systems (Limonov *et al.* 1993). The northeast–southwest trending Serrata Fault system extends offshore into Almeria Bay and beyond towards the Alboran Ridge from the Cabo de Gata. Recent movement on this system has been recorded onshore by dog-legged stream courses (Th. Roep, pers. comm.).

Though most of this slope is characterized by parallel, closely-spaced canyons with limited reach, the Almeria Bay margin is incised by one canyon and its tributaries, and extends far out across the shelf and down the continental slope. This canyon was first imaged on a GLORIA long-range sidescan survey of the western Mediterranean/Alboran Sea, collected in 1991 (Institute of Oceanographic Sciences, UK), where a sinuous system with further basinward extent than the rest of the canyons in the vicinity and a pronounced dog-leg ('boomerang' planform expression) was identified within the lower reaches of the channel.

Planform geometry of the Almeria Canyon

The canyon can be mapped for over 24 km, has an average width of 250 m, and a depth range of 40 m in the upper reaches to several metres in its lower reaches. It is deflected by almost 90° halfway along its course by an intrabasinal fault scarp (Fig. 14), and follows the trend of this fault into the basin. The channel has a variable sinuosity, one meander has a sinuosity of between 3.2 and 3.8 (meander loops go off the sonograph record on both sides), three meanders have a sinuosity of 2.0 (e.g. Fig. 15a), otherwise levels of sinuosity are more commonly in the order of 1.2–1.3. The relationship between sinuosity and gradient for the Almeria Canyon would appear anomalous. Work on a number of modern deep sea fan channel sinuosities has shown a direct relationship between sinousity and gradient (Clark *et al.* 1992), where steeper slopes generally result in straight channels and vice versa, whereas the Almeria Canyon has highest sinuosity values on the steeper slopes. Cronin *et al.* (1994) concluded that this was the result of a combination of: (1) the sinuosity being largely inherited when the gradient was lower in the area, and (2) local and major structural control on the channel course, accentuating meanders and deflecting the channel course entirely. The upper reaches of the channel are characterized by margin instability, with slump scars common near meander bends, and transparent seismic facies within the canyon covering the pelagic drape which characterizes the canyon floor elsewhere. The lower reaches (after channel deflection and a change in channel-floor gradient) are characterized by a thalweg channel, which is sometimes leveed (Fig. 15b & c).

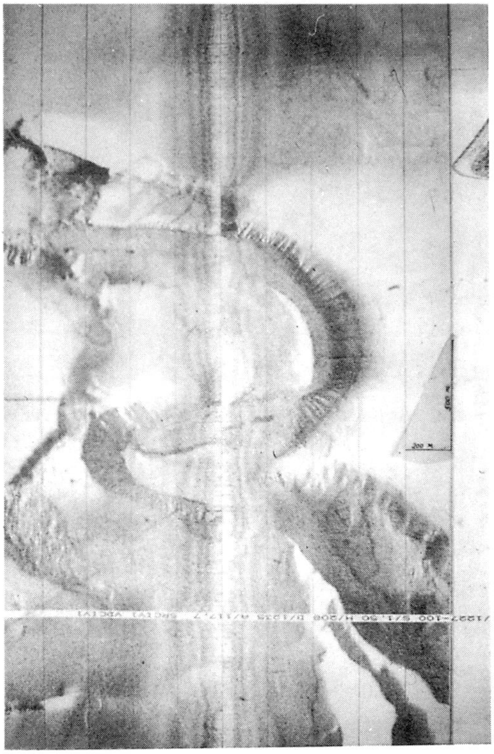

Fig. 15. (**a**) MAK-1 image of a meander in the Almeria Canyon immediately upslope of the canyon deflection (Fig. 14). Channel trend is from top to bottom. (**b**) 5.5 kHz sub-bottom profiler image of a cross-section of the canyon. Note the thalweg channel, which is leveed and elevated from the surrounding canyon floor. (**c**) Sonograph of lower reaches of the canyon, which is characterized by lower gradient, lower sinuosity, slumping at channel meanders and a thalweg channel which appears braided on the right-hand side of the diagram.

Discussion

The Almeria Canyon in the Alboran Sea and the Late Tortonian Solitary Channel in the Tabernas Basin are found in similar tectonic settings. The channels occur on the same scale, and have similar aspect (width-to-depth) ratios. In plan view their courses are comparable, and largely controlled by seafloor topography, which is inherited from structural elements. Both channel courses are interpreted to be deflected by emergent fault escarpments, and follow the dominant structural grain. This confinement has led to conduit stability which allows the channel to propagate further out into the basin than contemporary conduits. The stability of these channels allows coarse-grained material to be transported into mud-prone sectors of sedimentary basins. It also leads to increased channel longevity, as shown by biostratigraphic data from the ancient channel and seismic data for the modern one.

Deflection in channel course has been described from a number of modern settings where active faults or irregular topography are present [e.g. Baltimore and Wilmington Canyons (Twichell & Roberts 1982); Cape Breton Canyon (Kenyon *et al.* 1978); Guilvinec Canyon (Le Pichon & Renard 1982)]. It would appear that such a response should therefore be expected in channels and canyons in tectonically active deep sea settings. The similarity of planform characteristics of the studied channels may imply predictive planform development of

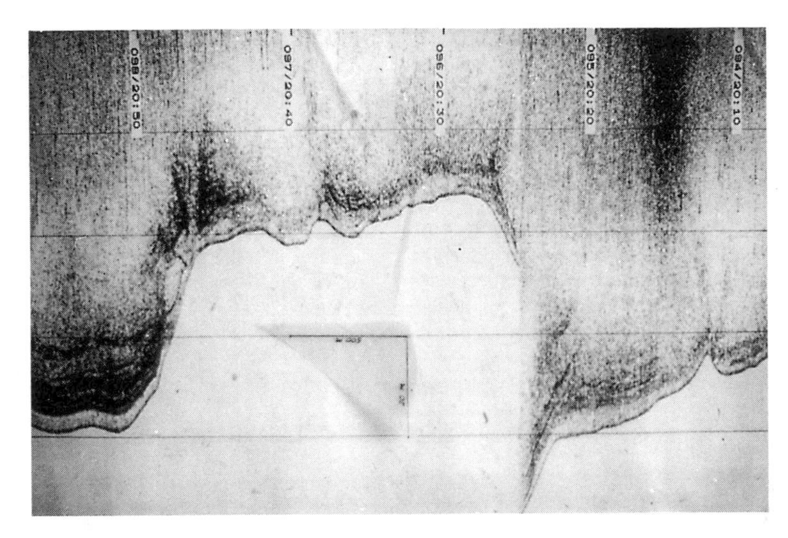

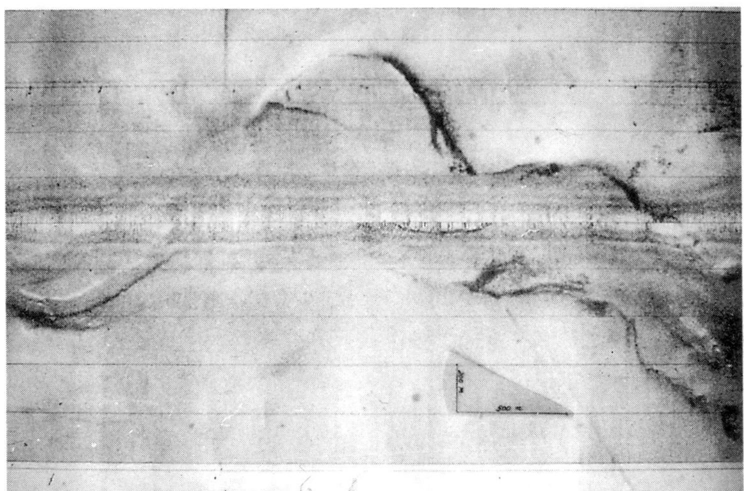

channels in tectonically active basins, resulting in long, linear bodies of coarse material, rather than sheet bodies of sand in proximal areas which characterize fans with mobile, migrating channels (Cronin 1994).

The implications of this analysis for reservoir characterization are threefold.

(1) The scale similarities in aspect ratio, planform geometry and length, are interpreted as having been controlled by tectonic setting. This indicates predictive effects of structural control on channel geometry and extent. In the stochastic modelling of analagous hydrocarbon reservoir data, it is therefore suggested that quantitative planform data of channels (e.g. sidescan sonographs) also be incorporated.

(2) The analysis shows that long, linear bodies of coarse material can develop in otherwise mud-prone areas of basins a relatively long distance from the basin margin. This is potentially a largely unexplored stratigraphic hydrocarbon trap.

(3) Conduit stability is interpreted to be controlled by seafloor topography, and this leads to an increase in channel longevity. The distinction can therefore be made between relatively long-lived sand-conduits in otherwise mud-prone sectors of basins, such as the two examples in this paper, and short-lived, sand-prone and mobile channels in more sandy areas.

Rob Kidd supervised the development of the project and encouraged my involvement with modern and ancient datasets. A. T. S. Ramsay undertook planning of coccolith sampling and subsequent biostratigraphical work. Jan Alexander and Adrian Cramp offered many useful discussions in the development of ideas and reading of manuscripts. Neil Kenyon and Dorrik Stow suggested ways of focusing important sections. John Woodside and Thom Roep were also very helpful from their knowledge of the study areas. Clive Bishop at BP Research offered many useful ideas and other support for this work over the past two years. I also wish to thank Adrian Hartley and Jeremy Prosser for pursuing the publication of this paper.

The work was sponsored by BP Research and BP Exploration, and cruise participation was supported by Cardiff University of Wales, Moscow State University, ESF and the UNESCO TREDMAR 'Training-through-research' program.

References

ADAMS, C. G., BENSON, R. H., KIDD, R. B., RYAN, W. B. F. & WRIGHT, R. C. 1977. The Messinian salinity crisis and the evidence of late Miocene eustatic changes in the world ocean. *Nature*, **269**, 383–386.

ALONSO, B. & MALDONADO, A. 1992. Plio-Quaternary margin growth patterns in a complex tectonic setting: Northeastern Alboran Sea. *Geo-Marine Letters*, **12**, 137–143.

BELDERSON, R. H., KENYON, N. H., STRIDE, A. H. & STUBBS, A. R. 1972. *Sonographs of the sea floor.* Elsevier, Amsterdam.

BOURROUILH, R. 1987. Evolutionary mass flow-megaturbidites in an interplate basin: example of the North Pyrenean Basin. *Geo-Marine Letters*, **7**, 69–81.

BOUSQUET, J.-C. 1979. Quaternary strike-slip faults in southeastern Spain. *Tectonophysics*, **52**, 277–286.

CHANNEL, J. E. T. & MARESCHAL, J. C. 1989. Delamination and asymmetric lithospheric thickening in the development of the Tyrrhenian Rift. *In:* COWARD, M. P. *et al.* (eds) *Alpine Tectonics*. Geological Society, London, Special Publication, **45**, 285–300.

CLARK, J. D., KENYON, N. H. & PICKERING, K. T. 1992. Quantitative analysis of the geometry of submarine channels: Implications for the classification of submarine fans. *Geology*, **20**, 633–636.

COSSEY, S. P. J. & KLEVERLAAN, K. 1994. Heterogeneity within a sand-rich submarine fan, Tabernas Basin, Spain. *In:* PICKERING, K. T., RICCI-LUCCHI, F., SMITH, R. D. A., HISCOTT, R.N. & KENYON, N. H. (eds) *Atlas of Deep Water Environments – Architectural Style in Turbidite Systems.* Chapman & Hall, London.

CRONIN, B. T. 1994. *Channel-fill architecture in deep-water sequences: variability, quantification and applications*. PhD thesis, University of Wales, UK.

——, KENYON, N. H., WOODSIDE, J. M., *ET AL.* 1994. Views of the Andarax submarine canyon: a meandering system on an active tectonic margin. *In:* PICKERING, K. T., RICCI-LUCCHI, F., SMITH, R. D. A., HISCOTT, R. N. & KENYON, N. H. (eds) *Atlas of Deep Water Environments – Architectural Style in Turbidite Systems*. Chapman & Hall, London.

DEWEY, J. F., PITMAN, W. C., RYAN, W. B. F. & BONNIN, J. 1973. Plate tectonics and the evolution of the Alpine system. *Bulletin of the Geological Society of America*, **84**, 3137–3180.

DROZ, L. & BELLAICHE, G. 1985. Rhone deep-sea fan: morphostructure and growth pattern. *American Association of Petroleum Geologists Bulletin*, **69**, 460–479.

DILLON, W. P., ROBB, S. M., GARY GREEN, M. & LUCENA, J. C. 1980. Evolution of the continental margin of southern Spain and the Alboran Sea. *Marine Geology*, **36**, 205–226.

KASTENS, K. 1984. Earthquakes as a triggering mechanism for debris flows and turbidites on the Calabrian Ridge. *Marine Geology*, **55**, 13–34.

KENYON, N. H. & BELDERSON, R. H. 1973. Bed forms of the Mediterranean undercurrent observed with side-scan sonar. *Sedimentary Geology*, **9**, 77–99.

——, BELDERSON, R. H. & STRIDE, A. H. 1978. Channels, canyons, and slump folds of the continental slope between SW Ireland and Spain. *Oceanologica Acta*, **1**, 369–380.

KLEVERLAAN, K. 1982. *Geologische onderzeokingen in het Tabernas bekken.* MSc thesis, University of Amsterdam, The Netherlands.

—— 1987. Gordo Megabed: a possible seismite in a Tortonian submarine fan, Tabernas Basin, Province Almeria, SE Spain. *Sedimentary Geology*, **51**, 165–180.

—— 1989. Three distinctive feeder-lobe systems within one time-slice of the Tortonian Tabernas fan, SE Spain. *Sedimentology*, **36**, 25–45.

KOSTASCHUK, R. A., LUTERMAUER, J. L., MCKENNA, G. T. & MOSLOW, T. F. 1992. Sediment transport in a submarine channel system: Fraser river delta, Canada. *Journal of Sedimentary Petrology*, **62**, 273–282.

LABAUME, P., MUTTI, E. & SEGURET, M. 1987. Megaturbidites: a depositional model from the Eocene of the SW-Pyrenean Foreland Basin, Spain. *Geo-Marine Letters*, **7**, 91–101.

LE PICHON, X. & RENARD, V. 1982. Avalanching: a major process of erosion and transport in deep-sea canyons: evidence from submersible and multi-narrow beam surveys. *In:* SCRUTON, R. A. & TALWANI, M. (eds) *The Ocean Floor.* John Wiley and Sons, New York, 113–128.

LIMONOV, A. F., WOODSIDE, J. M. & IVANOV, M. K. (eds) 1993. *Geological and Geophysical Investigations of Western Mediterranean Deep-sea Fans. Initial results of the UNESCO-ESF "Training-through-Research" cruise of the RV Gelendzhik in the Western Mediterranean (June–July 1992).* UNESCO Reports in marine science, **62**, 154p.

MALDONADO, A. (ed.) 1992. The Alboran Sea. *Geo-Marine Letters*, **12**, 186p.

MALDONADO, A. & ZAMARRENO, I. 1983. Modelos sedimentarios en la platformas continentales del Mediterraneo espanol: Factores de control y procesos que rigen su desarrollo. *In:* CASTELLVI J. (ed.) *Estudio oceanografico de la platforma continental.* Cadiz, Spain, 15–52.

MONTENAT, C., OTT D'ESTEVOU, P. & MASSE, P. 1987. Tectonic-sedimentary characters of the Betic Neogene basins evolving in a crustal transcurrent shear zone (SE Spain). *Bulletin de Compte Rendues Exploration et Production*, Elf-Aquitaine, **11**, 1–22.

MUTTI, E., RICCI-LUCCHI, F., SEGURET, M. & ZANZUCCHI, G. 1984. Seismoturbidites: a new group of resedimented deposits. *Marine Geology,* **55**, 103–116.

OKADA, H. & BUKRY, D. 1980. Supplementary modification and introduction of code numbers to the low latitude coccolith biostratigraphic zonation. *Marine Micropalaeontology,* **5**, 321–325.

PIPER, D. J. W. & NORMARK, W. R. 1983. Turbidite depositional patterns and flow characteristics, Navy submarine fan, California Borderland. *Sedimentology,* **30**, 181–199.

PLATT, J. P. & VISSERS, R. L. M. 1989. Extensional collapse of thickened continental lithosphere: A working hypothesis for the Alboran Sea and the Gibraltar Arc. *Geology,* **17**, 540–543.

——, VAN DEN EECKOUT, B., JANZEN, E., KONERT, G., SIMON, O.J. & WEIJERMARS, R. 1983. The structure and tectonic evolution of the Aguilon fold-nappe, Sierra Alhamilla, Betic Cordilleras, SE Spain. *Journal of Structural Geology,* **5**, 519–538.

PRIOR, D. B., BORNHOLD, B. D. & JOHNS, M. W. 1986. Active sand transport along a fjord-bottom channel, Bute Inlet, British Columbia. *Geology,* **14**, 581–584.

——, ——, WISEMAN JR. W. J. & LOWE, D. R. 1987. Turbidity current activity in a British Columbia fjord. *Science,* **237**, 1330–1333.

RIO, D., SPROVIERI, R. & CHANNELL, J. 1990. Pliocene to early Pleistocene chronostratigraphy and the Tyrrhenian deep-sea record from site 653. *In:* KASTENS, K. A., MASCLE, J. *et al.* (eds). *Proceedings of the ODP, Scientific Results,* **107**: College Station, Texas (Ocean Drilling Program), 705–714.

TWICHELL, D. C. & ROBERTS, D. G. 1982. Morphology, distribution and development of submarine canyons on the US Atlantic continental slope between Hudson and Baltimore. *Geology,* **10**, 408–412.

WEIJERMARS, R. 1991. Geology and tectonics of the Betic Zone, SE Spain. *Earth-Science Reviews,* **31**, 153–236.

——, ROEP, TH.B., VAN DEN EECKHOUT, B., POSTMA, G. & KLEVERLAAN, K. 1985. Uplift history of a Betic fold nappe inferred from Neogene-Quaternary sedimentation and tectonics (in the Sierra Alhamilla and Almeria, Sorbas and Tabernas Basins of the Betic Cordilleras, SE Spain). *Geologie en Mijnbouw,* **64**, 397–411.

WOODSIDE, J. M. & MALDONADO, A. 1992. Styles of compressional neotectonics of the Betic Zone, SE Spain. *Geo-Marine Letters,* **12**, 111–116.

ZIEGLER, P. A. 1982. *Geological Atlas of Western and Central Europe*. Elsevier, Amsterdam.

Sediment dispersal patterns in a deep marine back-arc basin: evidence from heavy mineral provenance studies

J. R. BROWNE & D. PIRRIE

Camborne School of Mines, University of Exeter, Redruth, Cornwall, TR15 3SE, UK

Abstract: Heavy mineral provenance studies can aid reservoir description in deep marine depositional systems by providing a clearer understanding of the lateral extent of discrete sediment bodies. In this study the lower Gustav Group (Lagrelius Point, Kotick Point and Whisky Bay formations) of James Ross Island, Antarctica, which was deposited in a deep marine slope-apron–submarine fan complex has been investigated. The Kotick Point and Whisky Bay formations crop out in two separate areas on the west coast of James Ross Island. The provenance of the group has been studied by both standard sandstone petrography and electron microprobe analysis of the heavy minerals' garnet, pyroxene and Fe–Ti oxides. The sandstones are lithic to arkosic arenites derived from the adjacent Antarctic Peninsula magmatic arc. Although the general source area can be recognized, individual stratigraphical or depositional units cannot be distinguished on the basis of sandstone petrography. However, electron microprobe analyses of the detrital heavy minerals allows a more detailed discrimination of sediment provenance. In particular, analyses of the detrital Fe–Ti oxide minerals show that in the Whisky Bay Formation, the southern outcrop area, contains ilmenite, whilst the northern outcrop area contains titanomagnetite. In conjunction with the available biostratigraphical and palaeocurrent evidence, this can be interpreted in two ways. Firstly, if the two outcrop areas are direct age equivalents then they were derived from separate point sources, and deposited as discrete sediment lobes. Alternatively, if they are not directly time-equivalent, the depositional system may have switched laterally with time, with sediment derived from source rocks of varying composition. Either interpretation has implications for the understanding of the deep marine depositional system. Detrital opaque minerals are commonly source specific and should not be ignored during provenance studies.

Sediment provenance studies aid the understanding of hydrocarbon reservoirs by constraining sediment source areas and sediment dispersal patterns within the depositional basin and by aiding lithostratigraphical correlation (e.g. Morton *et al.* 1989; Humphreys *et al.* 1991; Morton 1992). Many provenance studies rely upon the description and interpretation of broad framework grain compositions, and the discrimination of these petrographical data on the basis of a range of ternary diagrams (e.g. Pirrie 1991). Although this style of provenance study is important, it cannot provide a detailed understanding of provenance in basins where either the framework grain mineralogy has been extensively altered by diagenesis or the broad range of framework grains present do not allow detailed resolution of sediment provenance.

In recent years, the petrographical and geochemical analysis of heavy mineral suites, which are source diagnostic, has proven to be a valuable tool in evaluating sandstone provenance. However, the composition of a sandstone heavy mineral suite is not only controlled by source area geology, but also by the physical properties of the transporting fluid and subsequent diagenetic modification (Morton 1991). Sandstones derived from the same source area but transported by different mechanisms and affected by different diagenetic processes may be expected to have different resultant heavy mineral suites. These affects can, however, be minimized by detailed geochemical analysis of the compositional variation in a single heavy mineral species, because although the overall abundance of each heavy mineral may vary, the surviving sediment grains, unless diagenetically modified, will retain their provenance signature. This approach has been particularly effective in, for example, the evaluation of sandstone provenance in both the Middle Jurassic Brent Group and the Palaeocene Forties Formation in the North Sea (Morton 1991).

A wide range of heavy minerals may be of potential value in provenance studies. Ideally, the heavy minerals studied need to be relatively common throughout the sequence studied, compositionally variable allowing detailed source-

From Hartley, A. J. & Prosser, D. J. (eds), 1995, *Characterization of Deep Marine Clastic Systems*, Geological Society Special Publication No. 94, pp. 139–154.

area discrimination and diagenetically stable. Once a heavy mineral population has been separated, geochemical analysis of individual grains by electron microprobe is relatively routine, rapid and cost effective. In a recent review, Morton (1991) summarized the application of microprobe analyses of pyroxene, amphibole, epidote, staurolite, garnet, spinel, chloritoid, monazite, tourmaline and zircon in provenance studies. In particular, pyroxene microprobe analyses are commonly used to determine potential parent magma type and tectonic setting (e.g. Cawood 1991), and garnet has been found to be of considerable use in both lithostratigraphical correlation and understanding sediment dispersal in the Brent Group (Morton *et al.* 1989; Morton 1992).

Most heavy mineral provenance studies to date have focused upon the non-opaque heavy minerals. However, in recent years, the potential provenance signature of detrital opaque minerals has been examined (e.g. Darby 1984; Darby & Tsang 1987; Basu & Molinaroli 1989 1991; Grigsby 1990, 1992; Razjigaeva & Naumovo 1992). In particular these authors have examined the geochemistry and provenance signature of the Fe–Ti oxide minerals, primarily ilmenite and magnetite. Petrographical studies, combined with geochemical studies by electron microprobe, of Recent sands derived from discrete source areas have clearly shown that these Fe–Ti oxide minerals are useful provenance indicators (Basu & Molinaroli 1989; Grigsby 1990, 1992). However, there are few studies evaluating the use of Fe-Ti oxide minerals in ancient sediments, and the authors are unaware of any published studies utilizing these minerals in evaluating sediment provenance in hydrocarbon reservoirs.

In this study the aim is to show the potential use of microprobe geochemical studies of detrital Fe–Ti oxide minerals in constraining sediment provenance and dispersal within a Cretaceous deep marine submarine fan-slope-apron complex. The example studied is the lower part of the Gustav Group from the James Ross Basin, Antarctica. Although this region is not a hydrocarbon exploration area, the sedimentary system studied is comparable with other deep marine reservoir units (e.g. the Brae Formation, North Sea; Stow 1985), hence these results are applicable to reservoir studies elsewhere. In this paper the aim is to initially summarize the sedimentology of the Gustav Group, describe the sandstone petrography and provenance of these deep marine clastics, and describe the geochemistry of selected heavy mineral phases. Then the aim is to show that the heavy mineral geochemistry allows a more refined understanding of sediment provenance and dispersal, and therefore sediment body geometry, than otherwise possible on the basis of sandstone petrography.

Regional setting

Throughout the Mesozoic–Tertiary, the Antarctic Peninsula was an active magmatic arc terrane related to subduction of proto-Pacific oceanic crust beneath the southern margin of Gondwana. Major sedimentary basins developed in both the fore- and back-arc regions (Macdonald & Butterworth 1990). The Cretaceous–Tertiary fill of the back-arc basin, located to the east of the Antarctic Peninsula magmatic arc, is best exposed in the James Ross Island area (Fig. 1). These strata were deposited within the

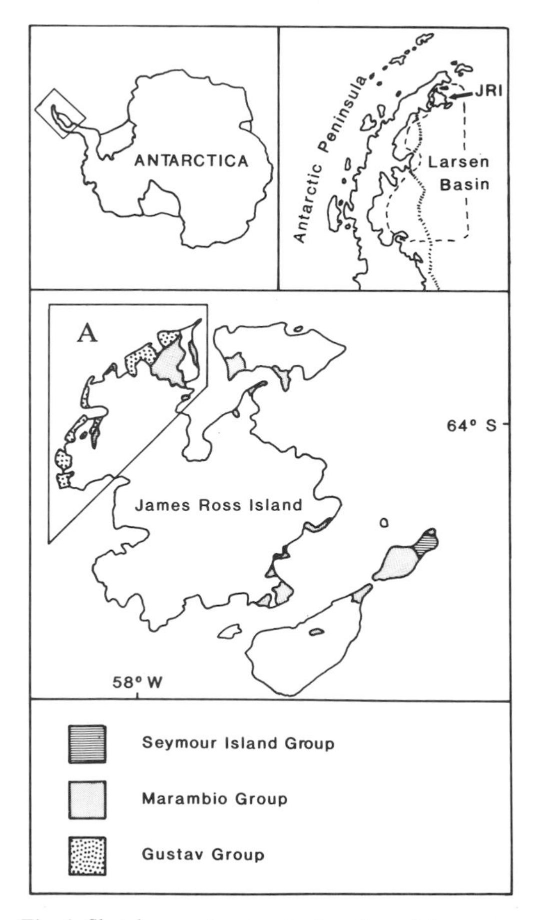

Fig. 1. Sketch map showing the location of James Ross Island, Antarctica. Area in box A is shown in more detail in Fig. 2.

James Ross Basin, a sub-basin of the larger Larsen Basin (Del Valle *et al.* 1992; Macdonald *et al.* 1988).

The tectonic setting of the basin is complex since both subduction-related processes and the opening of, and spreading within, the Weddell Sea affected its development. Storey & Nell (1988) suggested that the opening of the Weddell Sea may have led to strike-slip movement or oblique extension along the eastern margin of the Antarctic Peninsula, and thus may have controlled the early development of basins along this margin. However, during the early and mid-Cretaceous the dominant tectonic regime was broadly extensional, and geophysical and geological data from the James Ross Island region suggest that the basin developed as a half-graben that experienced repeated

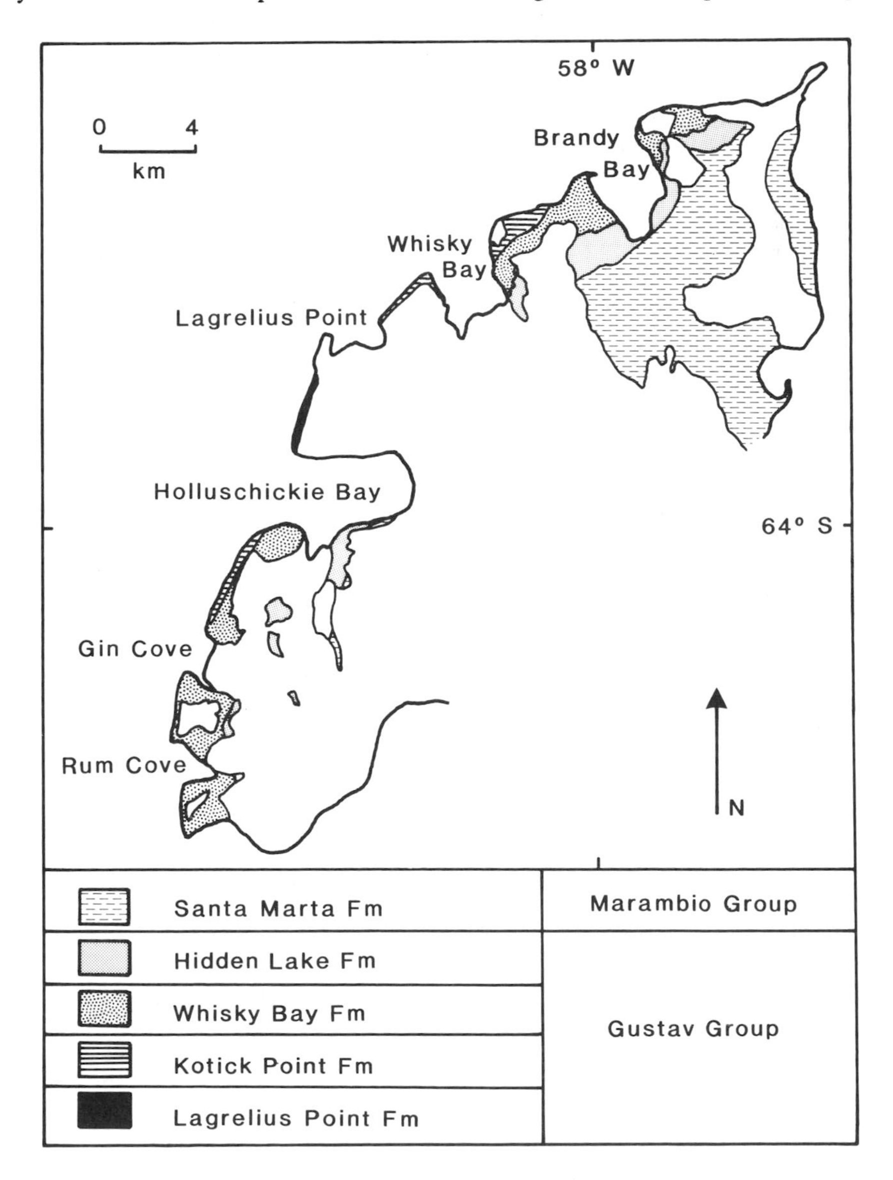

Fig. 2. Sketch geological map of western James Ross Island (after Ineson *et al.* 1986). Note that the Gustav Group crops out in two main areas, the Brandy Bay–Whisky Bay area to the northeast and the Gin Cove–Rum Cove area to the southwest.

movement along the basin-bounding fault (Farquharson *et al.* 1984; Whitham & Marshall 1988; Ineson 1989). Repeated movement along the basin bounding fault and syn-sedimentary deformation within the basin fill continued until the Coniacian, and partially controlled sediment dispersal (Whitham & Marshall 1988; Ineson 1989; Pirrie *et al.* 1991).

Approximately 5–6 km of clastic sediments derived from the Antarctic Peninsula magmatic arc terrane were deposited within the James Ross Basin between approximately the Barremian and the Eocene (Ineson *et al.* 1986; Crame *et al.* 1991). These strata are subdivided into three major lithostratigraphical units which define a regressive megasequence; the basal deep marine Gustav Group and the overlying shallow marine to deltaic Marambio and Seymour Island groups. The basal unit, the Gustav Group is thought to overlie a Late Jurassic–Early Cretaceous radiolarian mudstone–tuff sequence, the Nordenskjöld Formation (Macdonald *et al.* 1988). This mudstone–tuff sequence crops out at a number of isolated localities along the eastern margin of the Antarctic Peninsula and also occurs as derived blocks within the Cretaceous Gustav Group (Farquharson 1983; Ineson 1985*a*).

Samples examined in this study were collected from the lower part of the Gustav Group of approximately Barremian to Coniacian age, which crops out along the western coastline of James Ross Island (Fig. 2). Palaeocurrent, and previous sediment provenance studies, support sediment supply from the adjacent Antarctic Peninsula arc terrane (Ineson 1989; Pirrie 1991). The geology of the northern Antarctic Peninsula which represents the source area is reasonably well known. A Permian–Triassic accretionary complex comprising low grade metasediments (the Trinity Peninsula Group, TPG) is overlain by coarse alluvial sediments deposited in small fault bounded basins (the Botany Bay Group) which are in turn overlain

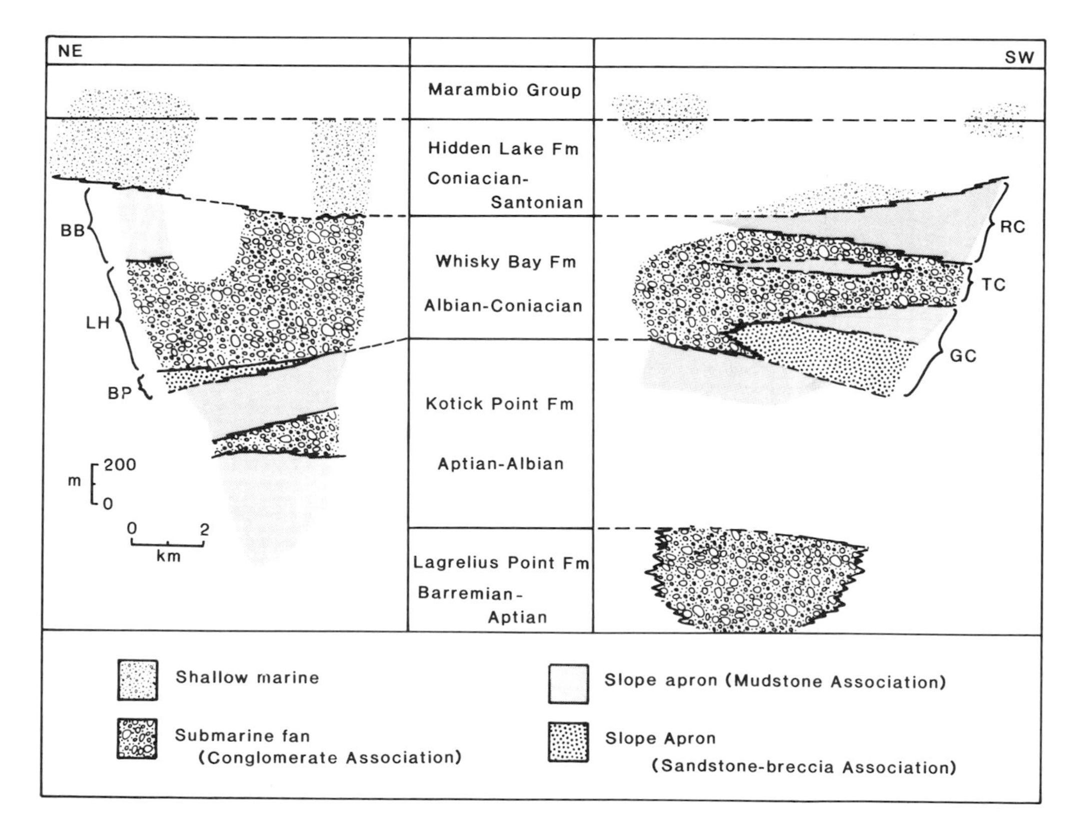

Fig. 3. NE–SW transect along the northwest coast of James Ross Island showing the lithostratigraphy and distribution of depositional environments (after Ineson *et al.* 1986; Ineson 1989). BP, Bibby Point Member; LH, Lewis Hill Member; BB, Brandy Bay Member; GC, Gin Cove Member; TC, Tumbledown Cliffs Member; RC, Rum Cove Member.

and intruded by calc-alkaline volcanics and cogenetic plutonics of the Antarctic Peninsula Volcanic Group (APVG)(Farquharson 1984).

Sedimentology of the Gustav Group

The Gustav Group is *c.* 2300 m-thick and comprises the deep marine Lagrelius Point, Kotick Point and Whisky Bay formations and the shallower marine Hidden Lake Formation (Ineson *et al.* 1986; Ineson 1989; Pirrie *et al.* 1991). Only the deep marine part of the Gustav Group is considered here. The sedimentology of the Lagrelius Point, Kotick Point and Whisky Bay formations has been described by Ineson (1989). The basal unit, the Lagrelius Point Formation is poorly known as it only crops out as a series of steep, inaccessable cliffs. The formation is conglomerate dominated and is estimated to be at least 500 m-thick (Ineson *et al.* 1986).

The Kotick Point and Whisky Bay Formations crop out in two main areas, the Brandy Bay–Whisky Bay region to the NE and the Gin Cove–Rum Cove area to the SW separated by an area of no exposure (Figs 2 & 3). The Kotick Point Formation is characterized by thinly interbedded sandstones and silty mudstones or claystones, with a 100–250 m thick graded conglomerate and pebbly sandstone unit in the middle of the formation (Ineson *et al.* 1986). Within the upper levels of the Kotick Point Formation a giant glide block of the Late Jurassic Nordenskjöld Formation, 200 × 800 m in size, is enclosed within mudstones (Ineson 1985*a*). The overlying Whisky Bay Formation is lithologically complex (Fig. 3). To the NE at Brandy Bay three members are recognised. The basal Bibby Point Member comprises bioturbated sandstones interbedded with lenticular conglomerates. The overlying Lewis Hill Member is 400–500 m thick and is dominated by graded channelized conglomerates and pebbly sandstones, in turn overlain by the fine grained Brandy Bay Member (Ineson *et al.* 1986). To the SW a further three members are recognized (Fig. 3). The Gin Cove Member comprises a basal sandstone-dominated unit, overlain by thin interbedded mudstones and fine-grained sandstones. This member is overlain by the Tumbledown Cliffs Member, which comprises graded sandstones, pebbly sandstones and conglomerates, along with two large Nordenskjöld Formation glide blocks (Ineson 1985*a*; Ineson *et al.* 1986). The overlying Rum Cove Member is mudstone dominated with subordinate thin sandstones.

Detailed facies analysis of the Kotick Point and Whisky Bay Formations allowed Ineson (1989) to recognize nine main facies which defined three broad facies associations: a Mudstone Association and Sandstone–Breccia Association, together interpreted as representing a slope-apron environment, and a Conglomerate Association interpreted as representing a submarine fan environment. The Mudstone Association, characterized by the Kotick Point Formation, was interpreted as representing a base of slope setting (Ineson 1989). The Sandstone–Breccia Association was interpreted as representing sediment deposition at the base of slope, as isolated or laterally coalescing sediment lobes, and is indicative of the tectonically active nature of the basin margin (Ineson 1985*a*, 1989). This association includes the Bibby Point Member and much of the Gin Cove Member of the Whisky Bay Formation (Ineson 1989). Stratigraphically the Conglomerate Association makes up most of the Whisky Bay Formation and the middle unit of the Kotick Point Formation. This association was interpreted as representing a submarine fan system (Ineson 1989; Buatois & Lopez Angriman 1992). Large-scale lenticularity shown by the major conglomerate bodies parallel to the basin margin was thought to suggest a localized point source for the sediment supply (Ineson 1989). Palaeocurrent data presented by Ineson (1989) revealed a dominant palaeoflow for the Conglomerate Association to the northeast in the northern outcrop area and to the south in the southern outcrop area (Fig. 4). This palaeocurrent pattern was interpreted by Ineson (1989) to represent a radial sediment dispersal system with a single major point source. However, palaeocurrent data from the slope-apron facies have a dominant easterly to south-easterly palaeoflow direction for both outcrop areas.

Biostratigraphical studies on both molluscan macrofossils and palynomorphs suggest a ?Barremian to ?Coniacian age for the lower Gustav Group (Ineson *et al.* 1986; Keating *et al.* 1992). Within the Whisky Bay Formation, precise biostratigraphical correlation between the northeast and southwest outcrop areas is not possible. However, the Gin Cove and Tumbledown Cliffs Members are approximately equivalent to the Bibby Point and Lewis Hill Members, and the Rum Cove and Brandy Bay Members are probably equivalent (Ineson *et al.* 1986; Keating *et al.* 1992) (see Fig. 3).

The uppermost unit in the Gustav Group, the Hidden Lake Formation of Coniacian–Santonian age reflects partial basin inversion with

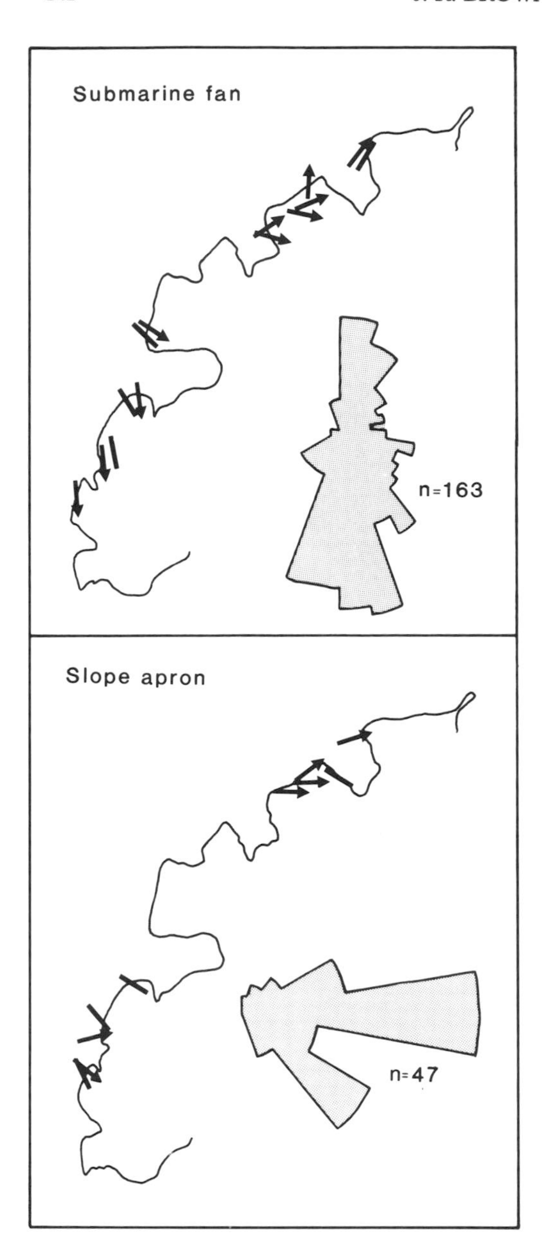

Fig. 4. Palaeocurrent data for the Lagrelius Point, Kotick Point and Whisky Bay Formations (after Ineson 1989).

deposition within a marine shelf setting below storm wave base (Pirrie *et al.* 1991). The formation is characterized by volcaniclastic sandstones, mudstones and conglomerates. The Hidden Lake Formation is conformably overlain by the Marambio Group which records the continued transition to predominantly shallow marine sedimentation (Crame *et al.* 1991).

Methods

The sandstone provenance of the lower Gustav Group was studied using both standard petrographic techniques and the separation of heavy mineral suites for geochemical analysis by electron microprobe. All samples studied were selected from the collections of the British Antarctic Survey, Cambridge. Forty-one sandstone samples were point counted following staining for plagioclase and K-feldspar (after Houghton 1980). The sandstones were point counted for all detrital grains using the 'QFL' method (Dickinson 1970) with a minimum of 500 grains counted per slide. Using this method all sand-sized monomineralic fragments within a thin section were counted as single grains, even if they occured within polymineralic lithic grains. Although the interdependence of grain size and mineralogy was reduced by using the QFL methodology (Ingersoll *et al.* 1984), medium-grained sandstones were selected for point counting whenever possible. To determine the relative proportions of lithic clasts, all samples were also point counted for lithic grains only, with a minimum of 200 grains counted per section. The point count data are summarized in Table 1; copies of the complete data set are available from the authors.

Heavy minerals were separated from 16 lightly crushed sandstones using standard gravity settling procedures. The heavy liquid di-idio methane stabilized with triethyl orthophosphate was used. The 63–180 μm heavy mineral grain-size fraction was selected for further analysis; this grain-size range contained most of the heavy mineral fraction. By selecting a restricted grain-size range, original effects of hydraulic differentiation of the heavy mineral suite were minimised. In addition, to limit potential facies-related variation in the heavy mineral suites, all of the sandstones selected for heavy mineral separation were from beds interpreted by Ineson (1989) as representing deposition from turbidity currents. Polished mounts of the heavy mineral concentrates were prepared for electron microprobe analysis. Microprobe analyses were carried out using an energy-dispersive SX 50 Electron Microprobe at the Department of Earth Sciences, University of Cambridge. The heavy minerals garnet, pyroxene, titanomagnetite and ilmenite were analysed for the elements Si, Al, Ti, Cr, Fe, Mn, Mg, Ca, Na and K. Results are presented as Wt% oxides. Garnet analyses with low totals (between 91 and 98%) were accepted providing stoichiometry showed Al to be close to four formula units, as this

Table 1. *Summary of sandstone modal data for each of the formations studied*

Stratigraphic unit	No. of samples	Qm	Qp	P	K	Lv	Lm	Ls	Lp	OPQ	CP
Lagrelius Point Formation	3	3.5–18.7 (8.8)	0–0.4 (0.2)	10.4–19.5 (14.7)	0–2.4 (1.2)	12.6–51.5 (30.3)	22–51.4 (39.3)	2.1–7.2 (4.3)	0 (0)	0 (0)	0–0 (0.1
Kotick Point Formation	12	0.4–26.5 (6.9)	0–1.8 (0.3)	10.5–87.7 (35.0)	0–2.1 (0.8)	3.5–74.2 (35.33)	0.4–32.8 (9.3)	0–61.9 (9.3)	0–0.2 (0)	0–2.0 (0.5)	0–0 (0.1
Whisky Bay Formation	26	0.2–11.4 (3.8)	0–2.0 (0.2)	3.4–77.4 (19.9)	0–8.4 (1.09)	1.7–72.9 (50.7)	0.3–38.6 (14.6)	0–30.5 (5.9)	0 (0)	0–1.5 (0.6)	0–0 (0.

All clasts were point-counted. Ranges of data are shown with averages (in brackets).
Explanation of parameters and abbreviations used in this study: Qm, monocrystalline quartz grain; Qp, polycrystalline quartz grain; P, plagioclase f potassium feldspar grain; Lv, lithic volcanic grain; Lp, lithic plutonic grain; Lm, lithic metamorphic grain; Ls, lithic sedimentary grain; OPQ, c minerals; CPX clinopyroxene grain; ACC, other accessory minerals.

indicates a lack of distortion in the Al:Fe balance precluding significant absorption effects (Morton 1985). For pyroxenes, low totals were accepted providing the Al + Si was approximately two formula units. The Fe–Ti oxides were only accepted where the totals were > 95%.

The polished mounts were subsequently prepared as thin sections and the relative abundance of the different heavy minerals was determined by point counting at least 200 grains.

Prior to this study, the only previous work on the provenance of the Gustav Group is that of Pirrie (1991), who examined sandstone samples spanning the entire Cretaceous–Tertiary James Ross Basin-fill. Pirrie (1991) point counted 22 sandstones from the Gustav Group and described pyroxene compositions on the basis of electron microprobe analyses.

Diagenesis

Vitrinite reflectivity and diagenetic studies imply limited post-depositional burial and diagenetic alteration of the samples studied. Whitham & Marshall (1988) showed that within the Lagrelius Point and Kotick Point Formations vitrinite reflectivity values average *c.* 0.4% Ro, decreasing to 0.3% Ro in the Hidden Lake Formation. This diagenetic work indicates that the dominant authigenic phases are carbonate cements (dominantly ferroan calcite), smectite and chlorite pore-rimming and pore-filling cements, the zeolite clinoptilolite–heulandite, along with minor quartz. Sandstone grains have variable degrees of alteration and replacement, although usually the original framework grains can be identified. Compaction-related brittle and ductile deformation with the formation of pseudomatrix is also observed. The absence of the diagenetic transformation of smectite to illite or the alteration of clinoptilolite to analcime, along with the vitrinite reflectivity results, suggests that the sequence has not been heated to > 80°C.

Sandstone petrography

The Gustav Group sandstones are arkosic and lithic arenites (*sensu* Pettijohn *et al.* 1972). The framework grains and accessory minerals are briefly described below.

Quartz

Both monocrystalline (Qm) and polycrystalline quartz (Qp) are present. Undulose and non-undulose Qm grains are present throughout the succession, undulose quartz being the most common. Rarely Qm grains contain inclusions. Qp grains (> 3 subcrystals) are scarce, but dominantly show sutured contacts and undulose extinction. Qp grains with 2–3 subcrystals occur very rarely in the Gustav Group.

Feldspar (F)

Plagioclase (P) is very abundant throughout the Gustav Group, comprising 4–85% of point-counted sandstones. The grains are commonly fresh and unaltered, and vary from large euhedral zoned crystals to broken twinned crystals. Compositionally, the plagioclase is mainly oligoclase to andesine, with rare grains of albite and labradorite. K-feldspar (K) is rare (average 0.5%) with euhedral to rounded, typically unaltered, microcline and orthoclase grains along with minor sanidine. Some feldspars have undergone either partial or total replacement by calcite.

Lithic volcanic grains (Lv)

Volcanic rock fragments, glass shards, pumice and accretionary lapilli are common throughout the Gustav Group ranging between 5 and 96% in abundance. The volcanic grains are mainly andesitic or rarely rhyolitic in composition. The groundmass and/or the plagioclase phenocrysts may be altered to calcite and clay minerals; grain replacement is occasionally extensive with only remnant grain outlines preserved. In some cases, relatively altered and unaltered lithic volcanic grains occur together within the same thin section. Opaque mineral-rich volcanic grains are common. The glass shards are usually cuspate and show varying degrees of diagenetic alteration. Pumice grains are usually partially replaced by calcite.

Lithic plutonic grains (Lp)

Lithic plutonic grains are rare within the Gustav Group, although grains showing either granophyric or myrmekitic textures do occur.

Lithic metamorphic grains (Lm)

Pelitic and psammitic metamorphic clasts are a common component within the Gustav Group. Grains typically comprise quartz–mica aggregates. Quartz subcrystals usually have sutured grain margins and show undulose extinction.

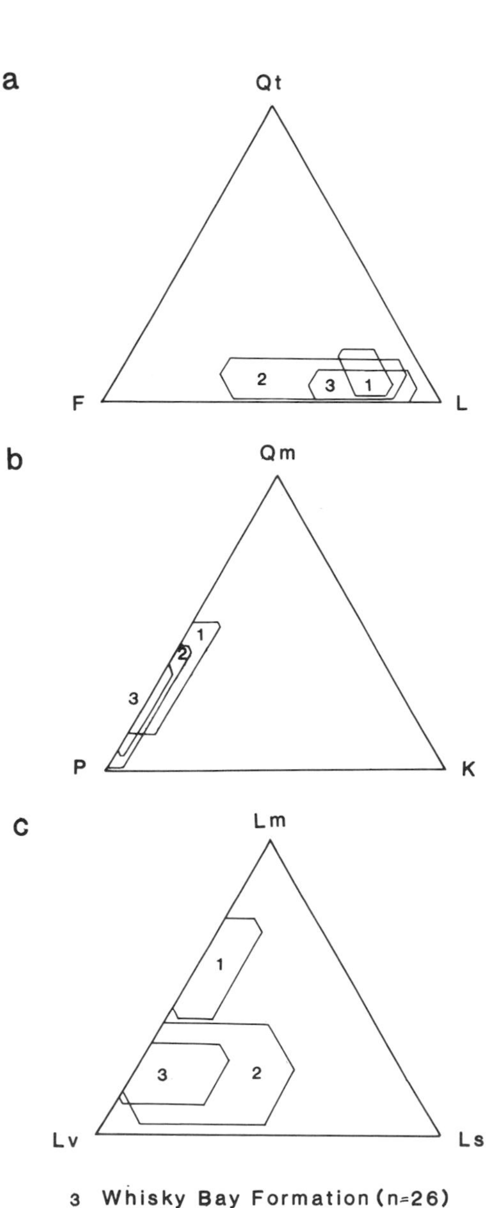

Fig. 5. Ternary diagrams showing the compositional trends for sandstones from the deep marine sequence. Data are shown as one standard deviation around the mean. Qt, Total quartz grains; F, feldspar grains; L, lithic grains; Qm, monocrystalline quartz grains; P, plagioclase feldspar grains; K, potassium feldspar grains; LV, lithic volcanic grains; LS, lithic sedimentary grains; LM, lithic metamorphic grains.

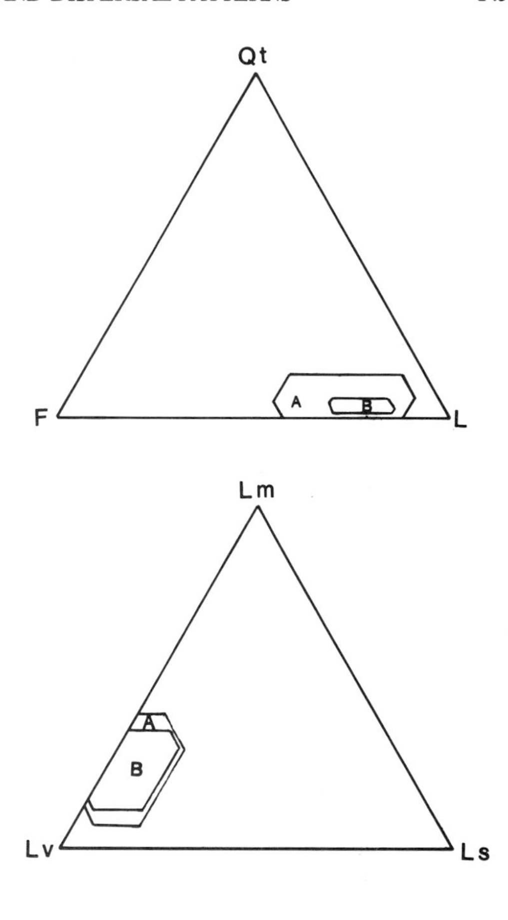

Fig. 6. Ternary plots showing the similarity in composition of sandstones between the slope-apron and submarine fan depositional environments in the Whisky Bay Formation for both outcrop areas. Data are shown as one standard deviation around the mean. Abbreviations as in Fig. 5.

Lithic sedimentary grains (Ls)

Organic-rich radiolarian mudstone and radiolarian-poor mudstone grains commonly occur. In all cases the grains are well rounded with either low or high sphericity.

The dominant accessory minerals present within the sandstones examined include biotite, muscovite, hornblende, pyroxene, zircon, apatite, garnet, rutile, pyrite, chalcopyrite, ilmenite and titanomagnetite.

Sandstone point-count data for all framework grains are shown on Qt–F–L and Qm–P–K ternary diagrams (Fig. 5a & 5b) and the point-count data for the lithic grains are shown on a Lv–Lm–Ls ternary diagram (Fig. 5c). Data are displayed as one standard deviation around the mean. As shown on both the Qt–F–L and Qm–P–K plots the lower three formations in the Gustav Group are lithic and plagioclase-rich, and cannot be clearly distinguished petrographically. In contrast, the lithics data show that the basal unit, the Lagrelius Point Formation, is petrographically distinct, with a greater abundance of Lm grains relative to the overlying formations. However, the Kotick Point and Whisky Bay Formations are petrographically comparable and dominated by lithic volcanic grains. In addition, the two major outcrop areas cannot be distinguished petrographically. Within the Whisky Bay Formation, sandstone samples assigned to either the slope-apron or submarine fan depositional environments cannot be distinguished petrographically (Fig. 6), although Ineson (1989), on the basis of field studies, suggested that large angular clasts of Nordenskjöld Formation are more abundant within the slope-apron deposits.

Sandstone source areas

On the basis of the petrographic data, the sandstone grains recognized can be matched with the known geology of the northern Antarctic Peninsula region (cf. Pirrie 1991); this is also consistent with the available palaeocurrent data (Ineson 1989). The Kotick Point and Whisky Bay Formations were predominantly sourced from calc–alkaline andesitic volcanic and cogenetic plutonic rocks of the APVG, along with a minor component from the metasedimentary rocks of the TPG and the Nordenskjöld Formation. Numerous pulses of lithic volcanic and euhedral plagioclase grains are indicative of coeval arc volcanism. These sandstones lie in the undissected to transitional arc fields of Dickinson *et al.* (1983). In contrast, the underlying Lagrelius Point Formation comprises a larger component of Lm grains derived from the TPG (see Fig. 5c). This implies that regional uplift of the TPG may have occurred prior to the development of an extensive cover of andesitic volcanics in the northern Antarctic Peninsula region.

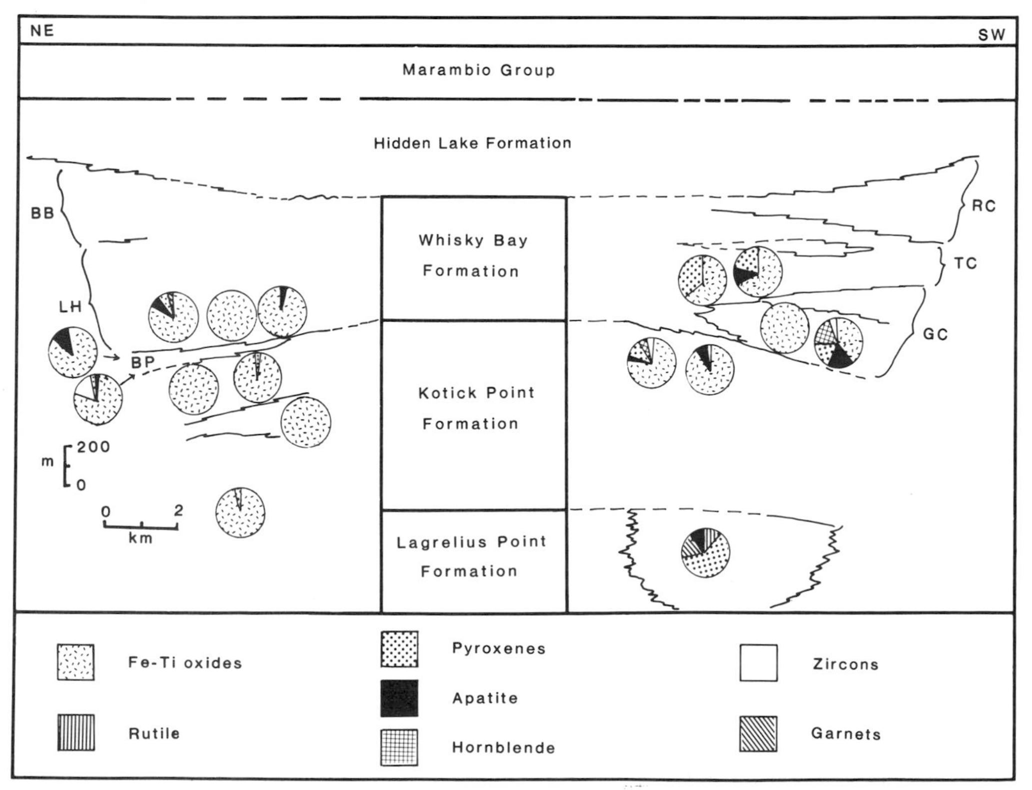

Fig. 7. Heavy mineral distribution within the Gustav Group. LH, Lewis Hill Member; BP, Bibby Point Member; RC, Rum Cove Member; TC, Tumbledown Cliffs Member; GC, Gin Cove Member; BB, Brandy Bay Member.

Table 2. *Summary of the microprobe data used in this study; ranges are shown as Wt% oxides for each element analysed*

Oxide	Ilmenite	Titanomagnetite	Clinopyroxene	Garnet
SiO_2	0–0.22	0.09–1.59	49.7–68.98	30.15–44.27
Al_2O_3	0–1.4	0.9–6.0	1.99–25.96	19.77–28.03
TiO_2	40–55.3	8–20.5	0–0.94	0.02–0.37
FeO	44.31–51.31	71.5–80.93	0.09–11.32	6.62–29.09
Cr_2O_3	0–0.4	0–0.24	0–0.45	0–0.06
MnO	0.5–4.5	0.3–3.5	0–1.26	0.06–0.62
MgO	0.3–5	0.1–3.8	0–16.6	0–12.39
CaO	0–0.02	0–0.29	0.03–20.31	8.1, 20.5–22.89
Na_2O	0–0.52	0–1.85	0.21–12.07	0–0.21

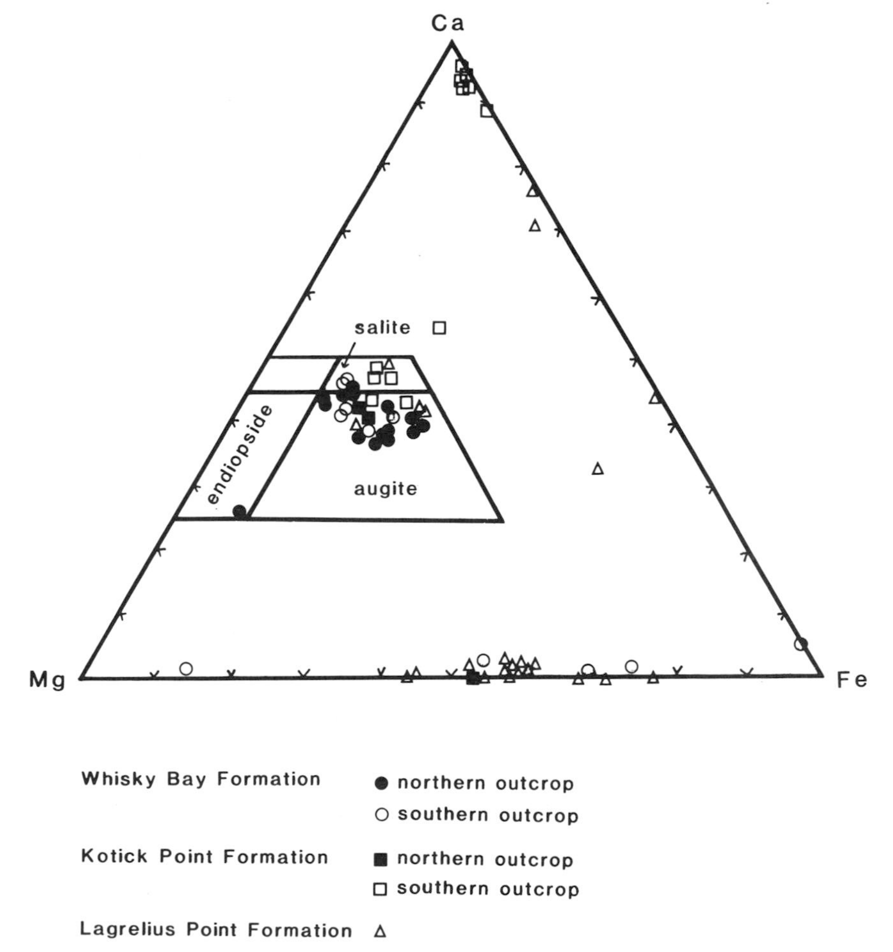

Fig. 8. Ca–Mg–Fe ternary plot for the clinopyroxene grains analysed.

In summary, although the traditional sandstone provenance data allow the sediment source area to be described in broad terms, it does not allow a detailed understanding of sediment dispersal within the submarine fan and slope-apron sequence. However, detailed electron microprobe analyses do allow a clearer understanding of sediment dispersal.

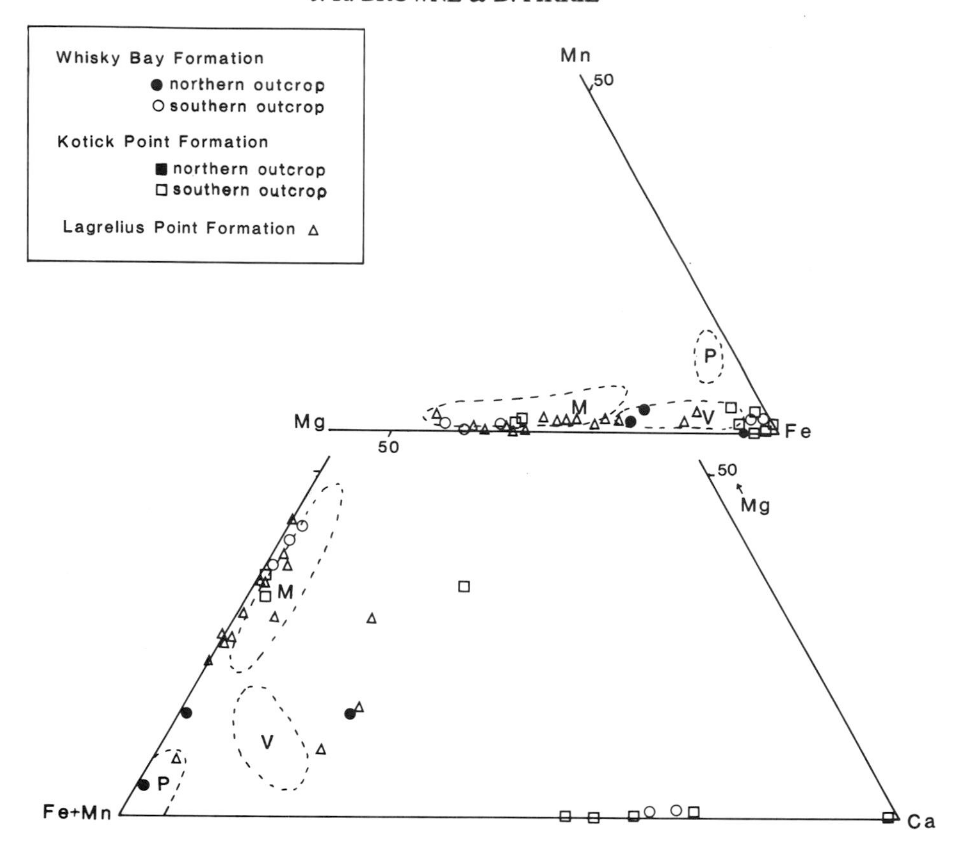

Fig. 9. Ca–Mg–Fe + Mn and Mg–Fe–Mn ternary diagrams of garnet composition in relation to the fields identified by Moyes & Hamer (1983). P, plutonic; V, volcanic; and M, metamorphic garnets.

Heavy mineral geochemistry

The results of heavy mineral point counts are shown schematically in Fig. 7. As can be seen, the most abundant heavy minerals are the Fe–Ti oxides and clinopyroxenes along with lesser amounts of apatite, zircon, garnet and hornblende. Clinopyroxene, garnet, ilmenite and titanomagnetite grains were analysed by electron microprobe for the elements Si, Ti, Al, Cr, Fe, Mn, Mg, Ca, Na and K. Results, reported as Wt% oxides, are summarized in Table 2. Pyroxene, although diagenetically unstable, is of importance as its chemistry is potentially indicative of magma type and tectonic setting of the source rocks (Cawood 1991; Morton 1991). Sixty-four clinopyroxene grains were analysed in this study, and were predominantly of augite composition with lesser amounts of jadeite, diopside, endiopside and salite (Fig. 8). These results are comparable with previous work from the James Ross Basin (Pirrie 1991). The abundance of unaltered augite grains with high Na_2O and MnO and low Cr_2O_3 contents supports derivation from an andesitic source, probably the APVG (cf. Pirrie 1991). Although variable, pyroxene grains from the Lagrelius Point Formation plot away from the augite field, implying a different source area for these grains.

Garnet is rare within the Gustav Group and only occurs as small unzoned spherical grains in the Lagrelius Point Formation and at the base of the Whisky Bay Formation. Electron microprobe results show that the grains are mostly almandine in composition (Fig. 9). Within the source region, garnet occurs within both the APVG and TPG. Hamer & Moyes (1982) analysed garnet from the APVG and recognized two main groups; Type A almandine-rich, primary igneous garnet and Type B, less almandine and more pyrope-rich garnet, interpreted as derived from xenocrysts or xenoliths within the volcanic rocks. As shown in Fig. 9, there is no apparent stratigraphical variation in the garnet composition with garnets derived from both volcanic and metamorphic sources. Recent

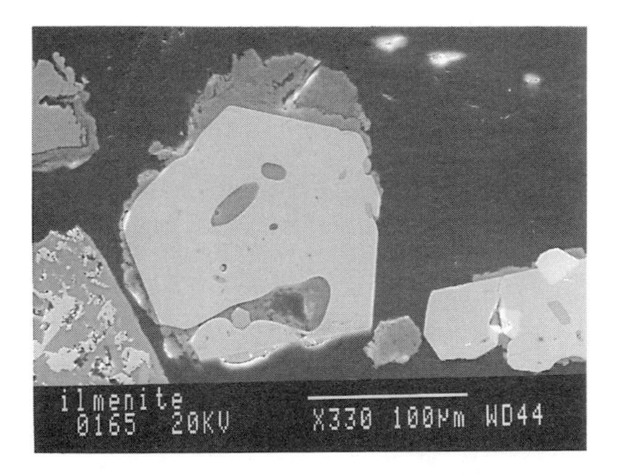

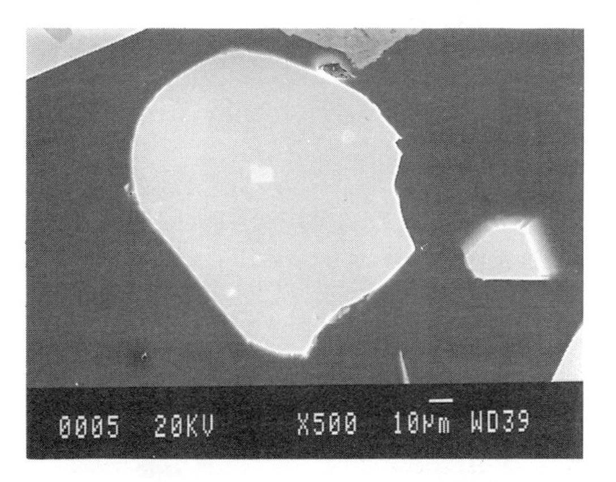

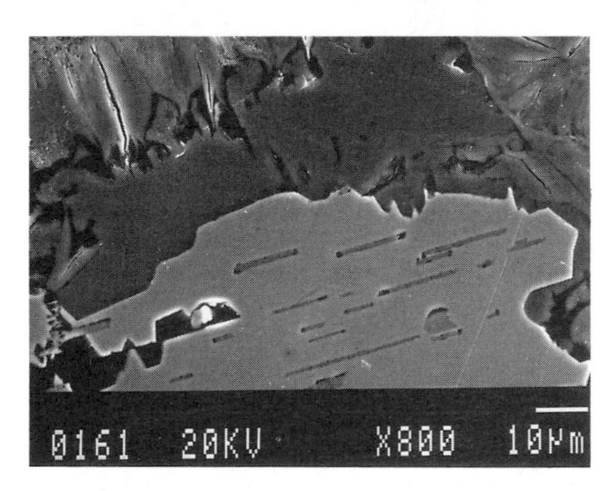

Fig. 10. Scanning electron microscope photomicrographs of (*a*) ilmenite; (*b*) titanomagnetite; and (*c*) exsolution lamellae within ilmenite.

studies by Elliot *et al.* (1992) on the provenance of Palaeocene strata within the James Ross Basin, suggest that garnet was also derived from a medium- to high-grade regional or contact metamorphic terrain, possibly to the south, although this source area is not recognized here.

Although Fe–Ti oxide minerals are present in a very wide range of lithologies (Force 1991), the ilmenite and titanomagnetite described here was probably derived from the APVG (Pirrie 1991). Previous studies on the use of Fe–Ti oxide minerals have focused on recent sands derived from either igneous or metamorphic source areas (e.g. Riezebos 1979; Darby 1984; Darby & Tsang 1987; Basu & Molinaroli 1989; Grigsby 1990, 1992) and Tertiary sandstones (Basu & Molinaroli 1991). Although commonly present as an accessory mineral in many sandstones, ilmenite has been shown to be diagenetically altered under a wide range of conditions (e.g. Reynolds 1982; Morad & Aldahan 1986). The alteration of ilmenite results in the removal of Fe, usually by leaching, resulting in an almost pure TiO_2 phase (Morad & Aldahan 1986). Diagenetic TiO_2 minerals (the polymorphs rutile, anatase and brookite) are precipitated during the diagenetic alteration of ilmenite, although the low solubility of Ti results in only local redistribution of these diagenetic phases (Morad 1986). Rutile is present throughout the Gustav Group, although it is most abundant within the Lagrelius Point Formation. Under SEM the rutile occurs as discrete rounded grains, which are interpreted as being detrital rather than diagenetic in origin. The textural and chemical features indicative of the diagenetic alteration of ilmenite (e.g. Ti enrichment and Fe depletion) were not observed in this study. Therefore, it is considered that this geochemical data was not significantly modified during diagenesis.

Previous workers have found that ilmenite grains may be homogeneous or have exsolution or intergrowth lamellae of other mineral phases present (e.g. Basu & Molinaroli 1989). However, in this work nearly all of the titanomagnetite and ilmenite grains examined were homogeneous uniform grains (Fig. 10a & b) with only very rare grains showing possible exsolution lamellae of haematite (Fig. 10c).

One hundred and four ilmenite and titanomagnetite grains have been analysed from 15 samples by electron microprobe. The ilmenite grains typically have TiO_2 and FeO values of 46–50%, MgO 0.46–5.02%, MnO 0.64–1.61% and low Cr_2O_3 and Al_2O_3 (<0.4 and 0.14%, respectively). Figure 11 shows TiO_2 plotted against MnO and MgO for the Fe–Ti grains from individual formations. As can be clearly seen on the MnO v. TiO_2 cross-plot, the ilmenite and titanomagnetite grains plot into

Whisky Bay Formation

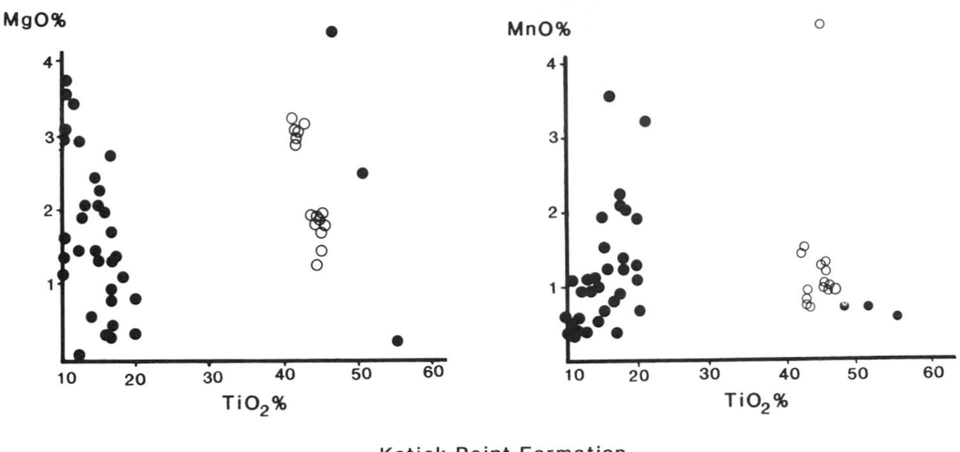

Kotick Point Formation

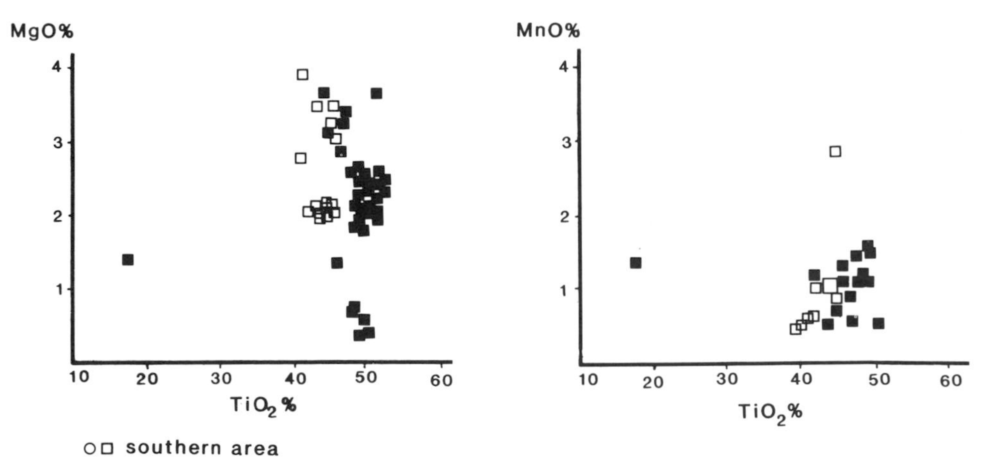

Fig. 11. Stratigraphical geochemical variablity within the Fe–Ti oxides for the Kotick Point and Whisky Bay Formations from the two outcrop areas. Note that overlapping analyses are not shown for the Kotick Point Formation. Open symbols, southern outcrop area; filled symbols, northern outcrop area. Ilmenite is characterized by TiO_2 contents ranging between 32 and 55% TiO_2. Titanomagnetite may have low concentrations of TiO_2 (5–20%).

two discrete fields. Ilmenite occurs within the Kotick Point and Whisky Bay Formations whilst titanomagnetite only occurs within the Whisky Bay Formation.

Figure 12 shows the variation in TiO_2 v. MnO both spatially and temporally within the lower Gustav Group. Ilmenite occurs in both outcrop areas of the Kotick Point Formation. However, in the overlying Whisky Bay Formation, ilmenite is only present in the southern outcrop area, within both the Gin Cove Member (slope-apron) and Tumbledown Cliffs Member (submarine fan). Titanomagnetite occurs within the Bibby Point Member (slope-apron) and Lewis Hill Member (submarine fan) in the northern outcrop area. These units are approximately equivalent biostratigraphically. It is interesting to note that the overlying Hidden Lake Formation contains ilmenite within both of the outcrop areas. Figure 13 shows MnO v. TiO_2 cross-plots for Whisky Bay Formation samples from both the submarine fan and slope-apron depositional environments for both the northern and southern outcrop areas. Irrespective of

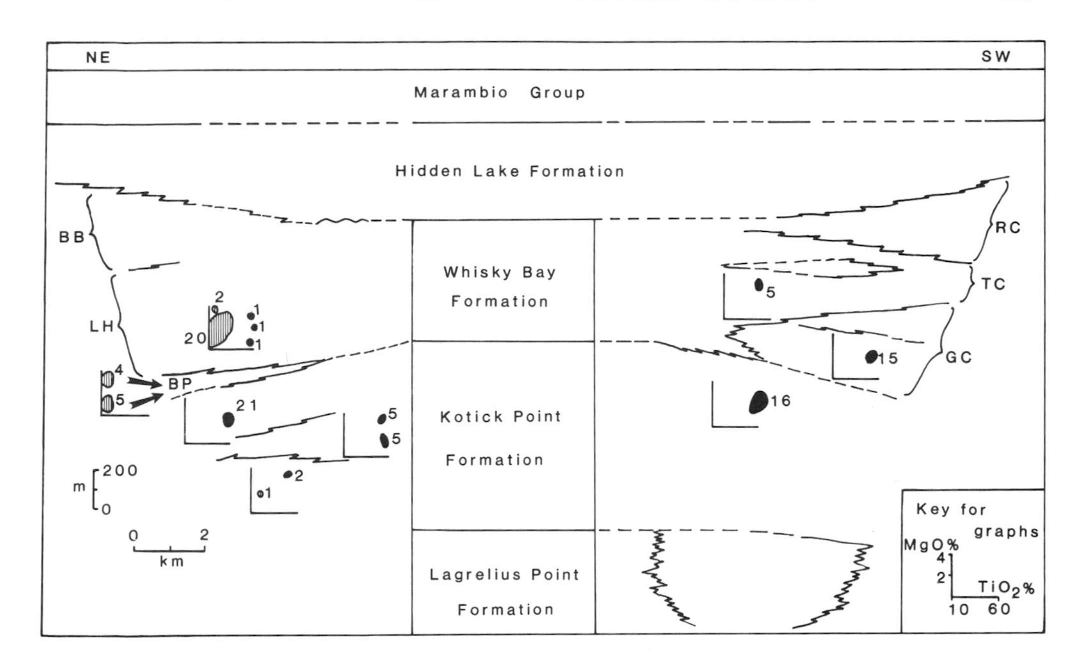

Fig. 12. Geochemical variability of the Fe–Ti oxides. LH, Lewis Hill Member; BP, Bibby Point Member; RC, Rum Cove Member; TC, Tumbledown Cliffs Member; GC, Gin Cove Member; BB, Brandy Bay Member. Data is shown as fields; the number of individual grains analysed for each field is shown. Ilmenite is shown by the solid symbol, titanomagnetite is shown by the vertical lines. Note that titanomagnetite only occurs within the northern outcrop area of the Whisky Bay Formation.

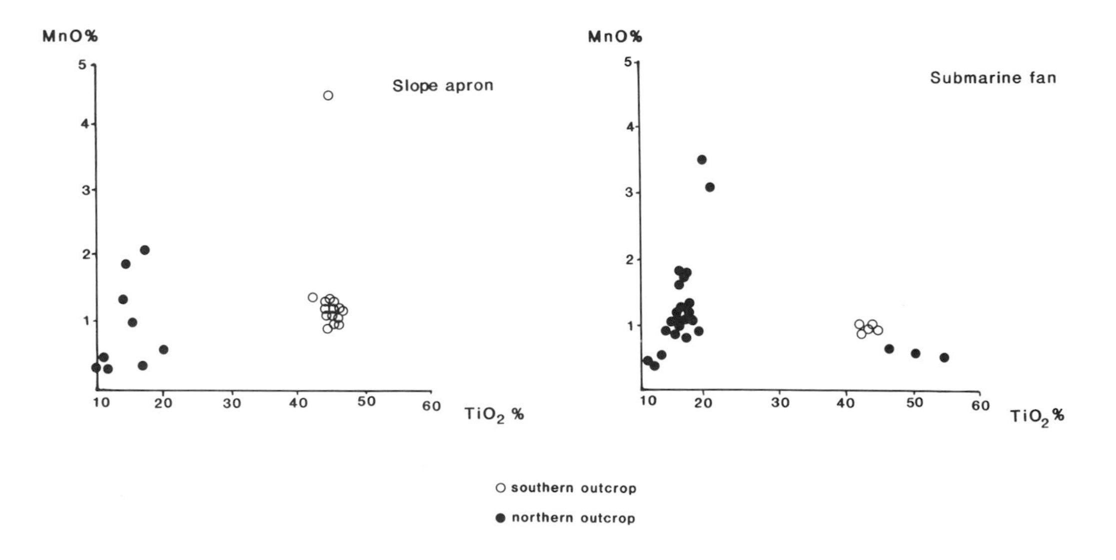

Fig. 13. Compositional similarity in the Fe–Ti oxides of the submarine fan and slope-apron deposits of the Whisky Bay Formation.

depositional setting, the northern outcrop area can still be characterized on the presence of titanomagnetite and absence of ilmenite. The southern outcrop area slope-apron and submarine fan sandstones contain ilmenite with rare titanomagnetite grains. Ilmenite and titanomagnetite have similar densities and occur within comparable facies and depositional environments. It is unlikely, therefore, that the apparent distribution of ilmenite and titanomagnetite has been controlled by sedimentary processes. In addition, there are no data to suggest differential diagenetic conditions within the two outcrop areas. The distribution of the heavy minerals is therefore considered to be controlled by source-rock geology.

Implications for sediment dispersal

During the deposition of the Kotick Point and Whisky Bay Formations, Ineson (1989) envisaged the basin margin to be faulted, with a mudstone-dominated slope-apron and laterally equivalent coarse-grained gravelly submarine fan systems. The approximately age-equivalent Tumbledown Cliffs Member to the southwest and Lewis Hill Member to the northeast are both conglomerate-dominated submarine fan sequences. The available palaeocurrent data for these units (Fig. 4) show diverging palaeoflow directions with the two outcrop areas separated by no exposure. The palaeocurrent data can either be interpreted as reflecting a radial sediment dispersal pattern within a single large fan body, that was fed from a point source, or two discrete sediment lobes flowing approximately oblique to the basin margin and fed from two separate source areas. On the basis of the sandstone petrographic data, the two outcrop areas cannot be distinguished. However, the Fe–Ti oxide data suggest that the two areas were supplied from different source areas and may therefore represent separate submarine fan lobes encased in slope-apron facies. In addition, heavy minerals from the slope-apron sediments in the two outcrop areas are also distinct, supporting the interpretation of sediment supply from two discrete volcanic source rocks. It is possible that sediment supply to the basin may have been from two or more point sources throughout the deposition of the Gustav Group, but if these areas were draining similar lithologies then petrographically they would be indistinguishable. The heavy mineral data can be interpreted as indicating that during Whisky Bay Formation times titanomagnetite-bearing volcanics were eroded in the north whilst ilmenite-bearing volcanics were derived from the southwest, allowing the recognition of the discrete sediment supply systems. Alternatively, despite the broadly comparable biostratigraphical ages for the two outcrop areas, it is possible that they are not directly contemporaneous. Ineson (1985*b*), suggested that the coarse-grained submarine fan system aggraded vertically, and switched laterally with time, so that the Lewis Hill Member is older, at least in part, than the Tumbledown Cliffs Member (Ineson pers. comm.). If this interpretation is correct then sediment supply may have been from a single point source, with a temporal change in source rock composition, from ilmenite-bearing to titanomagnetite-bearing with time.

In summary, this data either indicates two discrete depositional sediment lobes of equivalent age in the northern and southern outcrop areas, or that the outcrop areas are not stratigraphically equivalent. Either interpretation has implications for the understanding of the submarine fan depositional system.

Conclusion

On the basis of sandstone petrography data, the lower Gustav Group submarine fan–slope-apron complex can be interpreted as derived from rocks exposed on the northern Antarctic Peninsula. However, on the basis of standard petrographic studies the Kotick Point and Whisky Bay Formations are indistinguishable in both the northern and southern outcrop areas. The underlying Lagrelius Point Formation is petrographically distinct, and has a different heavy mineral suite, supporting the previous interpretation of this formation as a discrete lithostratigraphic unit (cf. Ineson *et al.* 1986). The geochemistry of Fe-Ti oxide minerals from the Kotick Point and Whisky Bay Formations provides a more detailed understanding of sediment supply. Kotick Point Formation sandstones in both outcrop areas were derived from ilmenite-bearing volcanics. However, in the overlying Whisky Bay Formation, the northern outcrop area is characterized by the presence of titanomagnetite whilst the southern outcrop area contains ilmenite. In conjunction with available biostratigraphical and palaeocurrent data this can be interpreted in two ways. Firstly, the two outcrop areas may be time-equivalent and represent two separate submarine fan lobes derived from discrete point sources. Secondly, if the outcrop areas are not time-equivalent the coarse-grained submarine fan sediments may have switched laterally with time and were derived from a single point-source area in which the composition of the opaque minerals present changed temporally. Either interpretation has implications for the understanding of the submarine fan depositional system.

Although the example studied here is based on outcrop samples from a non-prospective basin, these results do have implications for the description of hydrocarbon reservoirs. Although few hydrocarbon reservoirs are developed in volcaniclastic facies, the depositional system studied is comparable to a number of producing hydrocarbon reservoirs, and detrital opaque minerals are common in many sandstones. Detailed sediment provenance studies can aid the interpretation of deep marine reservoirs in subsurface by identifying discrete sediment lobes and aiding the interpretation of the lateral extent, and therefore connectivity of individual depositional bodies. In addition, high-resolution sediment provenance studies allow a clear understanding of sediment dispersal. Although varietal geochemical analysis of heavy minerals as an aid to understanding provenance is being used within reservoir description, most sedimentologists do not utilize the opaque minerals present. As shown in this study, detrital opaque phases are commonly source specific, and provide additional provenance data. Consequently, opaque minerals should not be ignored in provenance studies.

The work presented here was funded by a research studentship to J. R. B. from the Camborne School of Mines. The British Antarctic Survey is thanked for providing access to samples and for providing financial support. S. Pendray, J. Curnow, T. Ball (CSM) and T. Blesser (University of Cambridge) provided technical support. We are grateful to Drs. M. R. A. Thomson and J. R. Ineson for commenting on an early draft on this paper, and to journal referees Drs. A. G. Whitham and A.C. Morton for their constructive comments.

References

BASU, A. & MOLINAROLI, E. 1989. Provenance characteristics of detrital opaque Fe–Ti oxide minerals. *Journal of Sedimentary Petrology*, **59**, 922–934.

—— 1991. Reliability and application of detrital opaque Fe–Ti oxide minerals in provenance determination. *In:* MORTON, A. C., TODD, S. P. & HAUGHTON, P. D. W. (eds) *Developments in Sedimentary Provenance Studies,* Geological Society, London, Special Publication, **57**, 55–65.

BUATOIS, L. A. & LOPEZ ANGRIMAN, A. O. 1992. The ichnology of a submarine braided channel complex: the Whisky Bay Formation, Cretaceous of James Ross Island, Antarctica. *Palaeogeography, Palaeoclimatology, Palaeoecology,* **94**, 119–140.

CAWOOD, P. A. 1991. Nature and record of igneous activity in the Tonga arc, SW Pacific, deduced from the phase chemistry of derived detrital grains. *In:* MORTON, A. C., TODD, S. P. & HAUGHTON, P. D. W. (eds) *Developments in Sedimentary Provenance Studies,* Geological Society, London, Special Publication, **57**, 305–321.

CRAME, J. A., PIRRIE, D., RIDING, J. B. & THOMSON, M. R. A. 1991. Campanian–Maastrichtian (Cretaceous) stratigraphy of the James Ross Island area, Antarctica. *Journal of the Geological Society, London,* **148**, 1125–1140.

DARBY, D. A. 1984. Trace elements in ilmenite: A way to discriminate provenance or age in coastal sands. *Geological Society of America Bulletin,* **95**, 1208–1218.

—— & TSANG, Y. W. 1987. Variation in ilmenite element composition within and among drainage basins: Implications for provenance. *Journal of Sedimentary Petrology,* **57**, 831–838.

DEL VALLE, R. A., ELLIOT, D. H. & MACDONALD, D. I. M. 1992. Sedimentary basins on the east flank of the Antarctic Peninsula: proposed nomenclature. *Antarctic Science,* **4**, 477–478.

DICKINSON, W. R. 1970. Interpreting detrital modes of greywacke and arkose. *Journal of Sedimentary Petrology,* **40**, 695–707.

—— BEACH, L. S., BRACKENRIDGE, *et al.* 1983. Provenance of North American Phanerozoic sandstones in relation to tectonic setting. *Geological Society of America Bulletin,* **94**, 222–235.

ELLIOT, D. H., HOFFMAN, S. M. & RIESKE, D. E. 1992. Provenance of Palaeocene strata, Seymour Island. *In:* YOSHIDA, Y., KAMINUMA, K. & SHIRAISHI, K. (eds) *Recent Progress in Antarctic Earth Science.* Terrapub, Tokyo, 347–355.

FARQUHARSON, G. W. 1983. The Nordenskjöld Formation of the northern Antarctic Peninsula: An Upper Jurassic radiolarian mudstone and tuff sequence. *British Antarctic Survey Bulletin,* **60**, 1–22.

—— 1984. Late Mesozoic, non-marine conglomeratic sequences of northern Antarctic Peninsula (the Botany Bay Group). *British Antarctic Survey Bulletin,* **65**, 1–32.

—— HAMER, R. D. & INESON, J. R. 1984. Proximal volcaniclastic sedimentation in a Cretaceous back-arc basin, northern Antarctic Peninsula. *In:* KOKELAAR, B. P. & HOWELLS, M. F. (eds) *Marginal Basin Geology.* Geological Society, London, Special Publication, **16**, 219–229.

FORCE, E.R. 1991. *Geology of Titanium-mineral deposits.* Geological Society of America, Special Paper, **259**.

GRIGSBY, J. D. 1990. Detrital magnetite as a provenance indicator. *Journal of Sedimentary Petrology,* **60**, 940–951.

—— 1992. Chemical fingerprinting in detrital ilmenite: A viable alternative in provenance research? *Journal of Sedimentary Petrology,* **62**, 331–337.

HAMER, R. D. & MOYES, A. B. 1982. Composition and origin of garnet from the Antarctic Peninsula Volcanic Group of Trinity Peninsula. *Journal of the Geological Society, London,* **139**, 713–720.

HOUGHTON, H. F. 1980. Refined technique for staining plagioclase and alkali feldspars in thin section. *Journal of Sedimentary Petrology,* **50**, 629–631.

HUMPHREYS, B., MORTON, A. C., HALLSWORTH, C. R., GATLIFF, R. W. & RIDING, J. B. 1991. An integrated approach to provenance studies: a case example from the Upper Jurassic of the Central Graben, North Sea. *In:* MORTON, A. C., TODD, S. P. & HAUGHTON, P. D. W. (eds) *Developments in Sedimentary Provenance Studies,* Geological Society, London, Special Publication, **57**, 251–262.

INESON, J. R. 1985*a*. Submarine glide blocks from the Lower Cretaceous of the Antarctic Peninsula. *Sedimentology,* **32,** 659–670.

INESON, J. R. 1985*b*. A slope apron – submarine fan complex in the Lower Cretaceous of Antarctica. *Sixth European Meeting of the International Association of Sedimentologists,* LLeida 1985, Abstracts volume, 203–206.

—— 1989. Coarse-grained submarine fan and slope apron deposits in a Cretaceous back-arc basin, Antarctica. *Sedimentology,* **36,** 793–819.

—— CRAME, J. A. & THOMSON, M. R. A. 1986. Lithostratigraphy of the Cretaceous strata of west James Ross Island, Antarctica. *Cretaceous Research,* **7,** 141–159.

INGERSOLL, R. V., BULLARD, T. F., FORD, R. L., GRIMM, J. P., PICKLE, J. D. & SARES, S. W. 1984. The effect of grain size on detrital modes: A test of the Gazzi–Dickinson point-counting method. *Journal of Sedimentary Petrology,* **54,** 105–117.

KEATING, J. M., SPENCER-JONES, M. & NEWHAM, S. 1992. The stratigraphical palynology of the Kotick Point and Whisky Bay formations, Gustav Group (Cretaceous), James Ross Island. *In:* DUANE, A. M., PIRRIE, D. & RIDING, J. B. (eds) *Palynology of the James Ross Island Area, Antarctic Peninsula,* Antarctic Science, **4,** 279–292.

MACDONALD, D. I. M. & BUTTERWORTH, P. J. 1990. The stratigraphy, setting and hydrocarbon potential of the Mesozoic sedimentary basins of the Antarctic Peninsula. *In:* ST JOHN, B. (ed.) *Antarctica as an Exploration Frontier.* American Association of Petroleum Geologists, Studies in Geology **31,** 101–125.

—— BARKER, P. F., GARRETT, S. W., *et al.* 1988. A preliminary assessment of the hydrocarbon potential of the Larsen Basin, Antarctica. *Marine and Petroleum Geology,* **5,** 34–53.

MORAD, S. 1986. SEM study of authigenic rutile, anatase and brookite in Proterozoic sandstones from Sweden. *Sedimentary Geology*, **46,** 77–89.

—— & ALDAHAN, A. A. 1986. Alteration of detrital Fe–Ti oxides in sedimentary rocks. *Geological Society of America Bulletin,* **97,** 567–578.

MORTON, A. C. 1985. Heavy minerals in provenance studies. *In:* ZUFFA, G. G. (ed.) *Provenance of Arenites.* Reidel, Dordrecht, 249–277.

—— 1991. Geochemical studies of detrital heavy minerals and their application to provenance research. *In:* MORTON, A. C., TODD, S. P. & HAUGHTON, P. D. W. (eds) *Developments in Sedimentary Provenance Studies.* Geological Society, London, Special Publication, **57,** 31–45.

MORTON, A. C. 1992. Provenance of Brent Group sandstones: heavy mineral constraints. *In:* MORTON, A. C., HASZELDINE, R. S., GILES, M. R. & BROWN, S. (eds) *Geology of the Brent Group.* Geological Society, London, Special Publication, **61,** 227–244.

—— STIBERG, J. P., HURST, A. & QVALE, H. 1989. Use of heavy minerals in lithostratigraphic correlation, with examples from Brent sandstones of the northern North Sea. *In:* COLLINSON, J. D. (ed.) *Correlation in Hydrocarbon Exploration.* Norwegian Petroleum Society, 217–230.

MOYES, A. B. & HAMER, R. D. 1983. Contrasting origins and implications of garnet in rocks of the Antarctic Peninsula. *In:* OLIVER, R. L., JAMES, P. R. & JAGO, J. B. (eds) *Antarctic Earth Science.* Cambridge University Press, Cambridge, 358–362

PETTIJOHN, F. J., POTTER P. E. & SIEVER, R. 1972. *Sand and Sandstone.* Springer, Berlin.

PIRRIE, D. 1991. Controls on the petrographic evolution of an active margin sedimentary sequence: the Larsen Basin, Antarctica. *In:* MORTON, A. C., TODD, S. P. & HAUGHTON, P. D. W. (eds) *Developments in Sedimentary Provenance Studies.* Geological Society, London, Special Publication, **57,** 231–249.

—— WHITHAM, A. G. & INESON, J. R. 1991. The role of tectonics and eustasy in the evolution of a marginal basin: Cretaceous–Tertiary Larsen Basin, Antarctica. *In:* MACDONALD, D. I. M. (ed.) *Sedimentation, Tectonics and Eustasy*, Special Publication International Association of Sedimentologists, **12,** 293–305.

RAZJIGAEVA, N. G. & NAUMOVA, V. V. 1992. Trace element composition of detrital magnetite from coastal sediments of northwestern Japan Sea for provenance studies. *Journal of Sedimentary Petrology,* **62,** 802–809.

REYNOLDS, R. L. 1982. Post-depositional alteration of titanomagnetite in a Miocene sandstone, south Texas (U.S.A). *Earth and Planetary Science Letters,* **61,** 381–391.

RIEZEBOS, P. A. 1979. Compositional downstream variation of opaque and translucent heavy residues in some modern Rio Magdalena sands (Colombia). *Sedimentary Geology,* **24,** 197–225.

STOREY, B. C. & NELL, P. A. R. 1988. Role of strike-slip faulting in the tectonic evolution of the Antarctic Peninsula. *Journal of the Geological Society, London,* **145,** 333–337.

STOW, D. A. V. 1985. Brae oilfield turbidite system, North Sea. *In:* BOUMA, A. H., NORMARK, W. R. & BARNES, N. E. (eds) *Submarine Fans and Related Turbidite Systems.* Springer-Verlag, New York, 231–236.

WHITHAM, A. G. & MARSHALL, J. E. A. 1988. Syn-depositional deformation in a Cretaceous succession, James Ross Island, Antarctica: evidence from vitrinite reflectivity. *Geological Magazine,* **125,** 583–591.

The sedimentological record of a late Jurassic transgression: Rona Member (Kimmeridge Clay Formation equivalent), West Shetland Basin, UKCS

IVO VERSTRALEN, ADRIAN HARTLEY & ANDREW HURST

Production Geoscience Research Unit, Department of Geology & Petroleum Geology, King's College, University of Aberdeen, Aberdeen AB9 2UE, UK

Abstract: Late Jurassic sandstones form important potential reservoirs within the West Shetland Basin. These strata are thin (4–84 m) and broadly equivalent in age to the Kimmeridge Clay Formation. They were deposited predominantly in deep marine environments although subaerial and shallow marine facies are also present. The Rona Sandstone forms the lowermost of three sandstone units within the Upper Jurassic succession. Data from 23 exploration wells show, however, that it is a heterogeneous unit ranging from coarse-grained pebbly sandstones to mudstones and, because of this wide lithological variation, the Rona Member is considered a more appropriate name. The facies identified (cohesionless debris flow, high and low density turbidites, upper and lower shoreface) and their vertical transitions are considered typical of fan delta deposition, initially in a subaerial environment and later under progressively more marine conditions. Coarse-grained sediments (debris flow and high density turbidites) of the Rona Member are restricted to isolated topographic lows within Lewisianoid basement whereas finer-grained sediments (low density turbidites) have a wide distribution across the basin and cover much of the Rona Ridge. Previous models have suggested that the coarse-grained sediments were derived from active fault scarps. However, to explain the facies distribution it is suggested that footwall uplift on the Rona and Judd Faults (Rona Ridge and associated basement highs) was responsible for the supply of coarse-grained material to isolated fan deltas. Transgression eventually resulted in: flooding of the basement highs, cut off of sediment supply and deposition of finer-grained sediments. Deposition of the Rona Member is interpreted to have been controlled by the gradual transgression of residual islands of Lewisianoid basement.

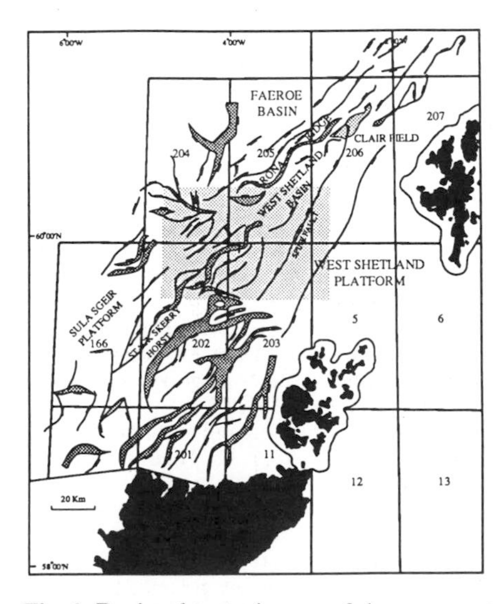

Fig. 1. Regional tectonic map of the area west of Shetland (after Duindam & van Hoorn 1987). Location of study area (Fig. 2) is shaded.

Clastic turbidite systems developed along the flanks of the West Shetland Basin (Fig. 1) during a regional late Jurassic transgression, provide one of a number of potential reservoirs in the largely frontier exploration acreage west of the Shetland Islands (Meadows *et al.* 1987). Previous studies of late Jurassic strata within the West Shetland Basin have necessarily been restricted by a limited dataset (e.g. Hitchen & Ritchie 1987; Meadows *et al.* 1987). Following an upturn in exploration interest in the late 1980s (see Verstralen & Hurst 1994 for a summary) an increased well database has allowed re-examination of potential late Jurassic reservoir facies and development of a new depositional model, which is presented here.

Previous depositional models have used late Jurassic rift-related depositional systems from the South Viking Graben of the North Sea as analogues for the West Shetland area (Hitchen & Ritchie 1987; Meadows *et al.* 1987). During late Jurassic sedimentation in the South Viking

From Hartley, A. J. & Prosser, D. J. (eds), 1995, *Characterization of Deep Marine Clastic Systems*, Geological Society Special Publication No. 94, pp. 155–176.

Graben, debris flow deposits and high density conglomeratic turbidites were deposited adjacent to the footwall scarp of the then active graben-bounding fault (Stow *et al.* 1982; Turner *et al.* 1987). However, facies analysis, particularly of the initial late Jurassic coarse clastic unit (the Rona Sandstone), suggests that this model is not applicable to the late Jurassic of the West Shetland Basin, and that a model involving transgression over an irregular topographic surface is more apt.

Geological setting

The West Shetland Basin was initially discovered from gravity and seismic refraction experiments in the 1970s (Browitt 1972; McLean 1978). McLean (1978) suggested that seismic refraction results were consistent with the presence of 3–4 km of Mesozoic-Tertiary rocks overlying Palaeozoic or Torridonian basement. Subsequent hydrocarbon exploration has further delineated the basin configuration and basin-fill.

Tectonic setting and basin-fill

The West Shetland continental margin is dominated by NE–SW trending faults that bound a number of horst and graben structures. The West Shetland Basin is a narrow NE–SW trending basin (Figs 1 & 2). It can be regarded simplistically as a half-graben with the basin-fill dipping southeastwards into the basin-bounding West Shetland–Spine Fault. To the southwest the basin can be divided into a number of smaller half-grabens known as the Papa, East and West Solan Basins (Booth *et al.* 1993; Fig. 2). The present-day structural configuration of the West Shetland, Papa and Solan Basins is attributed to early Cretaceous–Tertiary rifting as the Atlantic Rift System propagated northeastwards (Duindam & van Hoorn 1987). The West Shetland basin is bounded to the west by the Rona Ridge (Figs. 1 & 2) – a partly fault-bounded high composed of crystalline Lewisianoid basement (Ritchie & Darbyshire 1984) and Devono-Carboniferous sediments (Ridd 1981; Blackbourn 1987). The Rona Ridge is bounded to the west by the Rona Fault which forms the eastern margin to the major late Mesozoic to Tertiary Faeroe Basin (Fig. 2).

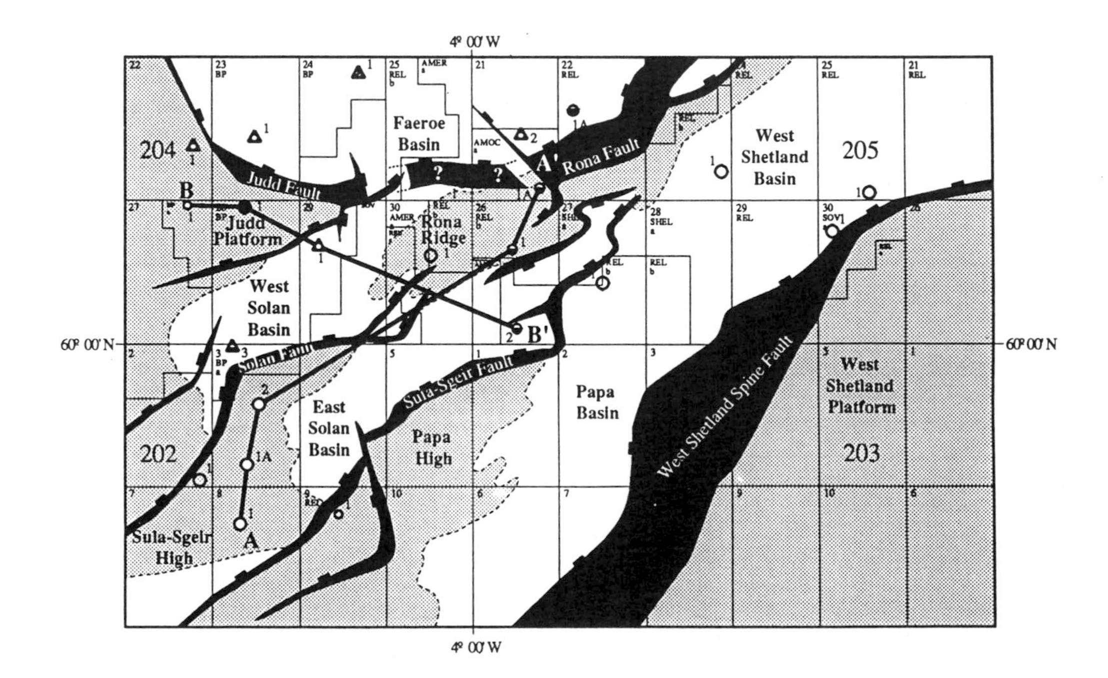

Fig. 2. Map of study area containing the main structural elements and well locations (after Booth *et al.* 1993).

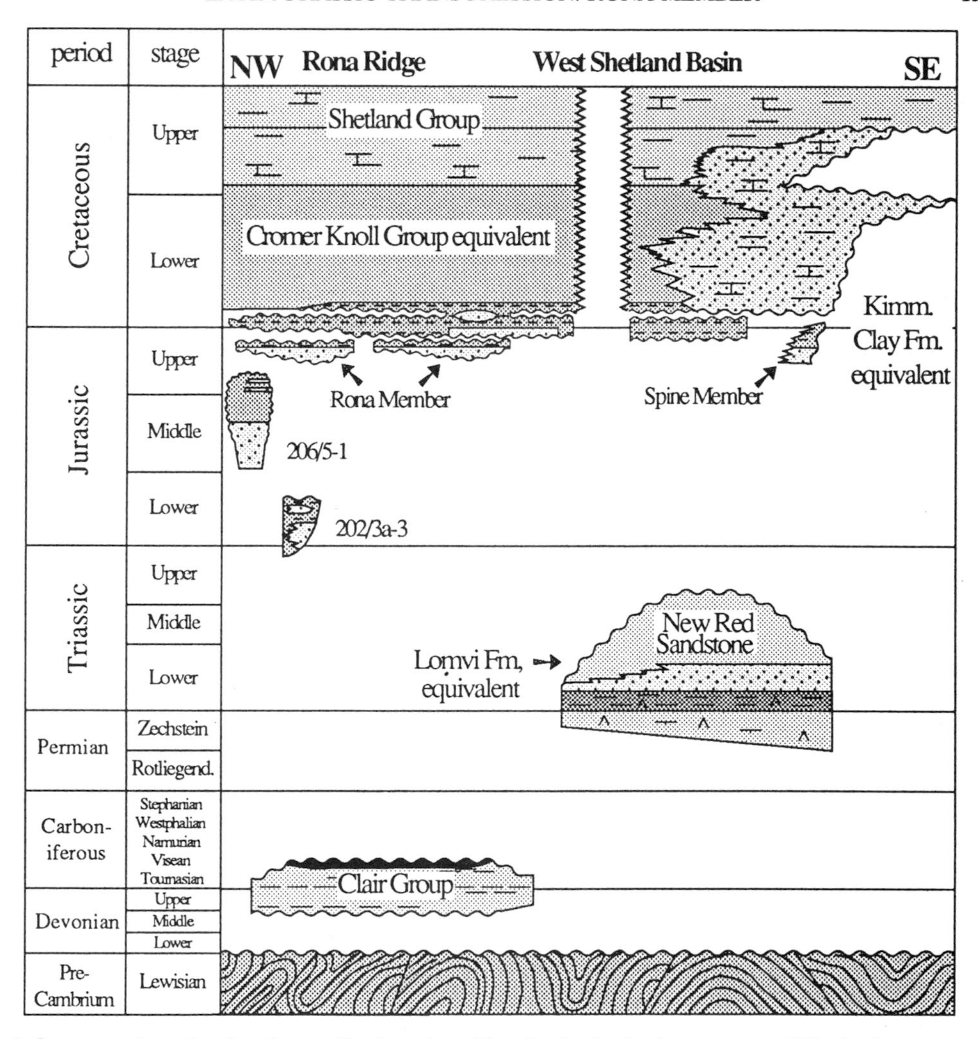

Fig. 3. Summary chart showing the pre-Tertiary depositional episodes in the area west of Shetland.

The stratigraphy of the west Shetland area is fragmentary and characterized by thick, stratigraphically isolated successions of strata (Fig. 3). Thick sections of Permo-Triassic strata are present in the West Shetland, Papa and East Solan Basins, whereas only a thin succession is preserved in the West Solan Basin. The thickness distribution of Permo-Triassic strata indicates a wedge-shaped geometry that is believed to have accumulated in an active half-graben (Booth *et al.* 1993) with thicknesses along the West Shetland–Spine Fault (Fig. 2) of at least 2.5 km and probably over 4 km (Browitt 1972; Haszeldine *et al.* 1987). Preserved thicknesses of Permo-Triassic strata have been modified by a substantial amount of late mid-Jurassic inversion and erosion which has exaggerated the wedge-shaped geometry. Because of the complex post-Triassic evolution of the area, it remains unclear whether Permo-Triassic deposition took place in a number of isolated basins or in one large single basin.

An apparently isolated section, at least 770 m thick, of Lower Jurassic (Dunlin Group equivalent) strata is present in the West Solan Basin (well 202/3a-3). Whether this succession represents Lower Jurassic deposition in a restricted part of the West Shetland area, or an erosional remnant of a once more widespread occurrence is uncertain. This latter hypothesis is favoured by: (1) the presence of reworked Lower Jurassic microfossils in several wells in the West Shetland area (Hitchen & Ritchie 1987); (2) the isolated occurrence of 770 m of marginal to open marine sandstones and mudstones; and (3) a mid–early late Jurassic (intra-Callovian) deformational event (Booth *et al.* 1993) that resulted in substantial uplift and erosion. Middle Jurassic strata have not been encountered in the West Shetland, Solan and Papa Basins, however, Lower and Middle Jurassic sediments are present in other wells in the region. Well 204/22–1, located on the Judd Platform (Fig. 2), penetrated 260 m of mid-Jurassic strata. In the northern part of the Faeroe Basin, well 206/5–1 penetrated > 700 m, including 450 m of coarse-grained clastics of mid–late Jurassic age (Haszeldine *et al.* 1987). Reworked Middle and Upper Jurassic dinocysts have been encountered in well 206/13–1 indicative of a previously more widespread occurrence of Middle Jurassic strata (Hitchen & Ritchie 1987).

Following late mid-Jurassic uplift and erosion Upper Jurassic sediments were deposited on Lewisianoid and Triassic strata. An angular (*c.* 20°) erosional unconformity is developed between Triassic and Upper Jurassic strata (Booth *et al.* 1993). Apatite fission track analysis from Triassic and Upper Jurassic rocks indicates that up to 1.5 km of sediment was removed prior to Upper Jurassic deposition (Booth *et al.* 1993). The Upper Jurassic is characterized by a thin, widespread cover consisting of a variety of clastic sediments, ranging from conglomeratic sandstones to basinal mudstones. Lower Cretaceous strata record the reactivation of normal faults and rejuvenation of source areas, resulting in a thick (1200 m), widespread cover of coarse-grained clastic sediments in the West Shetland Basin and a thinner (100–200 m), finer-grained section in the Solan Basin (Booth *et al.* 1993). Except for a short period of basin inversion during mid-Cretaceous times, due to initiation of the adjacent Faeroe Basin, deposition was continuous and resulted in a thick (1500 m) succession of mixed coarse- and fine-grained Upper Cretaceous sediments in the West Shetland and Solan Basins (Booth *et al.* 1993).

Transpressional reactivation of NE–SW trending faults during the early Palaeocene resulted in the formation of several localized, 'intense' inversion structures (Booth *et al.* 1993). During the Palaeocene 125–200 m of bryozoan-rich sandstones interbedded with mudstones were deposited on the flanks of the West Shetland Basin (Hitchen & Ritchie 1987). In the late Palaeocene a localized unconformity developed over parts of the basin which may be related to thermal uplift in response to Tertiary igneous activity. The bulk of Eocene and Lower Oligocene sediments comprise shallow marine glauconitic sandstones with minor mudstones and siltstones, although continental to paralic coal-bearing strata were developed on the southwestern edge of the Rona Ridge (Booth *et al.* 1993).

A major mid-Tertiary (?late Oligocene to early Miocene) erosional unconformity is developed across the West Shetland Basin. During this time period the Rona Ridge was uplifted and deeply eroded with as much as 1250 m of sediment removed in places (Booth *et al.* 1993). Minor reverse motion took place on NE–SW trending faults. Sediments overlying the Oligo-Miocene unconformity comprise coarse-grained sandstones and glacio-marine gravels of Pliocene to Quaternary age.

Upper Jurassic stratigraphy and sedimentology

Upper Jurassic sediments in the area west of Shetland have been penetrated by at least 17 wells. They unconformably overlie Lewisianoid basement, Triassic and early Jurassic rocks (Verstralen & Hurst 1994). Overall, the sequence consists of a number of sandstone units overlain by, or interbedded with, mudstones equivalent in age to the Kimmeridge Clay Formation (Fig. 3).

No formal litho-stratigraphic framework has been proposed for the area. The term Rona Sandstone has been informally used for the lowermost coarse clastic unit within the Upper Jurassic succession. The Rona Sandstone comprises a wide variety of lithologies, including conglomeratic sandstones, coarse- to very fine-grained sandstones, and siltstones interbedded with mudstones, considered to have been deposited in a number of different depositional environments. Therefore, the term Rona Member (of the Kimmeridge Clay Formation equivalent) is considered more appropriate and is introduced here.

At least two younger sandstone units (in wells 204/30–1 and 205/20–1) have been identified within the Upper Jurassic (Fig. 3). The position of these sandstone units relative to the Rona Member can be constrained by wireline log and biostratigraphic data within the Kimmeridge Clay Formation. The age of the Upper Jurassic succession ranges from early to late Volgian, although no reliable biostratigraphic data from the base of the Rona Member have been recovered. The Rona Member however, is considered to have been deposited during an overall transgressional event (see later), and is overlain by mudstones equivalent to the Kimmeridge Clay Formation. In similarity to the upper part of the Kimmeridge Clay Formation in the North Sea (Donovan *et al.* 1993; Price *et al.* 1993; Clark *et al.* 1993), these mudstones are

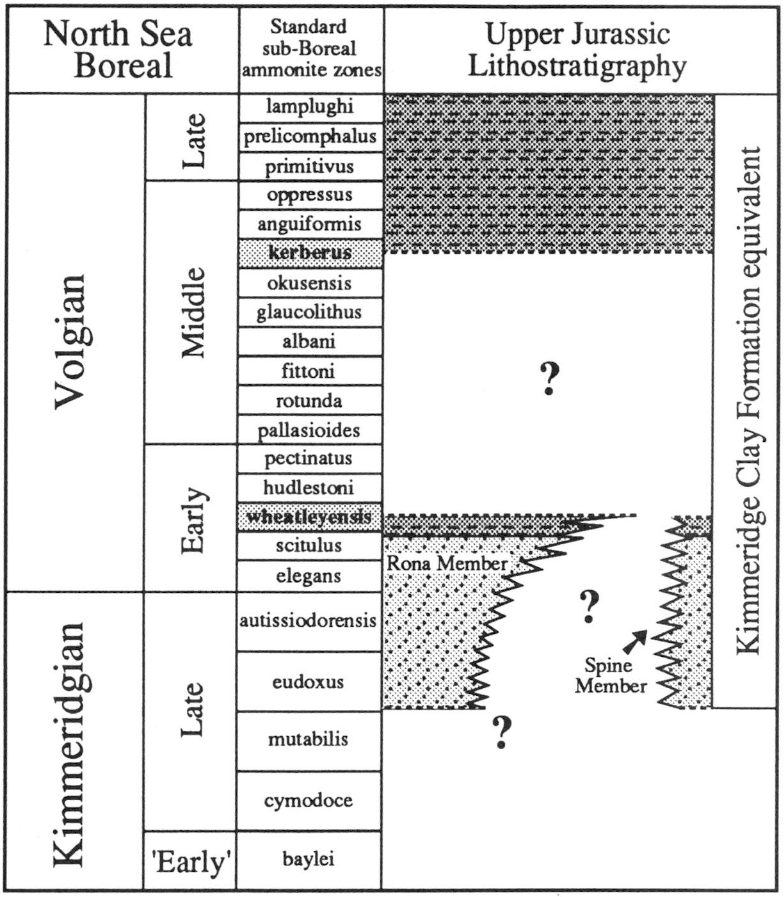

North Sea Boreal		Standard sub-Boreal ammonite zones
Volgian	Late	lamplughi
		prelicomphalus
		primitivus
	Middle	oppressus
		anguiformis
		kerberus
		okusensis
		glaucolithus
		albani
		fittoni
		rotunda
		pallasioides
	Early	pectinatus
		hudlestoni
		wheatleyensis
		scitulus
		elegans
Kimmeridgian	Late	autissiodorensis
		eudoxus
		mutabilis
		cymodoce
	'Early'	baylei

Upper Jurassic Lithostratigraphy
Kimmeridge Clay Formation equivalent
Rona Member
Spine Member
?
?
?

Fig. 4. Upper Jurassic biostratigraphic and lithostratigraphic framework for the study area. Biostratigraphically, the base and top of both the Rona and Spine Members are poorly constrained. For the same reason, the presence of a *'wheatleyensis-kerberus'* hiatus is uncertain.

typically characterized by a number of high gamma-ray peaks on the wireline logs, indicating the presence of 'hot shales'. These high gamma-ray shales are considered to represent maximum flooding surfaces and can be used for regional correlations (Price *et al.* 1993). In many wells, a distinctive high gamma-ray shale immediately overlying the Rona Member, often coincides with the presence of *wheatleyensis* ammonite zone palynomorphs. These palynomorphs may however have been reworked from unpreserved fine-grained sediments, deposited during the progressive marine transgression of the basement, as the first abundant biostratigraphic forms are from the middle Volgian ammonite zone (R. H. Poddubiuk, pers. comm.). The ambiguous nature of the biostratigraphic data, in particular whether the base of the Kimmeridge Clay Formation is as old as the *wheatleyensis* ammonite zone, is an element of uncertainty within the stratigraphic framework of the area. Consequently, it is debatable whether a small hiatus is present between the mudstones directly overlying the Rona Member and of possible *wheatleyensis* zone age, and younger mudstones of the Kimmeridge Clay Formation equivalent (Fig. 4). Stratigraphically, the Rona Member is therefore defined as the clastic sequence deposited on and adjacent to the present-day structural highs (Judd Platform, Sula-Sgeir High and Rona Ridge, see Fig. 2), on a varied pre-mid Jurassic subcrop. The Rona Member is directly overlain by mudstones in which *wheatleyensis* biozone palynomorphs have been identified.

Facies Description

The Rona Member has been studied in ten cored wells. Based on the core descriptions, a facies scheme has been developed. Figure 5

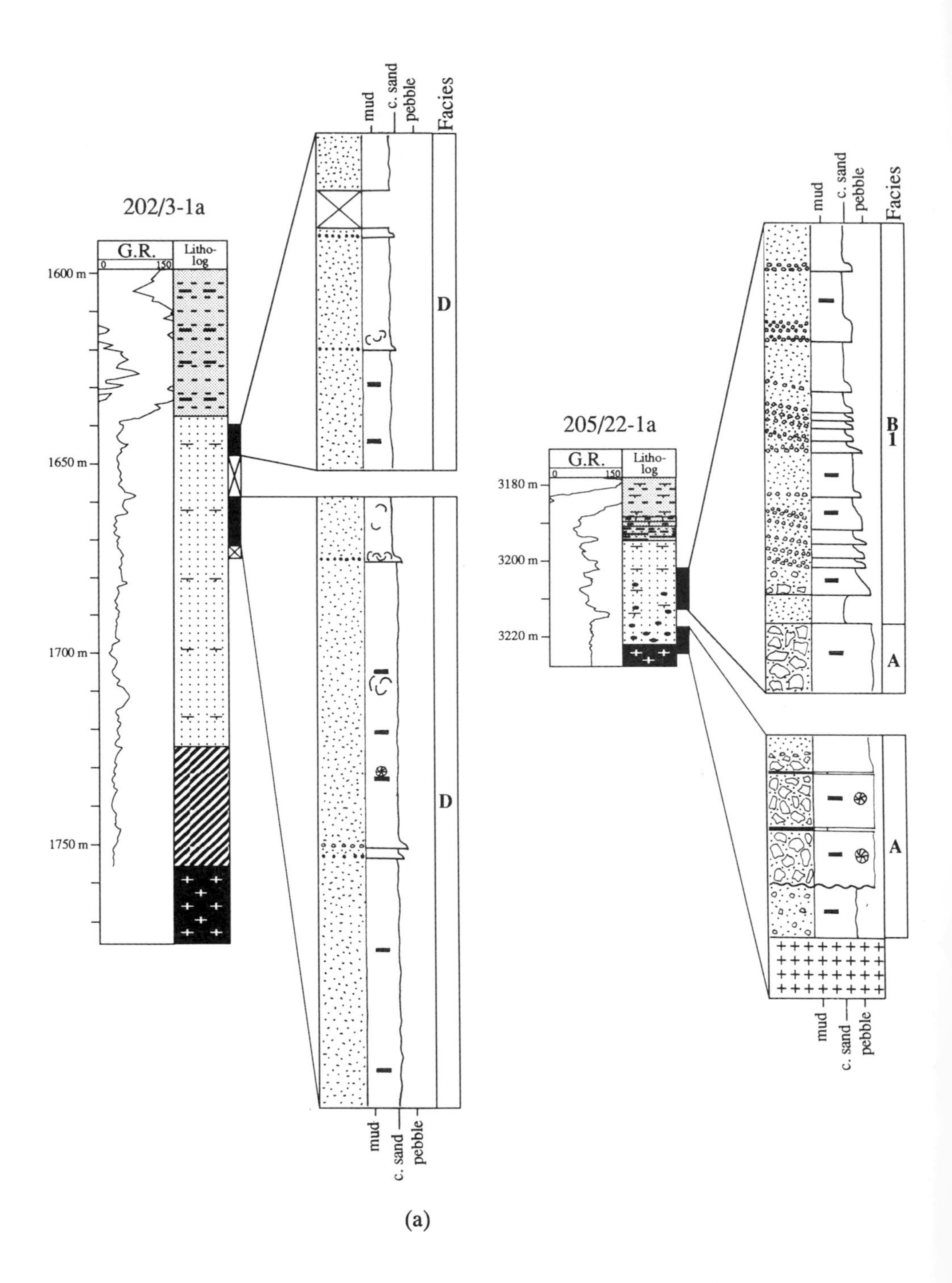

Fig. 5. Gamma-ray and litho-log of the Upper Jurassic intervals of wells **(a)** 202/3–1a and 205/22–1a, and **(b)** 204/27a-1. Descriptions of the cored sections show the lithological characteristics of each facies (A–G). The Upper Jurassic succession is underlain by Lewisianoid basement and is characterized by high gamma-ray responses of the 'hot shales' of the Kimmeridge Clay Formation in its top part.

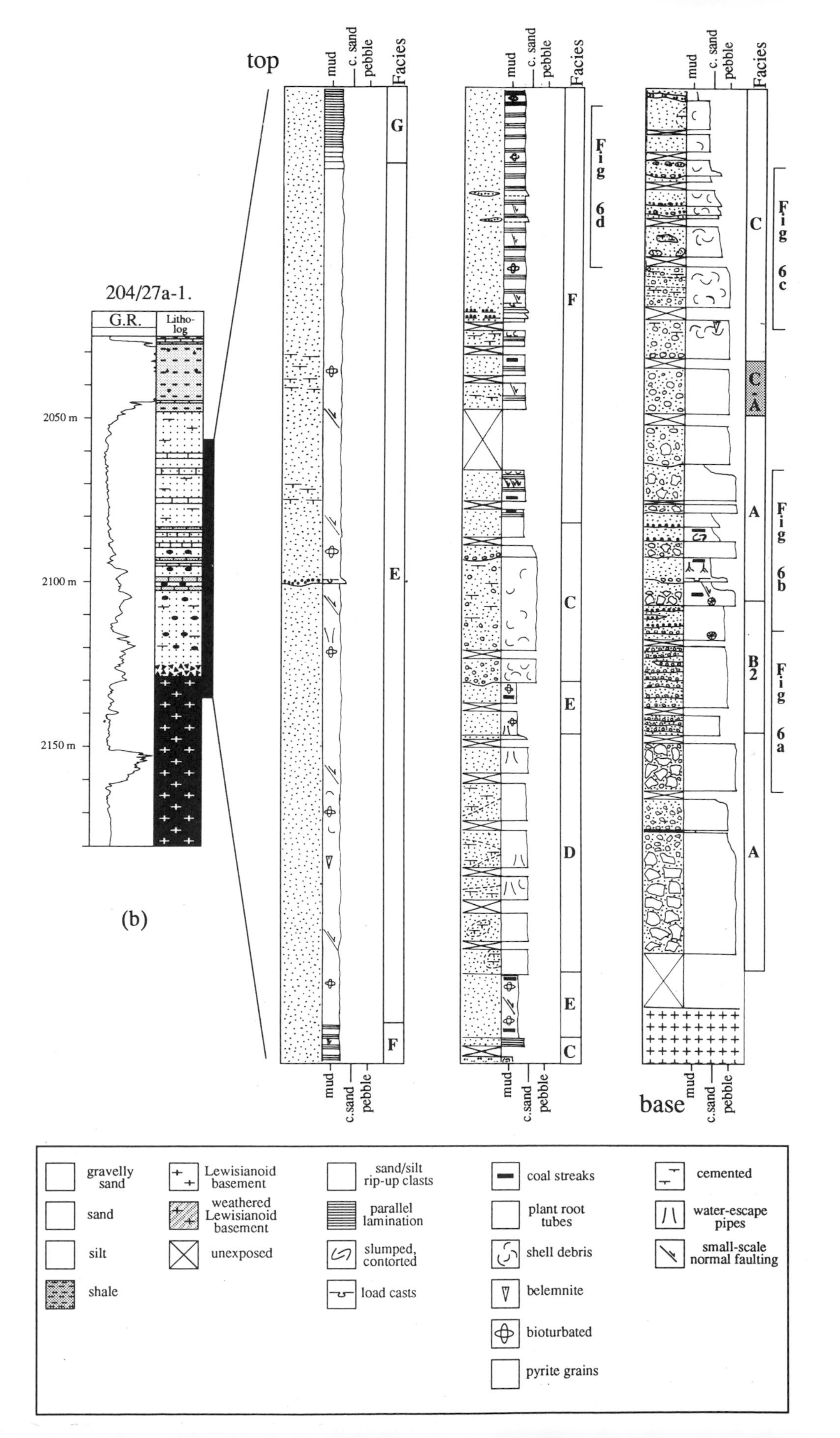

top
base
204/27a-1.
G.R.
Litho-log
2050 m
2100 m
2150 m
(b)
mud
c. sand
pebble
Facies
G
E
F
C
D
A
C-A
B2
Fig 6d
Fig 6c
Fig 6b
Fig 6a
gravelly sand
sand
silt
shale
Lewisianoid basement
weathered Lewisianoid basement
unexposed
sand/silt rip-up clasts
parallel lamination
slumped, contorted
load casts
coal streaks
plant root tubes
shell debris
belemnite
bioturbated
pyrite grains
cemented
water-escape pipes
small-scale normal faulting

contains lithological descriptions and an application of the facies scheme to the cored intervals of three wells.

Facies A. Matrix-supported conglomerates with clasts up to cobble size floating in a poorly sorted matrix (10–40%) of very fine- to very coarse-grained sandstone comprise this facies (Fig. 6a). Clasts are mainly angular to subangular but can be subrounded; they have a spherical to elongate shape and mainly comprise gneissose and granulitic material. Bed thicknesses range between 0.7 and 3 m. Commonly, the beds lack internal organization; clasts are randomly distributed and have no preferred orientation. Occasionally, bed tops are graded and fine upwards into pebbly sandstone. Conglomerate beds are separated by thin (0.02–0.1 m), graded sandstone, siltstone or mudstone layers. Coal streaks, black carbonaceous mud drapes and pyrite are common. In well 204/27a-1, a fine-grained, rooted palaeosol horizon is developed (Fig. 6b). Carbonate cementation and soft-sediment deformation are common. Slump and water-escape features characterized by flow structures and irregular contacts between fine- and coarse-grained sediment are present throughout the interval.

Facies A is interpreted as a cohesionless debris flow deposit. Both clasts and matrix are considered to have been derived from Lewisianoid basement. In well 204/27a-1 the rootlet horizon provides conclusive evidence of subaerial processes. In other wells positive distinction between subaerial and submarine environments cannot be made as the depositional processes active during transport and deposition of these flows are identical in both a subaerial and submarine environment. Absence of marine fauna within a limited dataset of coarse clastic deposits does not preclude a submarine origin, as has been proven from the Brae Field (Stow *et al.* 1982).

Facies B1. Facies B1 comprises normally graded pebbly sandstone with occasional intercalations of massive pebbly sandstone. Pebbly sandstones form fining upwards units, ranging in thickness from 0.1 to 1.3 m. They have a 50–80% clast-concentration within the basal parts (0.02–0.3 m thick) of fining upwards units, with a reduction in clast concentration to 20–30% in the top parts of fining upward units. Clasts comprise small pebbles and granules, and the matrix is composed of very coarse- to very fine-grained arkosic sandstone.

Contacts at the base of the fining upward units may be either diffuse or sharply erosive. The gravely clasts in the basal units are moderately well rounded and are generally aligned with long-axes parallel to bedding. In a number of cases the fining upward units become increasingly better stratified and also show an upward decrease in bed thickness.

Structureless units vary in thickness from 0.2 to 1.6 m and consist of small pebbles and granules (10–20%) floating in a very coarse- to very fine-grained sandstone matrix. Black, carbonaceous mud films and comminuted bivalve debris are common, and occasional shale rip-up clasts are also present. The amount of shell debris increases upwards. Calcite-cemented zones are particularly common around high concentrations of shell debris.

Facies B1 is interpreted to have been transported by high density turbidity currents, with sedimentation occuring from suspension via grain-by-grain deposition (Lowe 1982). The clast distribution together with the generally upward decrease in grain size indicate a combination of content and coarse-tail grading. Clast lithology suggests original derivation from Lewisianoid basement, whereas abundant shell debris suggests input from a shallow marine source. The composition (arkosic), sphericity and angularity of the sandstone matrix are also indicative of derivation from (weathered) Lewisianoid basement. Deposition is inferred to have taken place below storm wave base as wave reworked structures are not observed on bed tops. It should be noted, however, that bed tops may well have been eroded by subsequent flows or that sedimentation was so rapid as to preclude wave reworking.

Facies B2. Facies B2 is only developed in well 204/27a-1, where it occurs intercalated with the subaerial cohesionless debris flow deposits of facies A (Fig. 6b). It is poorly sorted and comprises angular to subrounded clasts ranging from granules to small pebbles floating in a matrix of coarse to very coarse sandstone at the base. Clasts are commonly aligned with their long axes parallel to the bedding plane. Pebbly beds (0.1–0.2 m) alternate irregularly with medium- to coarse-grained sandstone beds that display horizontal and occasional low angle cross-lamination (0.05 m). As the pebbly beds commonly comprise small scour-fills, bed thicknesses vary laterally (even at core scale). Towards the top of beds, an upward decrease in both pebble and matrix grain size (medium to very fine sandstone matrix) can be recognized. Fossil debris is notably absent.

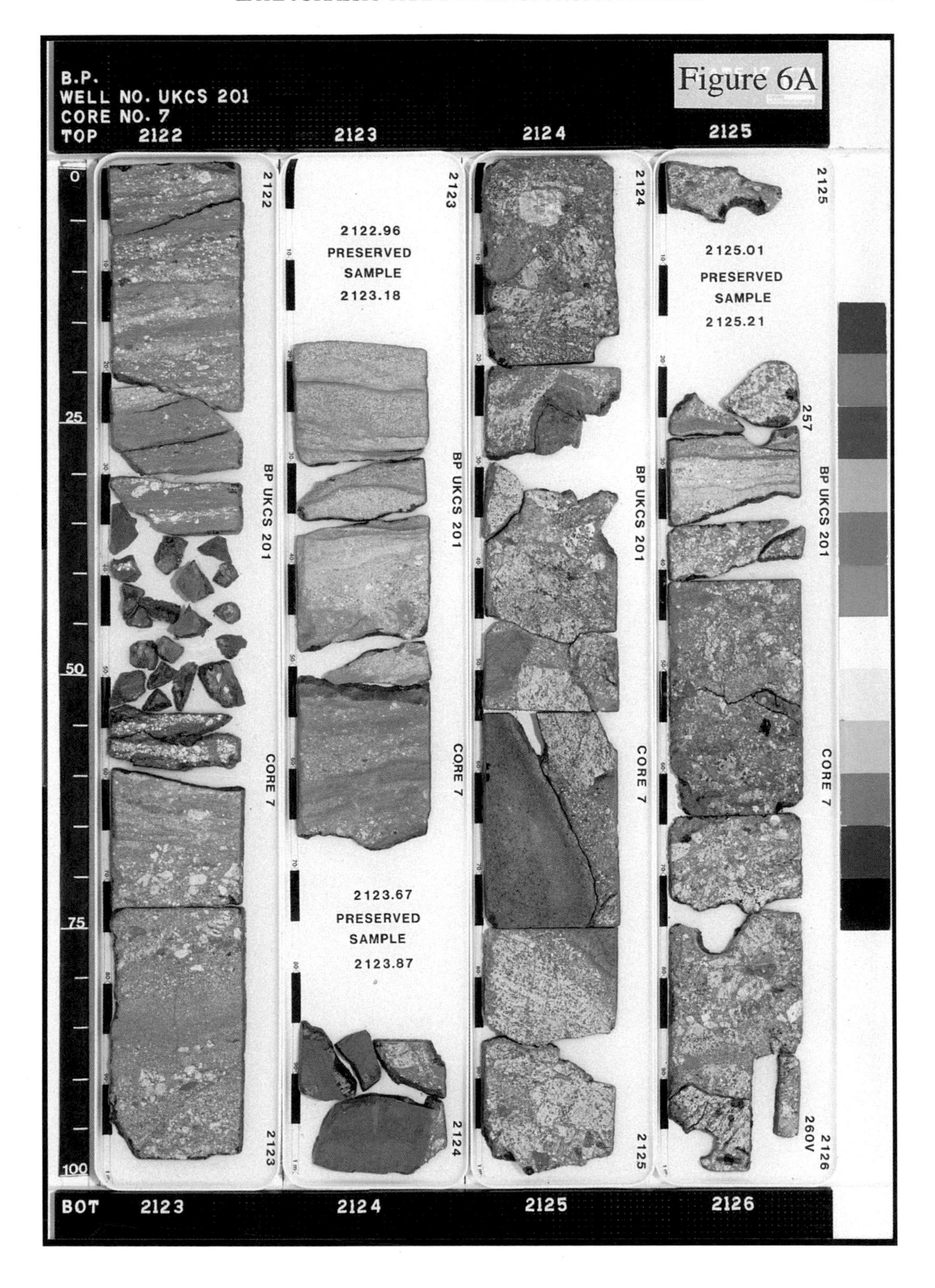
B.P.
WELL NO. UKCS 201
CORE NO. 7
TOP 2122
2123
2124
2125
Figure 6A
2122.96
PRESERVED
SAMPLE
2123.18
2123.67
PRESERVED
SAMPLE
2123.87
2125.01
PRESERVED
SAMPLE
2125.21
BP UKCS 201
CORE 7
257
260V
BOT 2123
2124
2125
2126

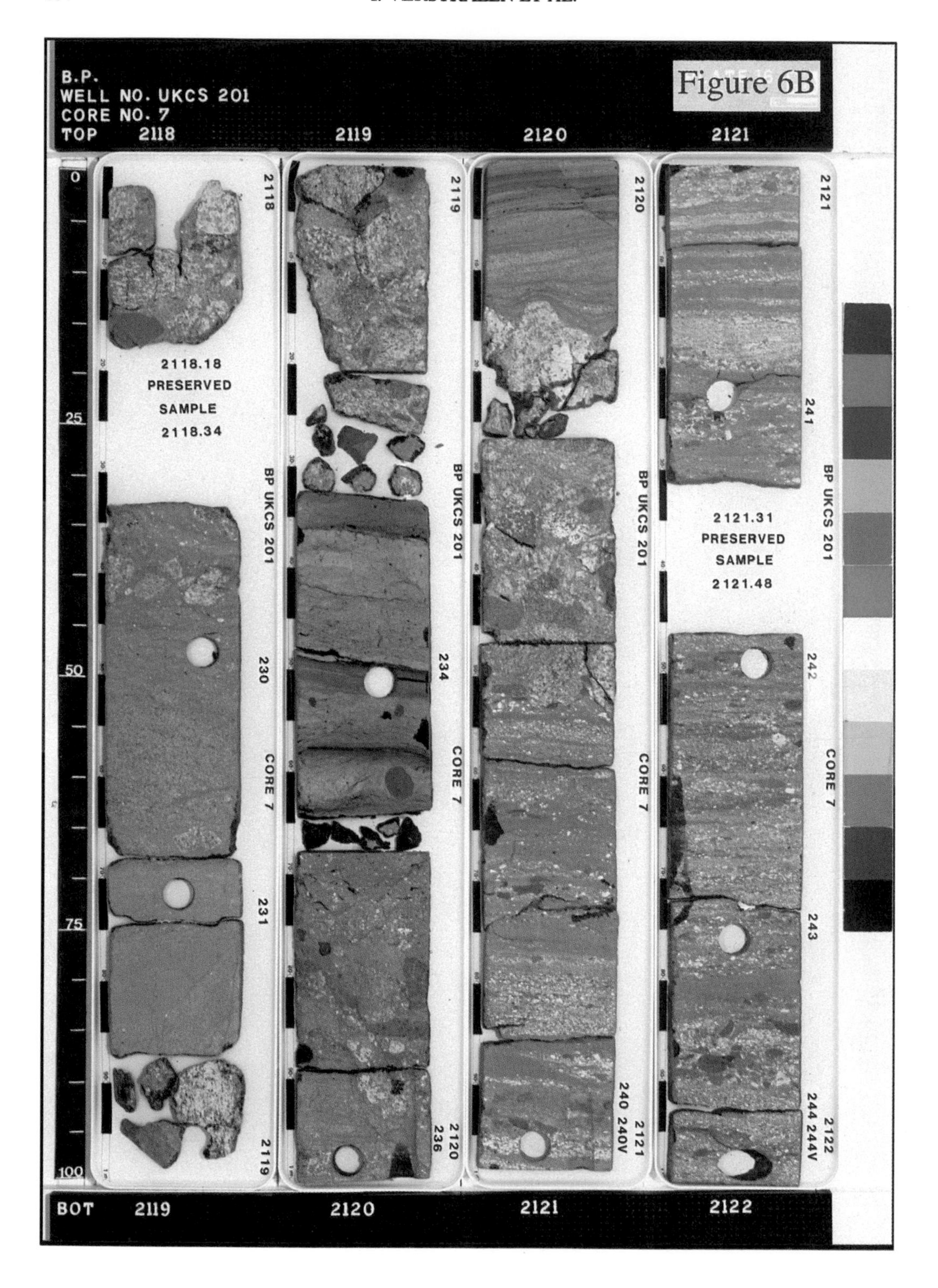
B.P.
WELL NO. UKCS 201
CORE NO. 7
TOP 2118
2119
2120
2121
Figure 6B
2118.18
PRESERVED
SAMPLE
2118.34
2121.31
PRESERVED
SAMPLE
2121.48
BP UKCS 201
CORE 7
230
231
234
236
240 240V
241
242
243
244 244V
BOT 2119
2120
2121
2122

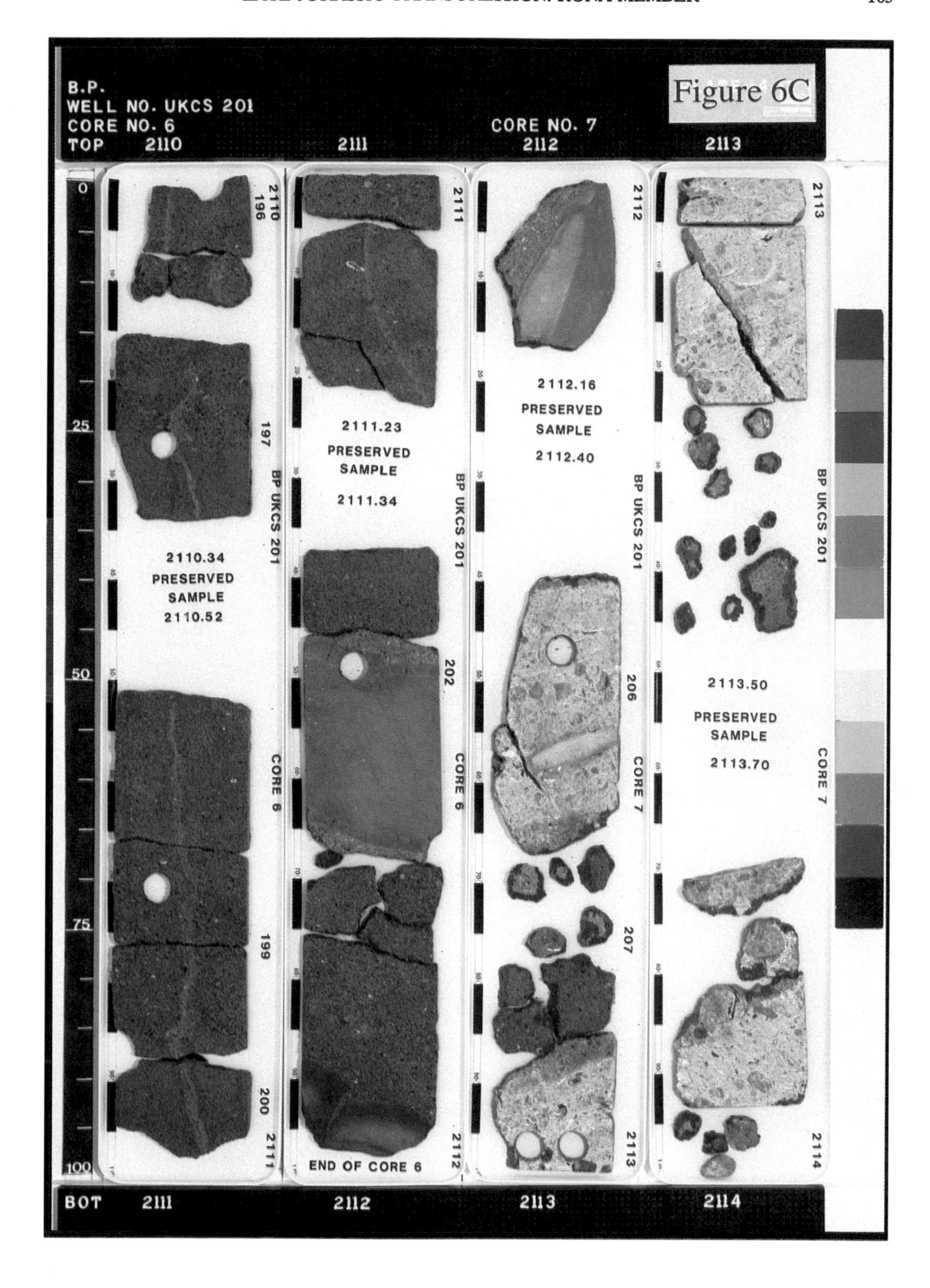
B.P.
WELL NO. UKCS 201
CORE NO. 6
TOP 2110
2111
CORE NO. 7
2112
2113
Figure 6C
2110.34
PRESERVED
SAMPLE
2110.52
2111.23
PRESERVED
SAMPLE
2111.34
2112.16
PRESERVED
SAMPLE
2112.40
2113.50
PRESERVED
SAMPLE
2113.70
END OF CORE 6
BP UKCS 201
CORE 6
CORE 7
BOT 2111
2112
2113
2114

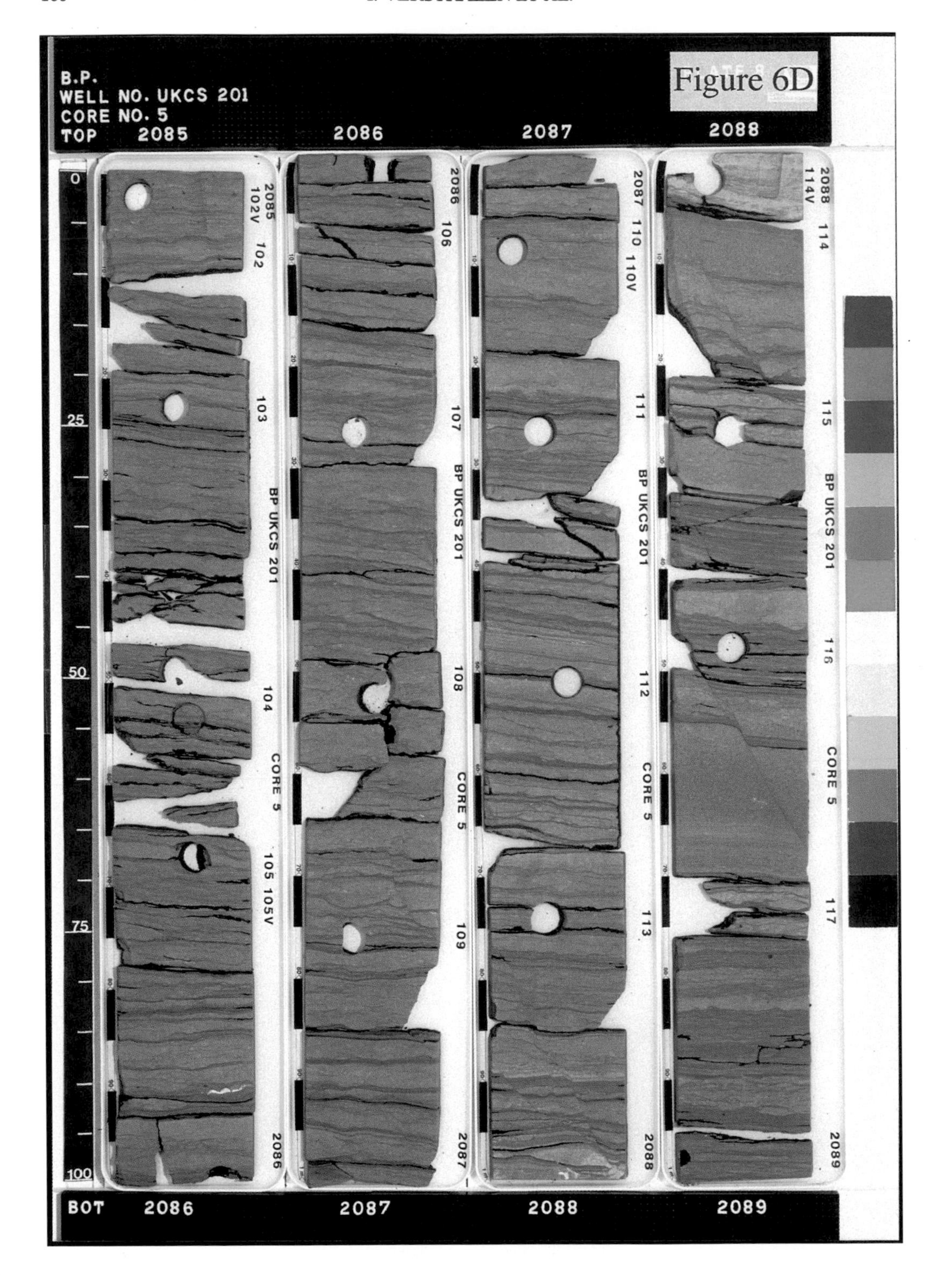
B.P.
WELL NO. UKCS 201
CORE NO. 5
TOP 2085
2086
2087
2088
Figure 6D
BP UKCS 201
CORE 5
BOT 2086
2087
2088
2089

Fig. 6. Core photographs of well 204/27a-1 (see Fig. 5 for location). (**a**) Disorganized, matrix-supported conglomerates of facies A and poorly sorted pebbly sandstones of facies B2. Clasts are composed of Lewisianoid gneiss. (**b**) Poorly sorted pebbly sandstones of facies B2 and conglomerates of facies A. A palaeosol is present between 2119.35 and 2119.65 m. (**c**) Subrounded to well-rounded, calcite-cemented, pebbly sandstones and very coarse-grained sandstones of facies C. Bivalve debris is present throughout this interval. (**d**) Laminated argillaceous sandstones alternating with clean, structureless sandstones (facies F).

The association of facies B2 with a rootleted horizon proves evidence for a subaerial depositional environment. Clast alignment together with horizontal and occasional low angle cross-lamination are generated by transport and deposition from traction currents. The upward fining nature of individual beds is indicative of waning flow. The above features together with the poor sorting, particularly in the lower parts of beds, is suggestive of rapid transport and deposition from poorly confined sheet flows.

Facies C. Facies C is known only from well 204/27a-1 where it overlies subaerial deposits of facies A (Fig. 6c). At its base, this facies comprises 0.8 m of well-rounded clasts (up to small cobble in size) floating in a very coarse sandstone matrix. Clast size decreases in the next 1.6 m to small subrounded to well-rounded pebbles and granules within a very coarse- to coarse-grained sandstone matrix which comprises 60% of the rock. The pebbly sandstones are ungraded and bedding is only occasionally observed (defined by small grain size changes). A 4.6 m thick coarse- to very coarse-grained sandstone unit, containing granules (up to 10% of the rock) overlies the pebbly sandstones. Large pebble-sized rip-up clasts of very fine- to fine-grained sandstone are present at the base of the coarse-grained sandstones unit. The sandstone is poorly sorted and structureless at its base, fines upward from very coarse- to medium-grained sandstone, and is better sorted towards its top. Bedding surfaces identified comprise one scour filled with granules, and two thin, brown silty layers with surrounding rip-up clasts. Comminuted bivalve debris is abundant above the first 0.8 m of this facies but decreases towards the top of the largely structureless sandstone unit. The sandstones contain scattered calcite-cemented veins.

The well-rounded pebbly sandstones and coarse-grained sandstones containing abundant large bivalve fragments were deposited in a high energy marine environment, possibly upper shoreface. The well-rounded, basement-derived clasts of facies C contrast markedly with the more angular gravels within underlying facies A and B. The shell material is indicative of a marine environment with the well-rounded coarse grain size being suggestive of beach deposits. Overall, facies C fines upward, which is taken to reflect an increase in water depth.

Facies D. This facies comprises disorganized, poor to moderately well-sorted, very coarse- to very fine-grained sandstone. Grains are subangular to subrounded. Bedding and grading are rare. Facies D has been cored in two wells, 202/3a-1 and 204/27a-1. In well 202/3a-1 a number of thin (0.02–0.1 m) fining-upward units occur within otherwise structureless, poorly sorted sandstone. The units have sharp bases overlain by very coarse sandstone and occasional granules. Beds of mudstone rip-up clasts are present but rare. Black carbonaceous films and comminuted bivalve debris are abundant.

In well 204/27a-1, thin fining upward units are absent, and facies D comprises fine- to coarse-grained, moderately well-sorted structureless sandstone. Occasional bivalve debris is present, but the black carbonaceous films observed within 203/3a-1 are absent. Water-escape structures, soft-sediment deformation structures and calcite-cemented zones are only present in well 204/27a-1. Observations indicate that calcite cementation is often concentrated adjacent to water-escape structures.

Facies D is interpreted as the deposits of high concentration turbidity currents. Rapid mass deposition of sand-trapped pore water and later fluidization is interpreted to have initiated secondary water-escape and soft-sediment deformation processes (Lowe & LoPiccolo 1974; Lowe 1975). Water-escape structures have largely overprinted any primary structures such as bedding, lamination and grading.

Facies E. This facies comprises very fine- to fine-grained, well-sorted sandstone. Intensive bioturbation has resulted in a mottled appearance and has overprinted primary sedimentary structures. *Planolites, Chondrites, Terebellina, Teichichnus, Helminthopsis* and *Skolithos* burrow forms have been identified. Grading is absent. In one example (well 204/27a-1) a thin, coarse-grained sandstone lens was observed. Macrofossils comprise comminuted unidentified bivalve species and an 8 cm long belemnite was observed in well 204/27a-1. Calcite-cemented zones are randomly distributed throughout the facies. Small-scale faulting is abundant and associated with soft sediment deformation. Water-escape features (pipes) are present, but rare.

The development of an extensive and diverse ichnofauna is interpreted to represent deposition below fair-weather wave base. The coarse-grained sandstone bed may represent rapid deposition from a storm-induced flow. A lower shoreface to inner-shelf environment is proposed.

Facies F. Laminated argillaceous sandstones alternating with structureless sandstone beds comprise facies F (Fig. 6d). The argillaceous sandstone beds are very fine- to fine-grained and range from 0.05 to 0.1 m in thickness. The massive, clean sandstone beds fine upwards, are fine- to medium-grained and up to 0.3 m thick. Thin (*c.* 0.01 m) very coarse-grained sandstone lenses occasionally overlie the laminated argillaceous sandstones. Overall, beds of the clean sandstone become increasingly rare upwards, through facies F with beds of the very fine-grained more argillaceous sandstone predominating. Coal streaks and comminuted shell debris are occasionally present. Burrowing is abundant in both the argillaceous and clean sandstones, with *Planolites* and *Chondrites* representing the dominant forms. Loading at the base of the structureless sandstone beds and small-scale faulting are common and occur associated with soft-sediment deformation.

The clean, massive sandstone beds are interpreted to represent transportation by episodic high concentration turbidity currents. The laminated argillaceous sandstones are considered to record deposition from dilute, low concentration turbidity currents. Petrographically, the clean and argillaceous sandstones are very similar, suggesting a common source area. The thin, discontinuous, very coarse-grained sandstone lenses may represent deposition from storm-induced currents. The overall grain size and trace fossil assemblage suggest a relatively lower energy environment than that in which facies E was deposited and a shallow marine shelf environment is envisaged. The general upward increase in low density turbidite deposits may reflect a relative increase in water depth.

Facies G. Facies G comprises black, very fine-grained argillaceous sandstones and siltstones interbedded with very thin, 1–5 mm fine-grained sandstone laminae. The fine-grained sandstone is well sorted and planar laminated. Bioturbation (burrow forms are not identifiable) is common in the lower part of this facies where it partly obliterates primary lamination. Bioturbation decreases upwards and planar lamination predominates towards the top of facies G.

The argillaceous sandstones and siltstones are interpreted to have been deposited by low concentration turbidity currents. Deposition is believed to have taken place below storm wave base in a shelf environment.

The Rona Member succession is overlain by a Kimmeridge Clay Formation equivalent which comprises black, parallel laminated, micaceous mudstones, with very thin intercalated siltstone laminae. The mudstones and siltstones were deposited from suspension and hemipelagic fall-out in an anoxic environment.

Regional facies distribution

Application of the detailed core-based facies scheme to non-cored wells is not possible due to the limited resolution of wireline log responses. However, non-cored wells can be broadly subdivided into two facies associations: (1) comprising the basal coarse-grained facies A, B1, B2, C and D (termed FA1); and (2) comprising the finer-grained facies E, F and G (termed FA2). Overall, vertical facies transitions within the Rona Member (from facies A at the base to facies G at the top) together with the transition into the overlying basinal mudstones of the Kimmeridge Clay Formation equivalent, reflect both a fining- and deepening-upward succession (see facies transitions in cored sections in Fig. 7). A complete vertical sequence in which all facies transitions can be recognized is only developed in well 204/27a-1. The transition (from subaerial conglomeratic sandstones into shoreface deposits and finally into basinal mudstones) observed in this well, plays a crucial role in development of the depositional model. In the other wells, the sequence is incomplete or the absence of cored sections prohibits further facies subdivision (Fig. 7). Although the overall succession of the Rona Member is fining upward and reflects a deepening environment, repetitions of facies E are recognized within well 204/27a-1. These repetitions are either simply related to autocyclicity, or suggest small relative sea-level fluctuations during deposition. High gamma ray peaks of 'hot shales' within the overlying Kimmeridge Clay Formation equivalent indicate the presence of high frequency sea-level changes.

Wells 205/25-1 and 205/30-1 have a different log response to those adjacent to the Rona Ridge. These wells are located adjacent to the West Shetland–Spine Fault and contain 40.8 m and 39.0 m respectively of strata which have previously been assigned to the Rona

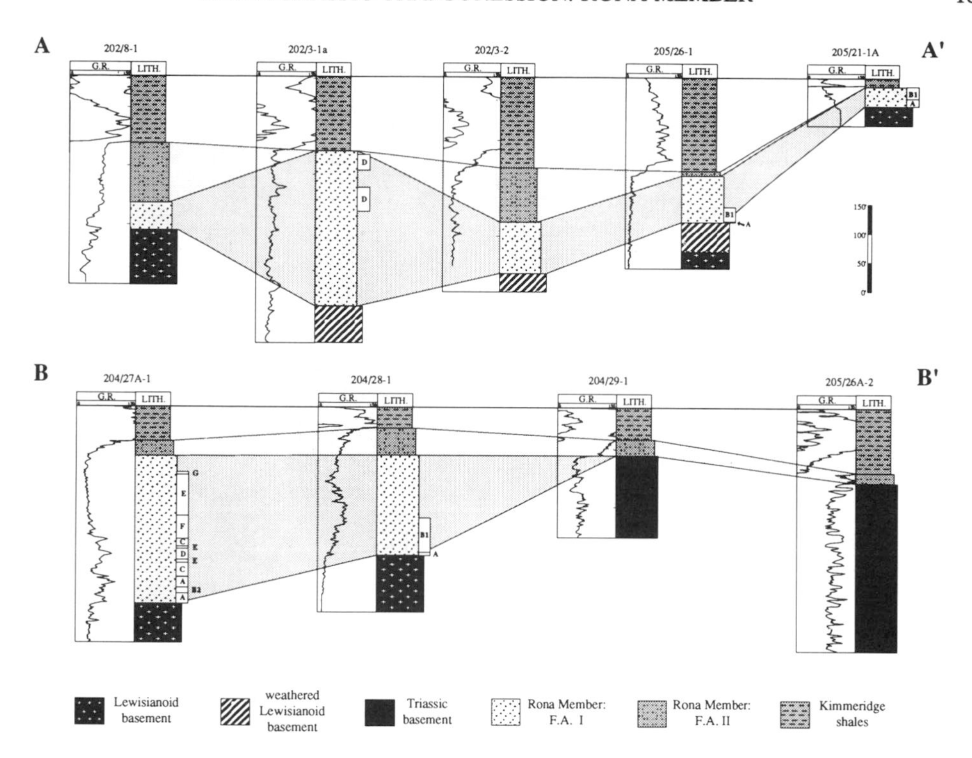

Fig. 7. Wireline correlation panels AA′ and BB′ (see Fig. 2). Panel AA′ shows the presence of both FA1 and FA2 in areas underlain by Lewisianoid basement whereas in panel BB′, FA1 pinches out away from the Rona Ridge. On the Permo-Triassic basement of the Solan Basin only FA2 is present. For wells in which the Rona Member has been cored, facies transitions are shown.

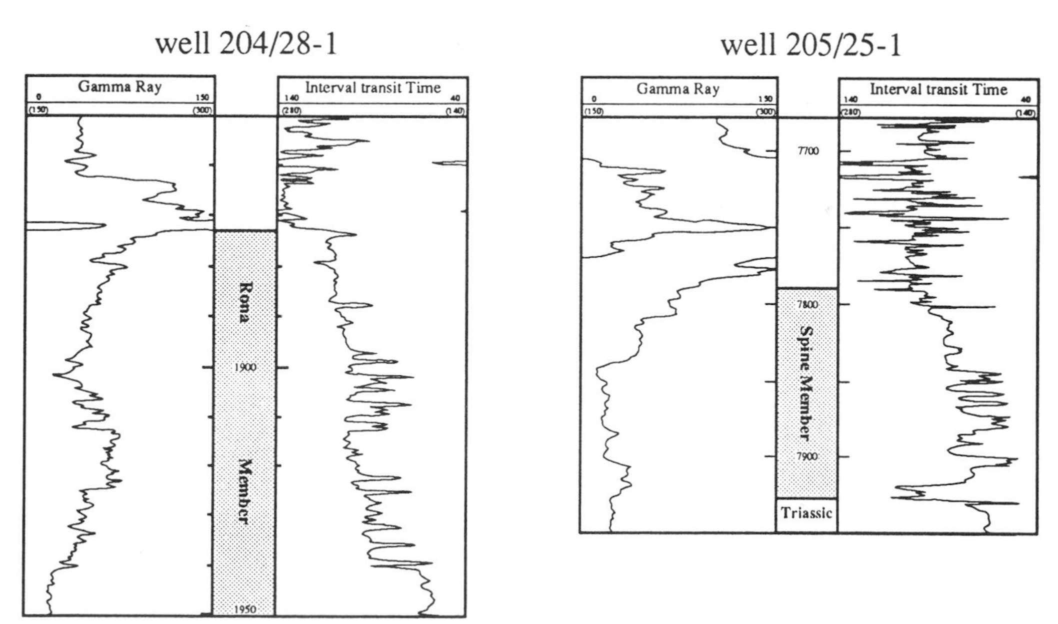

Fig. 8. Typical gamma-ray and sonic-log responses for the Rona Member (well 204/28–1) and Spine Member (well 205/25–1).

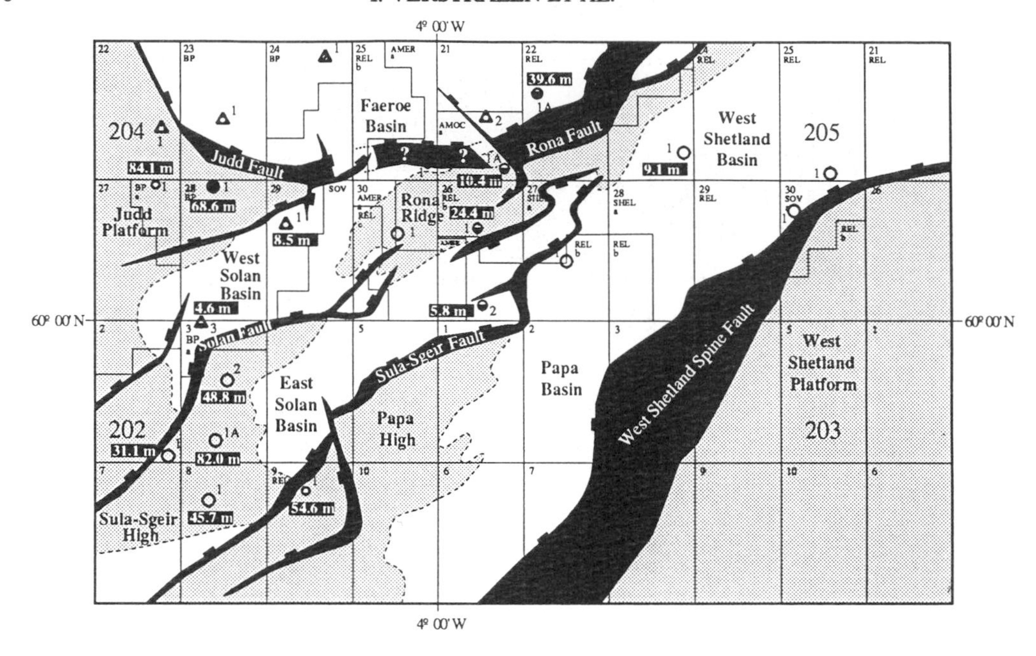

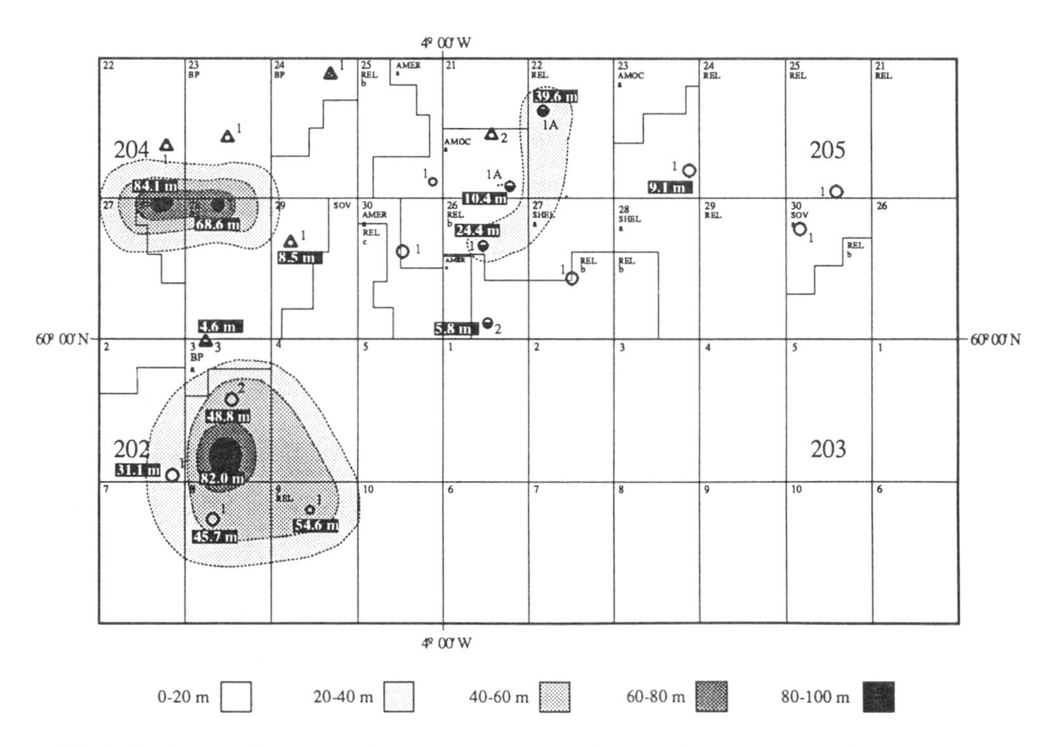

Fig. 9. (**a**) Distribution and thickness of the Rona Member based on completion log data. (**b**) Rona Member isopach map. Despite being based on limited data, the map shows the thickest successions of Rona Member being preserved on the present day highs (i.e. Sula Sgeir High, Judd Platform and Rona Ridge).

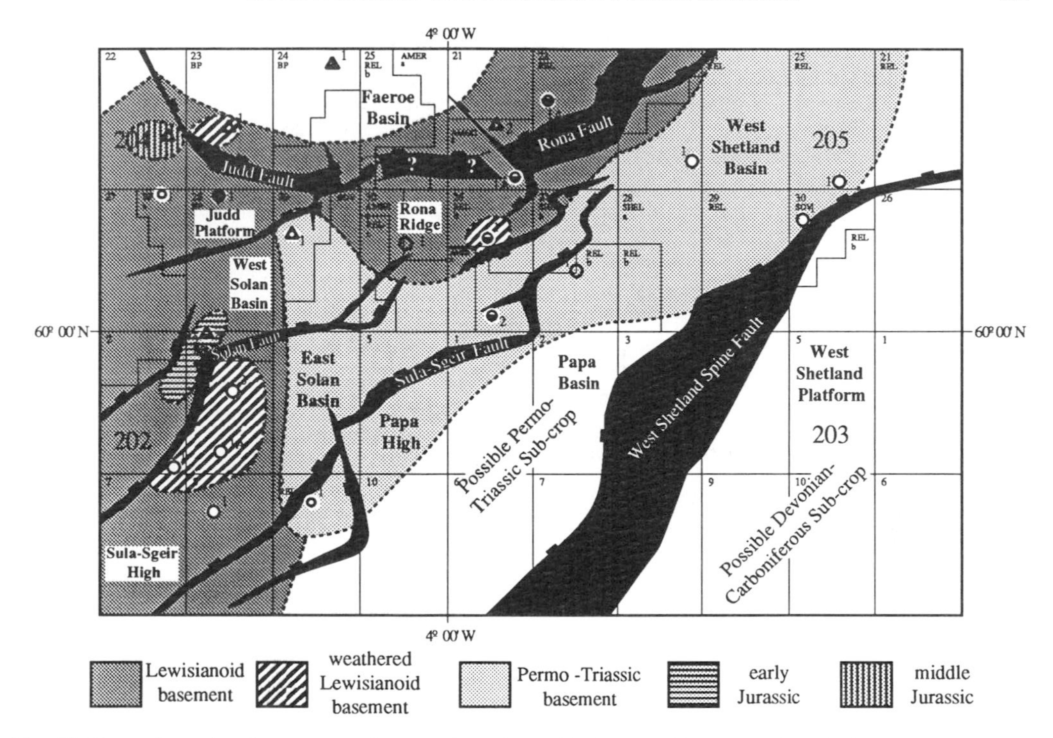

Fig. 10. Upper Jurassic subcrop map, showing the relationship between Rona Member thicknesses and the underlying subcrop. Note that subcrop (particularly the Permo-Triassic) bears little resemblance to the present-day fault pattern.

Sandstone. Wireline log responses within this interval are dissimilar to those from the Rona Member sandstones elsewhere. Although the sandstone present in these wells is considered to be time-equivalent to the Rona Member (i.e. it occurs between basement and is conformably overlain by mudstones equivalent to the Kimmeridge Clay Formation), from the wireline log responses the sandstone appears to be cleaner and much better sorted and is therefore considered unlikely to have been derived from Lewisianoid basement (Fig. 8). Henceforth, this sandstone is referred to as the Spine Member (Fig. 3). A potential source for the Spine Member sandstones are the possible Devonian-Carboniferous strata located on the West Shetland Platform (see subcrop maps from Stoker *et al.* 1993).

Two well correlation panels have been constructed (Fig. 7). Panel AA′ is oriented SW–NE and shows the Rona Member distribution on the Sula Sgeir High and Rona Ridge. Panel BB′ is oriented WNW–ESE and shows the relationship between facies associations FA1 and FA2, and the underlying subcrop (Fig. 2). In both panels, the top of the Kimmeridge Clay Formation equivalent is used as a horizontal datum. The boundary between the Kimmeridge Clay Formation equivalent and overlying Lower Cretaceous sediments is well defined by the change from high gamma-ray log responses for the 'hot shales' to low gamma-ray log responses for Lower Cretaceous mudstones and limestones. For wells which contain cored sections in the Upper Jurassic the facies interpretations are given (Fig. 7). Within all studied wells a continuous section is present from basement to the *wheatleyensis* biozone (except for well 205/21–1a where the *wheatleyensis* zone and possibly the upper, finer-grained part of the Rona Member are eroded). The sections are therefore considered not to have been reduced by erosion and to represent the depositional thicknesses of the Rona Member.

The present-day thickness of the Rona Member in the study area varies from 4.6 m to 84.1 m (Fig. 9) and thicknesses from the basement to the top of the Kimmeridge Clay Formation equivalent range from 14.3 to 121.1 m. This thickness variation within the Rona Member is taken to reflect deposition on an irregular topography. However, while the Rona Member is relatively thin, it is widespread across much of the study area (Fig. 9).

Facies-association distributions within the study area show a relationship with the Upper Jurassic subcrop (Figs. 9 & 10). FA1 is largely restricted to areas of Lewisianoid basement subcrop, whereas FA2 has a basinwide distribution overlying Lewisianoid, Permo-Triassic and early Jurassic strata. Also, the thickest sections of Rona Member contain FA1 at the base. This indicates that FA1 was deposited in topographic lows or valleys developed on a weathered Lewisianoid surface over a wide area. The relatively deeper water facies and more uniform widespread distribution of FA2 indicates that the study area was subject to a widespread transgression which culminated in a flooding surface. Consequently, any depositional model must take into account: (1) a variable thickness of FA1, developed across an irregular Lewisianoid subcrop; (2) the widespread distribution of FA2 over both the basement highs of the Rona Ridge and basinal areas; (3) the increase in water depth represented by FA2.

Depositional model

Facies A and B1 in well 204/27a-1 are interpreted to have been deposited in a subaerial part of a fan delta, probably an alluvial fan environment. From the available data the subaerial parts of fan deltas seem only to be locally developed and preserved. The overlying facies C–G were deposited in a marine environment and reflect an overall deepening from beach through shoreface to a shelf setting. A fluctuation in relative sea level is considered likely where repetition of facies E occurs (e.g. in 204/27a-1; Fig. 7). In wells 204/28–1, 205/21–1a, 205/22–1a and 205/26–1 upper shoreface sandstones and lower shoreface/inner shelf mudstones are absent. There is no evidence for subaerial deposition in these wells and as facies A is conformably overlain by turbiditic shelly pebbly sandstones of facies B1 (Fig. 7) it is believed that deposition took place in a submarine fan delta environment. Deposition of both subaerial and submarine fan delta deposits was largely confined to a subcrop of weathered and unweathered Lewisianoid basement.

The limited thicknesses of the fan delta deposits suggests that the amount of source area relief and consequently drainage basin area were restricted, and that fan delta systems were short-lived. It is envisaged that, as the subaerially exposed areas of Lewisianoid basement were progressively flooded, source areas for fan delta systems became increasingly restricted. Fan deltas were flooded and shallow marine sedimentation prevailed. Increased water depth and subsequent reworking of fan delta sediments led to initiation of turbidity currents. Initially, coarse-grained high density turbidites prevailed. With waning sediment supply the flow regime changed, and finer- grained sandstones were deposited from low density turbidity currents. During the final stage of transgression, sedimentation spread basinwide with deposition of FA2 on early Jurassic and Permo-Triassic subcrop. Finally, basinal mudstones were deposited marking the maximum extent of transgression.

Discussion

The late Jurassic palaeogeographic setting of the area west of Shetland is still poorly known. Booth *et al.* (1993) suggested that a mid–early late Jurassic (intra-Callovian) tectonic event was responsible for inversion and erosion of *c.* 1500 m of Permo-Triassic, and probably early–mid-Jurassic, strata in the West Shetland Basin and adjacent basins. Following inversion, the area was peneplaned prior to the onset of the early Volgian transgression, except for residual relief of Lewisianoid basement along the line of the present-day Rona Ridge. Localized topographic lows within the basement were infilled by subaerial and shallow marine fan delta deposits. Subsequent transgression resulted in flooding of the basin and large parts of the Rona Ridge producing isolated islands of Lewisianoid basement (Fig. 11). Well 204/30–1, located on the Rona Ridge, does not contain any pre-*kerberus* biozone sediments and Kimmeridge Clay Formation equivalent strata directly overlie basement. This could indicate that: (1) the ridge was never fully submerged during pre-*kerberus* time as the post-*kerberus* sediments that overlie the basement are basinal mudstones and conformably overlie the *kerberus* biozone; (2) Pre-*kerberus* sediments were deposited and eroded/reworked. The thin and incomplete Rona Member succession in well 205/21–1a supports the latter scenario. In both cases the Rona Ridge was passively transgressed during a period of tectonic quiescence, with local erosion and reworking of Rona Member sediments. However, a problem with this hypothesis is the necessity to generate relief on the Rona Ridge in order to supply coarse-grained sediment to deposit the Rona Member.

Previous models for deposition of the Rona Member have invoked the presence of an active fault scarp (Ridd 1981; Meadows *et al.* 1987; Haszeldine *et al.* 1987; Hitchen & Ritchie 1987). These models have been drawn from analogies taken from fault scarp-derived proximal conglomerates in the Upper Jurassic of the Viking

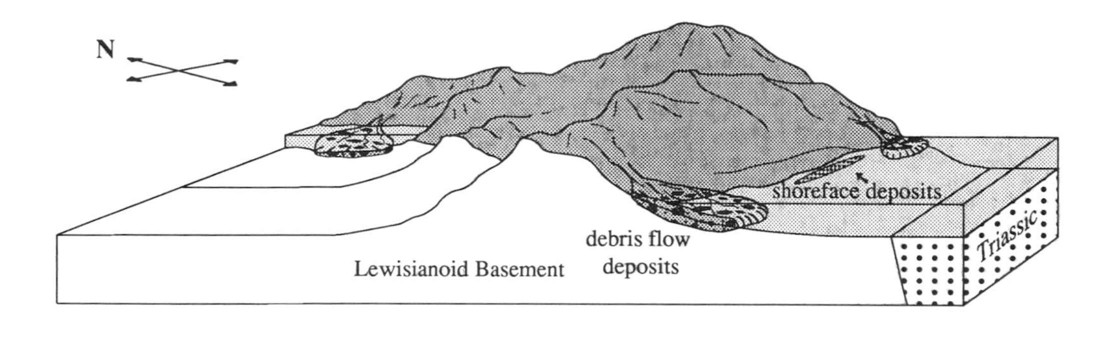

(a)

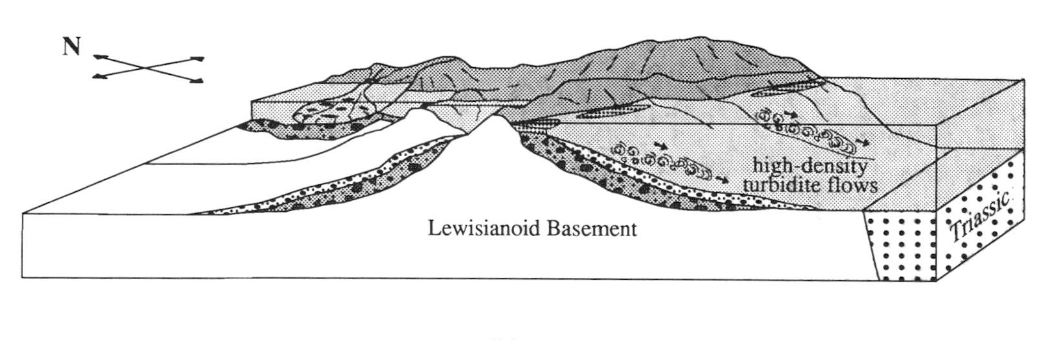

(b)

Fig. 11. The genesis of the Rona Member. (**a**) Subaerial and submarine fan delta deposition dominate. Locally, the coarse-grained deposits are reworked in a shoreface environment. (**b**) Fan-deltas are gradually flooded and their deposits reworked and subsequently resedimented by high concentration turbidity currents. (**c**) Sediment supply decreases as subaerially exposed Lewisianoid basement is flooded. Low concentration turbidite flows predominate and sedimentation is spread basinwide. (**d**) Basinwide deposition of the dark mudstones of the Kimmeridge Clay Formation equivalent.

Graben and Moray Firth (e.g. Stow *et al.* 1982; Pickering 1984) and have the advantage of generating relief for production of coarse-grained fan delta deposits via active faulting. Unfortunately, the current dataset does not support these models.

Present-day fault patterns in the study area cannot be assumed to have been active in the late Jurassic as they are largely a result of early Cretaceous extension associated with the main phase of rifting in the Faeroe Basin. Consequently, with the exception of the West Shetland–Spine Fault, the Rona Fault and the Judd Fault, the present-day fault pattern is unlikely to bear any resemblance to that of the late Jurassic. The West Shetland–Spine, Rona and Judd Faults have long-lived histories. The West Shetland–Spine Fault has been periodically active since the Permo–Triassic and the Rona and Judd Faults are likely to have been periodically active since at least the early Jurassic (if not before) when the initial opening of the Faeroe Basin took place (Haszeldine *et al.* 1987; Stoker *et al.* 1993 fig. 24). It is

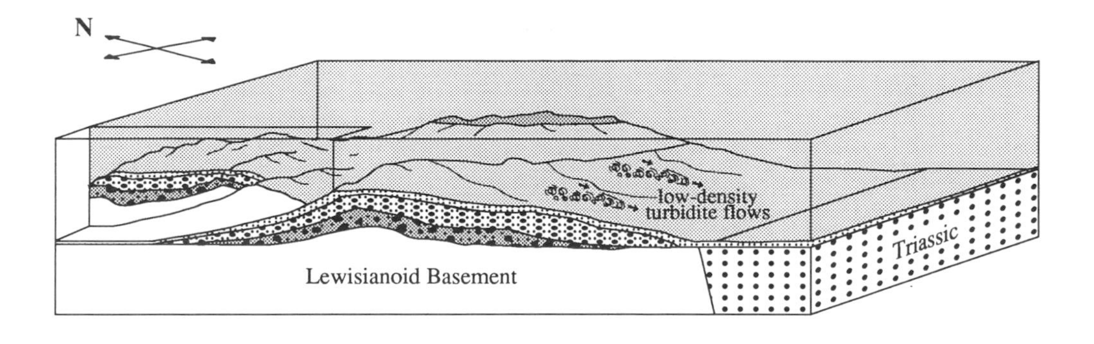

(c)

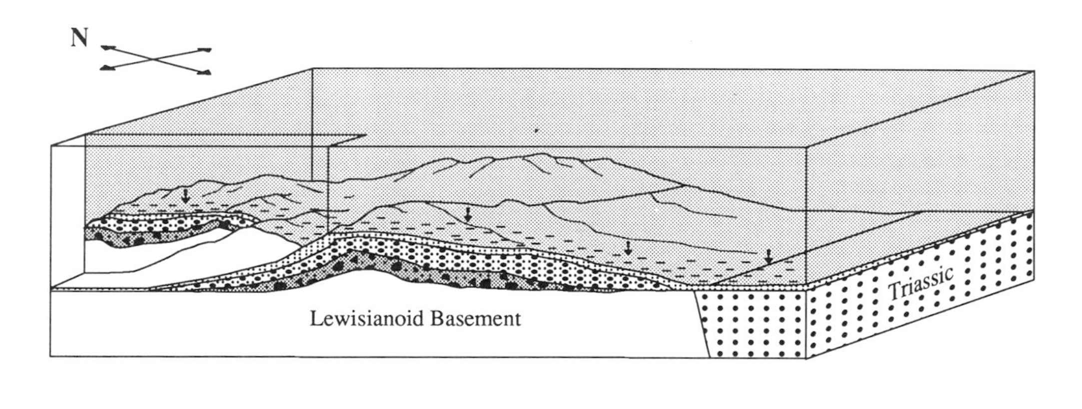

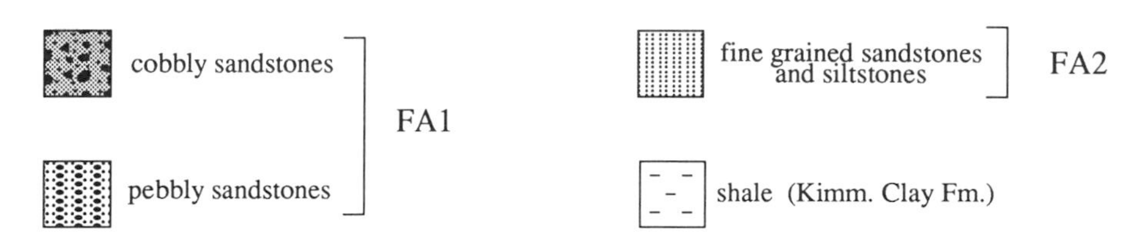

(d)

therefore postulated that the West Shetland–Spine, Rona and Judd Faults are the only likely faults, observed today, to have influenced late Jurassic sedimentation. It is, of course, possible that the postulated fault scarps from which the Rona Member was derived are beyond seismic resolution and have not been imaged. However, because of the widespread distribution of FA1 within the Rona Member, this scenario would require a number of small faults to be ideally situated close to the pockets of FA1 across the study area. In addition, facies and thickness distributions of the Rona Member do not imply deposition controlled by an active fault scarp. Time-equivalent, submarine fault-scarp successions from the North Sea are typically composed of a sequence of stacked fining-upwards units, related to renewed phases of fault activity (Turner *et al.* 1987; Pickering 1984). The Rona Member, however, comprises just one overall fining- upward sequence, with a relatively thin, partially subaerially deposited conglomeratic unit at the base.

An alternative mechanism to explain the emergence of the Rona Ridge and associated basement highs is that of footwall uplift associated with movement on the Rona and Judd Faults. Footwall uplift of weathered Lewisianoid strata would account for the generation of source area relief for deposition of FA1. In addition, the need for a number of ideally distributed small faults across the study area is not required. In this scenario transgression of the Rona and Judd fault's footwalls (or West Shetland–Spine Fault hanging-wall dip-slope) would result in trapping of FA1 and FA2 sediments on the Rona Ridge and associated basement highs. This hypothesis is supported by the observation of Haszeldine *et al.* (1987) from well 206/5–1 situated north of the study area in the Faeroe Basin just west of the Rona Fault. These authors noted the presence of conglome rates containing pebbles of gneiss and Devonian sediments within late Jurassic submarine fan deposits which are inferred to have been derived from an adjacent active fault scarp, which we suggest was the Rona Fault.

Conclusions

Recently released well data allow modifications for existing depositional models for the Rona Member to be made. Facies transitions and facies distribution indicate that Rona Member deposition was mainly controlled by marine transgression of residual islands of Lewisianoid basement probably elevated by footwall uplift along the Rona and Judd Faults. Transitions from coarse-grained subaerial fan delta deposits into shoreface/shallow marine, and finally deep marine deposits reflect an overall transgressive succession. Basal coarse-grained facies were both sourced by, and deposited on, a subcrop of Lewisianoid basement. As transgression continued sedimentation spread and finer-grained marine sediments were deposited basinwide.

The authors would like to acknowledge Amerada Hess for supporting this research. A. J. H. and A. H. acknowledge funding from Mobil North Sea Ltd and Shell Expro UK, respectively. ARCO British, BP Exploration and Production, ESSO Expro UK, and Shell Expro UK are kindly thanked for permission to examine their core data. Constructive comments from referees S. Haszeldine and S. Brown helped to clarify ideas and improve the manuscript.

References

BLACKBOURN, G. A. 1987. Sedimentary environments and stratigraphy of the Late Devonian–Early Carboniferous Clair Basin, West of Shetland. *In:* MILLER, J., ADAMS A. E. & WRIGHT V. P. (eds) *European Dinantian Environments.* Wiley, 75–91.

BOOTH, J., SWIECICKI, T. & WILCOCKSON, P. 1993. The tectono-stratigraphy of the Solan Basin, west of Shetland. *In:* PARKER, J. R. (eds) *Petroleum Geology of Northwest Europe: Proceedings of the 4th Conference.* Geological Society, London, 987–998.

BROWITT, C. W. A. 1972. Seismic refraction investigation of deep sedimentary basins in the continental shelf West of Shetland. *Nature,* **236,** 161–163.

CLARK, D. N., RILEY, L. A. & AINSWORTH, N. R. 1993. Stratigraphic, structural and depositional history of the Jurassic in the Fisher Bank Basin, UK North Sea. *In:* PARKER, J. R. (ed.) *Petroleum Geology of Northwest Europe: Proceedings of the 4th Conference.* Geological Society, London, 415–424.

DONOVAN, A. D., DJAKIC, A. W., IOANNIDES, N. S., GARFIELD, T. R. & JONES, C. R. 1993. Sequence stratigraphic control on Middle and Upper Jurassic reservoir distribution within the UK Central North Sea. *In:* PARKER, J. R. (ed.) *Petroleum Geology of Northwest Europe: Proceedings of the 4th Conference.* Geological Society, London, 251–269.

DUINDAM, P. & VAN HOORN, B. 1987. Structural evolution of the West Shetland continental margin. *In:* BROOKS J. & GLENNIE, K. (eds) *Petroleum Geology of Northwestern Europe.* Graham & Trotman, London, 765–773.

HASZELDINE, R. S., RITCHIE, J. D. & HITCHEN, K. 1987. Seismic and well evidence for the early development of the Faeroe-Shetland Basin. *Scottish Journal of Geology* **23,** 283–300.

HITCHEN, K. & RITCHIE, J. D. 1987. Geological review of the West Shetland area. *In:* BROOKS, J. & GLENNIE, K. (eds) *Petroleum Geology of Northwestern Europe.* Graham & Trotman, London, 737–749.

LOWE, D. R. 1975. Water escape structures in coarse-grained sediments. *Sedimentology* **22,** 157–204.

—— 1982. Sediment gravity flows: II. Depositional models with special reference to the deposits of high-density turbidity currents. *Journal of Sedimentary Petrology* **52,** 279–297.

—— & LOPICCOLO, R. D. 1974. The characteristics and origins of dish and pillar structures. *Journal of Sedimentary Petrology* **44,** 484–501.

MCLEAN, A. C. 1978. Evolution of fault-controlled ensialic basins in NW Britain. *In:* BOWES, D. R. & LEAKE, B. E. (eds) *Crustal evolution in NW Britain and adjacent regions.* Seel House Press, Liverpool, 325–346.

MEADOWS, N. S., MACCHI, L., CUBITT, J. M. & JOHNSON, B. 1987. Sedimentology and reservoir potential in the west of Shetland, UK, exploration area. *In:* BROOKS, J. & GLENNIE, K. (eds) *Petroleum Geology of Northwestern Europe.* Graham & Trotman, London, 723–736.

PICKERING, K. T. 1984. The Upper Jurassic 'Boulder Beds' and related deposits: a fault-controlled submarine slope, NE Scotland. *Journal of the Geological Society, London,* **141,** 357–374.

PRICE, J., DYER, R., GOODALL, I., McKIE, T., WATSON, P. & WILLIAMS, G. 1993. Effective stratigraphical subdivision of the Humber Group and the Late Jurassic evolution of the UK Central Graben. *In:* PARKER, J. R. (ed.) *Petroleum Geology of Northwest Europe: Proceedings of the 4th Conference.* Geological Society, 443–458.

RIDD, M. F. 1981. Petroleum geology West of Shetland. *In:* ILLING L. V. & HOBSON, G. D. (eds) *Petroleum Geology of the Continental Shelf of North-West Europe,* Heyden, 414–425.

RITCHIE, J. D. & DARBYSHIRE, D. P. F. 1984. Rb–Sr dates on Precambrium rocks for marine exploration wells in and around the West Shetland Basin. *Scottish Journal of Geology* **20,** 31–36.

STOKER, M. S., HITCHEN, K. & GRAHAM, C. C. 1993. *The geology of the Hebrides and West Shetland shelves, and adjacent deep-water areas.* British Geological Survey, United Kingdom Offshore Regional Report.

STOW, D. A. V., BISHOP, C. D. & MILLS, S. J. 1982. Sedimentology of the Brae oilfield, North Sea: fan models and controls. *Journal of Petroleum Geology* **5,** 129–148.

TURNER, C. C., COHEN, J. M., CONNELL, E. R. & COOPER, D. M. 1987. A depositional model for the South Brae Oilfield. *In:* BROOKS J. & GLENNIE, K. (eds) *Petroleum Geology of Northwestern Europe* Graham & Trotman, London, 853–864.

VERSTRALEN, I. R. M. J. & HURST, A. 1994. The sedimentology, reservoir characteristics and exploration potential of the Rona Sandstone (Upper Jurassic), West of Shetland, UKCS. *First Break,* **12,** 11–20.

The formation of carbonate cements in the Forth and Balmoral Fields, northern North Sea: a case for biodegradation, carbonate cementation and oil leakage during early burial

ROSELEEN S. WATSON[1], NIGEL H. TREWIN[2] & ANTHONY E. FALLICK[3]

[1] *Department of Geology and Petroleum Geology, Meston Building, Kings College, University of Aberdeen, Aberdeen AB9 2UE, UK; Present address: Nederlandse Aardolie Maatschappij B.V., Postbus 28000, 9400 HH Assen, The Netherlands*

[2] *Department of Geology and Petroleum Geology, Meston Building, Kings College, University of Aberdeen, Aberdeen AB9 2UE, UK*

[3] *Scottish Universities Research and Reactor Centre, East Kilbride, Glasgow G75 0QU, UK*

Abstract: The Tertiary reservoirs of the Forth and Balmoral Fields are sand-rich turbidites, which were deposited in a deep marine setting in the vicinity of the Fladen Ground Spur and East Shetland Platform. The Forth Field reservoir contains biodegraded oil and consists of Palaeocene and Eocene unconsolidated to semi-consolidated, well sorted, medium- to fine-grained sub-arkoses at a maximum present burial depth of 1.8 km. Pervasive ferroan and non-ferroan calcite cements dominate the Forth paragenetic sequence. Bitumen-filled inclusions within these cements indicate that oil emplacement and carbonate cementation occurred simultaneously, probably within the first 1000 m of burial. Carbonate $\delta^{18}O$ data suggest that meteoric water from the East Shetland Platform flushed the reservoir, biodegraded the migrated oil and displaced the original sea water. Biodegradation of oil took place at the palaeo-oil–water contact, producing a complex sequence of laterally extensive non-ferroan and ferroan calcite concretions. Frequent oil leakage during shallow burial may have produced a large number of concretionary cement layers. The cemented layers have the potential to act as major barriers to vertical fluid movement and to significantly reduce the net to gross ratio of the reservoir. Pore-filling ankerite post-dates non-ferroan and ferroan calcite. Ankerite precipitation marks the final stages of bacterial influence within the sediment column.

The Balmoral Field is located *c.* 100 km to the south of Forth. The reservoir consists of a series of poorly sorted subfeldspathic arenites with infrequent spherical carbonate concretions. These non-ferroan and ferroan calcitic concretions preferentially precipitated around localized accumulations of organic matter at < 500 m burial depth. Pore fluids were meteoric, and carbonate was provided by both bacterial oxidation and sulphate reduction of organic matter. Meteoric water is thought to have been derived from the East Shetland Platform to the north of Balmoral. Oil migrated into Balmoral during the Oligocene, post-dating meteoric flooding.

Laterally extensive carbonate-cemented sand horizons, formed in association with oil biodegradation, have the potential to compartmentalize a reservoir. The distribution of these cements within Tertiary deep water sandstone reservoirs adjacent to the East Shetland Platform is likely to be controlled by the relative timing of meteoric flushing and oil migration.

The Palaeocene and Eocene reservoirs of the northern North Sea have measured porosities of 25–45% and permeabilities up to 10 D, in moderately well sorted sublith/subfeldspathic arenites (Tonkin 1990; Newman *et al.* 1993). However, carbonate cemented sand horizons and concretions have been documented in a number of Tertiary Fields e.g. Balmoral, Forth, Gryphon and Frigg (Conort 1986; Hertier *et al.* 1981; Newman *et al.* 1993; Watson 1993). Carbonate cements are common within Jurassic North Sea oil reservoirs (e.g. Saigal &

From Hartley, A. J. & Prosser, D. J. (eds), 1995, *Characterization of Deep Marine Clastic Systems*, Geological Society Special Publication No. 94, pp. 177–200.

Bjørlykke 1987; Walderhaug & Bjørkum 1992; Giles *et al.* 1995). Carbonate-cemented intervals in reservoirs can form impermeable barriers to fluid flow during hydrocarbon production, thus an understanding of their formation is important. The aim of this paper is to describe the morphology, distribution and isotope geochemistry of carbonate concretions from the Forth and Balmoral Fields, UK Sector, North Sea. This study largely focuses on four Forth Field wells which cored Palaeocene and Eocene sands and these are compared with material from the Palaeocene of the Balmoral Field. Oxygen isotopic data have been used to estimate the isotopic composition of the pore fluid from which carbonates precipitated. Potential cement sources have been described with reference to carbon isotopes and Sr isotopic compositions of selected carbonate-cemented sands. The data are used to propose two separate models by which the carbonate concretions in the Forth and Balmoral Fields may have formed. These models may then be extended to predict carbonate cement formation and distribution in Tertiary reservoirs adjacent to the East Shetland Platform.

Regional setting

The Cenozoic sediments of the North Sea were deposited in a rapidly subsiding, deep water basin initiated within the post-rift or 'sag' phase of extensional basin formation. However, in the northern North Sea, a final tensional period affected the region during the Danian–Middle Palaeocene resulting in the reactivation of some of the graben margin faults. Associated with this tectonism was uplift and tilting of the Orkney–Shetland landmass and Scottish Highlands, giving rise to an easterly-directed drainage pattern, which shed large quantities of deep water

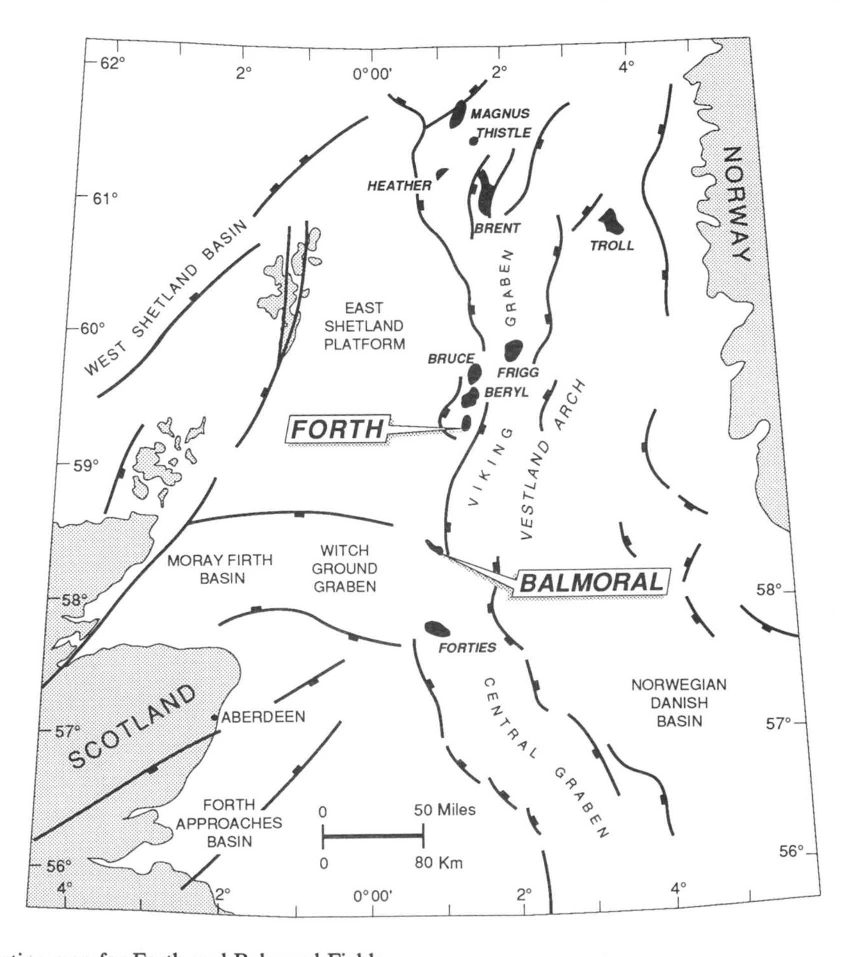

Fig. 1. Location map for Forth and Balmoral Fields.

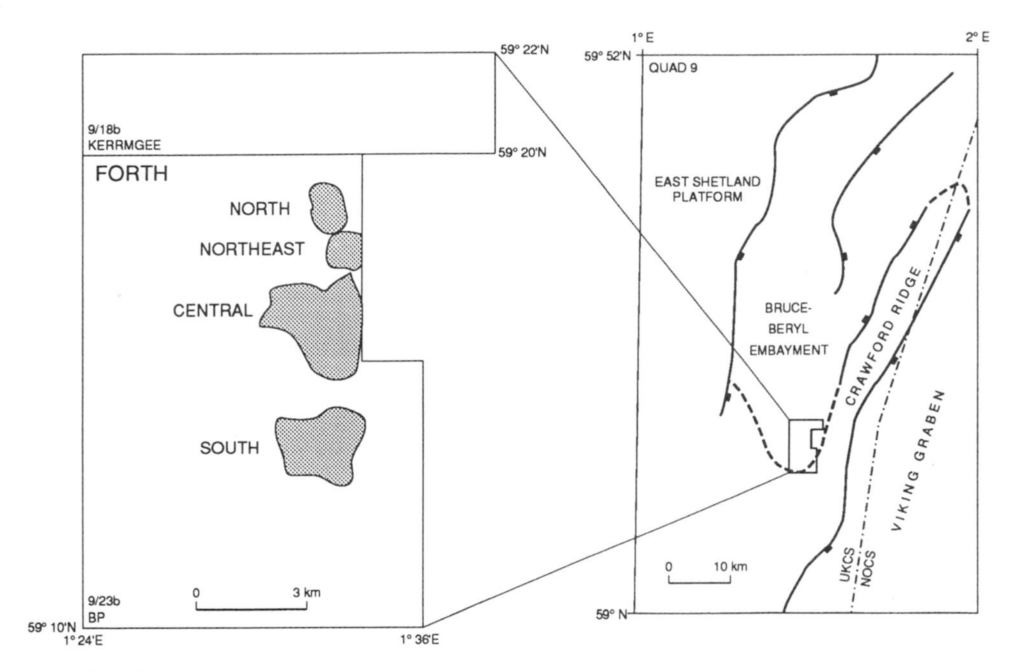

Fig. 2. Forth Field tectonic setting.

clastics into the rapidly subsiding Viking and Outer Moray Firth Basins (Ziegler 1981). These deep water clastics form the reservoirs of Forth and Balmoral Fields which are located at the western margin of the northern North Sea basin within UK continental shelf licence blocks 9/23b and 16/21a/b/c, respectively (Fig. 1).

Forth Field

The Forth Field is situated on the southeastern flank of the Bruce–Beryl Embayment, an arcuate sub-basin on the western footwall of the South Viking Graben (Fig. 1). The embayment plunges to the northeast, opening into the South Viking Graben *c.* 10 km north of the field. The southeastern margin of the embayment is defined by the Crawford Ridge, a structurally high Mesozoic basement feature lying below and immediately east of the field (Fig. 2). During the late Palaeocene and early Eocene, the Bruce–Beryl Embayment was situated in a basinal setting with water depths in the order of 400–600 m. The sands of the Forth Field are thought to have been deposited by turbidity currents, originating from a large delta complex on the East Shetland Platform, which was active during the Palaeocene and Eocene (Dixon & Pearce 1995). Post-depositional sliding initiated by tectonic activity related to growth of the Crawford anticline, is thought to have modified the original depositional geometries of these submarine fans (Alexander *et al.* 1991; Dixon *et al.* 1995). These sands are enclosed by the tuffs and shales of the Balder and Sele Formations.

Four wells were selected from the Forth Field area, in the vicinity of Forth South (Fig. 2):

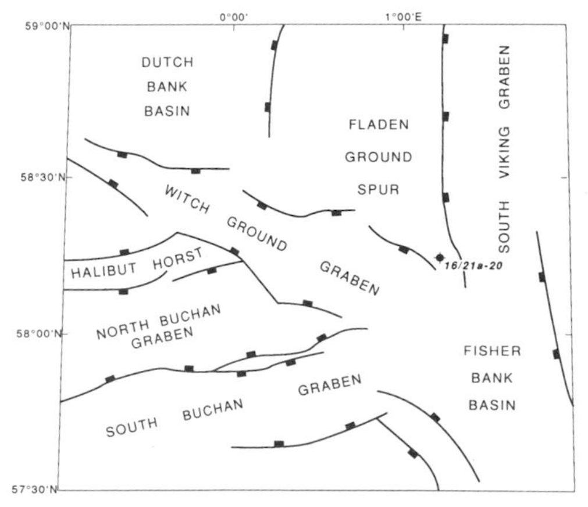

Fig. 3. Balmoral Field tectonic setting and well location.

UKCS 184, UKCS 255, UKCS 255Y, and UKCS 255Z. These wells contain a series of shales, tuffs, unconsolidated, semi-consolidated and carbonate-cemented subarkoses of Lowermost Ypresian and uppermost Thanetian age (T40–T50, Dixon & Pearce 1995) buried to a

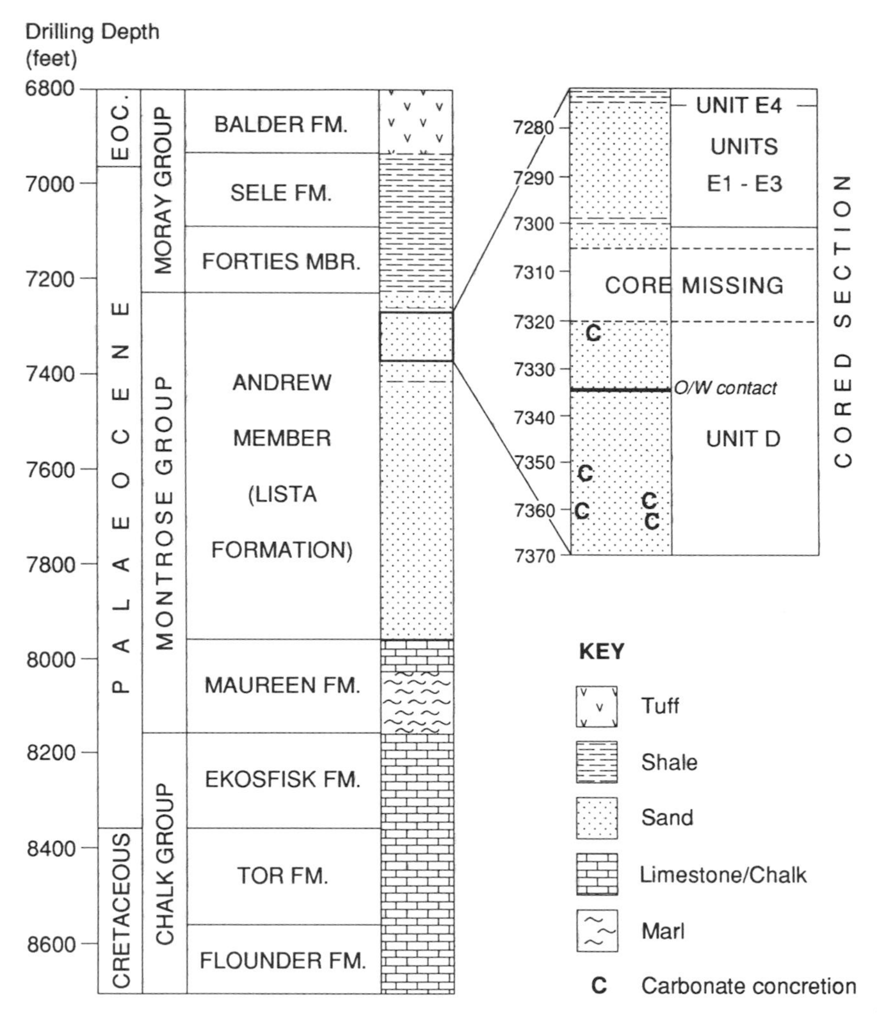

Fig. 4 . Stratigraphy and reservoir units of 16/21a-20, Balmoral Field.

maximum depth of 1.8 km. All 4 wells contain a similar Kimmeridge Clay Formation sourced, biodegraded oil (19–23° API).

Balmoral Field

Well 16/21a-20 from the Palaeocene Balmoral Field was also sampled. The well is located on the tip of the Fladen Ground Spur – a structural high on the western footwall of the most southerly part of the South Viking Graben (Fig. 3). The turbidite sands which constitute the Balmoral reservoir were sourced from areas to the north and west and locally from intra-basinal highs (Tonkin 1990). These sandstones belong to the Late Palaeocene Andrew Member of the Lista Formation and overlie the reworked chalk and limestone deposits of the Maureen Formation (Fig. 4).

Five reservoir units can be defined in the Balmoral Field based on log characteristics, degree of consolidation, depositional environment, biostratigraphy and shale content. From oldest to youngest these are Units A–E. Units A–D frequently consist of unconsolidated sands with occasional carbonate concretions (S. Wright, *pers. comm.*). Units D and E are present in 16/21a-20 (Fig. 4). These form a series of poorly-consolidated, consolidated and carbonate-cemented subfeldspathic arenites at a depth of *c.* 2200 m. The transition from Unit E into Unit D is marked by increased core friability and an increase in the abundance of lignite fragments. An oil-water contact is observed within the sand sequence. Oil is thought to have migrated into the reservoir during the Oligocene.

Samples and methodology

Carbonate-bearing sandstone samples were analysed from available Forth and Balmoral wells. Mineralogical, fabric and isotopic variation were investigated by systematic sampling throughout selected concretions. Each sample was examined using standard petrography, cathodoluminescence (CL) and carbonate staining. Selected cemented sands were etched overnight using Na-acetate buffer solution. Each sample was then examined by SEM and carbonate mineralogies were substantiated by EDS microprobe analysis. Oil was removed from all samples by soxhlet extraction using xylene and methylated spirits (Trewin 1988). Isotopic analyses used standard procedures (Hamilton *et al.* 1987) which have been documented in detail in Watson (1993). Average reproduciblity, based on pooled mean deviation, was determined as better than 0.05 $‰_{PDB}$ for $\delta^{13}C$ and 0.08 $‰_{PDB}$ for $\delta^{18}O$.

Concretion morphology and distribution

Forth Field

Carbonate concretions in the Forth Field sands can be grouped into four distinct morphologies at core-width scale: non-tapered, laterally tapered, irregular and bed cemented (Fig. 5). Non-tapered cemented layers and cemented beds may be laterally extensive but concretions with an obvious rounded, tapered outline are likely to be non-continuous and limited in lateral extent. Irregularly-shaped concretions are likely to be the result of concretion merging. Non-tapered concretions display the greatest thickness, typically ranging from 1 to 4 m. All other concretions are typically 0.5 m thick but can be up to 1 m thick. The margins of non-ferroan calcite concretions are typically sharp, but ferroan calcite concretions frequently display diffuse edges. Non-ferroan calcite concretions are more common than ferroan calcite concretions in all four wells (approximate ratio of 3:2). Non-ferroan and ferroan calcite cements have precipitated in UKCS 255, 255Y and 255Z, but only non-ferroan calcite concretions are present in UKCS 184. Ankerite does not form concretions but occurs as a late pore-filling cement within the porous sands of UKCS 255Y and 255Z. Carbonate concretions occlude *c.* 15, 8, 30 and 10 % porosity in the cored sections of UKCS 184, 255, 255Y and 255Z, respectively.

Balmoral Field

Five spherical non-ferroan and ferroan calcite concretions up to 0.6 m in thickness are present

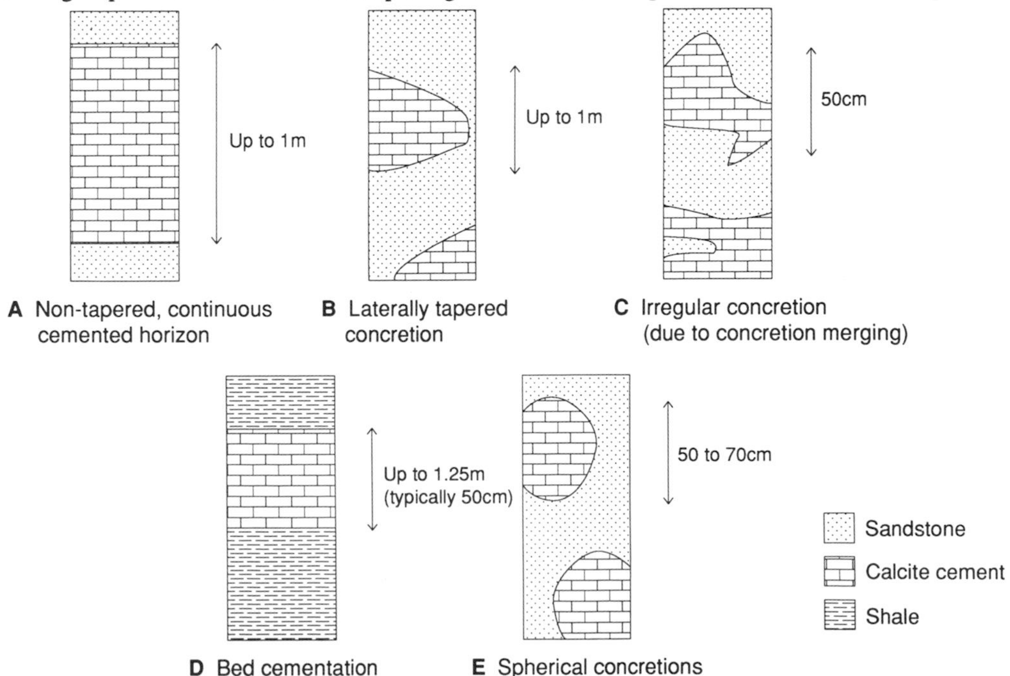

Fig. 5. Concretion morphologies observed in the Forth and Balmoral Fields.

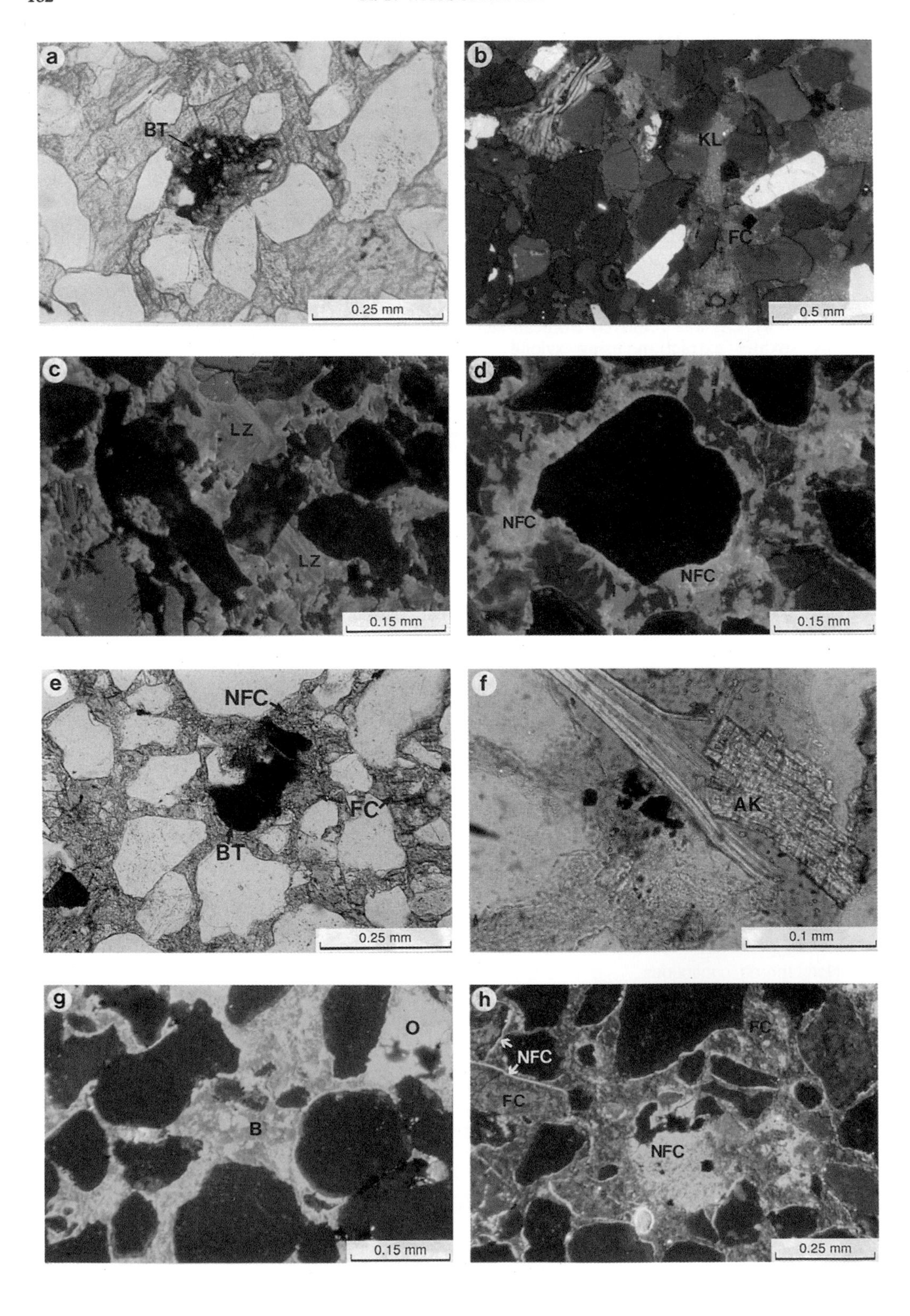
a
BT
0.25 mm
b
KL
FC
0.5 mm
c
LZ
LZ
0.15 mm
d
NFC
NFC
0.15 mm
e
NFC
FC
BT
0.25 mm
f
AK
0.1 mm
g
O
B
0.15 mm
h
FC
NFC
FC
NFC
0.25 mm

in Unit D (Fig. 4). Each concretion has a distinct rust-coloured Fe-enriched rim. These concretions are more spherical than the tapered concretions of the Forth Field (Fig. 5), and cement up to 4 % of the total sand in the well. Their spherical outlines suggest they are limited in lateral extent. There are no concretions in Unit E.

Carbonate mineralogy and petrography

Forth Field

Non-ferroan calcite, ferroan calcite and ankerite are present in the Forth Field. Mineralogical analysis and point-count data are presented in Table 1, and a representative sample of each carbonate phase is illustrated in Fig. 6. Petrographic studies illustrate that non-ferroan calcite was the earliest phase to precipitate followed by ferroan calcite and ankerite (Fig. 7).

Non-ferroan calcite occurs as large equant, poikilotopic crystals 1–2 mm in size, with straight, curved and sutured crystal boundaries. The majority of these calcite cements are homogeneous under CL, lacking any zoning or significant variation in their bright orange luminescence, either within or between cemented horizons. Grain packing within the non-ferrous calcite cement displays an open fabric, and cement crystals frequently enclose patches of bitumen (Fig. 6a).

The abundance of ferroan calcite ranges from 7 % (at diffuse concretion boundaries) up to 40 % (concretion centres) of point-count totals (Table 1). On the basis of EDS microprobe and CL analyses, three different phases of ferroan calcite have been differentiated (see below). All ferroan calcite phases contain bitumen inclusions:

(1) The majority of ferroan calcites contain > 1.5 mol% $FeCO_3$. The ferroan calcite forms 1–2 mm sized poikilotopic crystals with straight, curved or irregular crystal contacts, and patches of up to 1 mm of intergranular microcrystalline carbonate. The microcrystalline carbonate is frequently intergrown with kaolinite (Fig. 6b), and forms in optical continuity with, and appears identical in composition to the poikilotopic crystals. Under CL, both microcrystalline and poikilotopic ferroan calcite exhibit homogeneous, dull orange–brown luminescence.

(2) A smaller proportion of poikilotopic ferroan calcites contain < 1.5 mol% $FeCO_3$, displaying orange luminescence and subtle dull orange–brown lamellar zonation resembling, the sector-specific zonation described by Hendry & Marshall (1991) (Fig. 6c). Elsewhere a complex, irregular sector zonation/replacement texture is present, consisting of irregular domains of different orange and dull orange–brown CL which are in optical continuity within individual crystals. This zonation cannot be seen in stained thin sections, and compositional variations associated with it are insufficient to be characterized by EDS microprobe.

(3) Two ferroan calcite-cemented samples (Y25 and Y26) from one cemented horizon exhibit complex replacement or sector zonation textures consisting of irregular domains

Fig. 6. CL and PPL photomicrographs of Forth and Balmoral carbonates. (**a**) Bitumen (BT) enclosed by non-ferroan calcite cement. Sample F, 1727.2 m, UKCS 184, Forth Field. (**b**) Detrital quartz, K-feldspars and plagioclase feldspar are partially enclosed by dull orange–brown luminescent ferroan calcite cement (FC). Kaolinite (KL) is intergrown with, but largely post-dates the carbonate. Sample Y1, 2254.5 m, UKCS 225, Forth Field. (**c**) Quartz and K-feldspar are enclosed by orange ferroan calcite cement with dull orange–brown lamellar zonation (LZ). Sample Y22, 2281.1 m, UKCS 255, Forth Field. (**d**) The carbonate cement consists of a complex, interlocking, irregular sector zoned pattern (I) of dull orange–brown, ferroan calcite (FC) and bright orange, non-ferroan calcite cement (NFC). The brighter yellow luminescent, non-ferroan calcite coats quartz suggesting that non ferroan calcite was the earlier phase. Sample Y26, 2283.5 m, UKCS 255, Forth Field. (**e**) Bitumen (BT) enclosed by microcrystalline non-ferroan calcite (NFC) and poikilotopic ferroan calcite (FC). Both phases form in optical continuity. The non-ferroan calcite coats grains and is likely to predate ferroan calcite. Sample Y26, 2283.5 m, UKCS 255, Forth Field. (**f**) Ankerite associated with muscovite, and forcing a part muscovite cleavage flakes during crystallization. Sample Z1, 1957.5 m, UKCS 255Z, Forth Field. (**g**) Detrital quartz and minor K-feldspar enclosed by ferroan calcite cement. The cement consists of an optically continuous, complex sector zoned/replacement texture of orange (O) and dull orange–brown cement (B). Sample A6, 7352.5, 16/21a-20, Unit D, Balmoral Field. (**h**) The carbonate cement consists of two distinct mineralogies: a bright orange luminescent non-ferroan calcite (NFC) which coats grains and appears to predate the dull orange–brown ferroan calcite cement (FC). Sample A4, 7322 m, 16/21a-20, Unit D Balmoral Field.

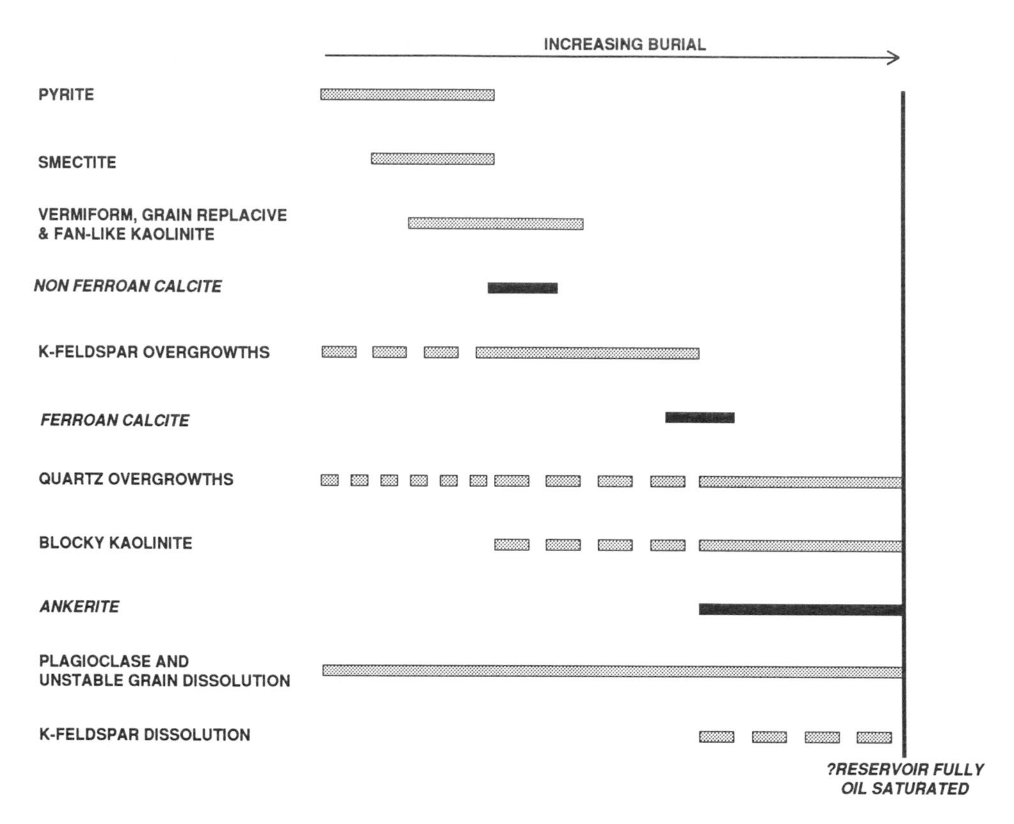

Fig. 7. Paragenetic sequence for Forth Field.

of microcrystalline non-ferroan calcite (bright orange luminescence) and coarsely crystalline, poikilotopic ferroan calcite (dull orange–brown luminescence) (Fig. 6d). Both ferroan and non-ferroan phases occur in optical continuity. The non-ferroan calcite tends to coat grains and probably pre-dates ferroan calcite (Fig. 6e). Non-ferroan calcite dominates within both samples, forming up to 75% of the total carbonate present.

Ankerite (Fig. 6f) only occurs in the sands of UKCS 255Y and 255Z. Average carbonate composition from EDS microprobe analyses are as follows (mol%): 57.8 $CaCO_3$, 28.5 $MgCO_3$, 12.6 $FeCO_3$ and 1.1 $MnCO_3$. The term ankerite is adopted here as these carbonates have Mg:Fe < 4:1 (Deer *et al.* 1966). These ankerites are non-stoichiometric, exhibiting 10 mol% Ca enrichment over ideal compositions.

Balmoral Field

Samples A4, A8, A9a and A10 were removed from the outer edges of separate concretions, whereas A6 sampled a concretion centre. The calcites sampled were non-ferroan (A9 and A10), ferroan (A6 and A8) and mixed non-ferroan and ferroan (A4) (Table 2). All the carbonates form large poikilotopic crystals 0.125–1.0 mm in size, with straight or slightly curved to jagged, highly irregular intercrystalline boundaries. Grain packing within these cements is very low, and lithoclasts observed in SEM appear uncompacted. Carbonate concretions appear to have formed early in the paragenetic sequence (Fig. 8). The non-ferroan calcites of the Balmoral Field are similar to the Forth Field non-ferroan calcites. Individual crystals of ferroan calcite consist of irregular domains of different orange and dull orange–brown CL (Fig. 6g), which are in optical continuity within individual crystals. Compositional variations between the orange and dull–brown orange luminescent phases were insufficient to be characterized by EDS microprobe. In CL, individual optically continuous crystals of mixed ferroan calcite/non-ferroan calcite consist of complex and irregular domains, within these mixed ferroan/non-ferroan calcites, the non-ferroan calcite luminesces bright orange and coats grains, probably pre-dating a

Table 1. *Point count % of carbonate, EDS microprobe and* $\delta^{13}C_{PDB}$ and $\delta^{18}O_{SMOW}$ results for UKCS 184, 255, 255Y and 255Z, Forth Field

Sample	Depth	Lithology	%Ank*	%Cal*	%Fe-Cal*	Mol% Ca	Mol% Mg	Mol% Mn	Mol% Fe	$\delta^{13}C_{PDB}$	$\delta^{18}O_{SMOW}$
UKCS 184											
A	1726.50	Carb.cem.sst.		53		97.0	2.5	0.5	0.0	−29.78	21.74
E	1727.10	Carb.cem.sst.		45		96.2	3.5	0.3	0.0	−31.01	21.72
F	1727.20	Carb.cem.sst.		49		96.7	2.8	0.4	0.0	−29.35	22.41
H	1736.00	Carb.cem.sst.		51		97.4	2.3	0.3	0.0	−26.90	21.06
I	1736.10	Carb.cem.sst.		46		97.1	2.7	0.2	0.0	−25.10	19.92
J	1748.20	Carb.cem.sst.		43		96.4	2.6	1.1	0.0	−27.48	20.56
UKCS 255											
XI	1727.68	Carb.cem.sst.		51		96.6	2.7	0.5	0.2	−26.67	19.89
X2	1729.90	Carb.cem.sst.		53		97.8	1.3	0.9	0.0	−21.76	20.06
X4	1735.20	Carb.cem.sst., oil bearing			20	91.5	3.3	2.0	3.0	3.14	21.72
X13	1749.75	Carb.cem.sst., oil bearing			28	92.1	2.7	2.2	2.8	4.42	20.40
X26	1764.50	Pyr. & carb.cem.sst.			36	94.1	1.2	2.4	2.2	−6.08	20.12
X27	1765.10	Pyr. & carb.cem.sst.			27	93.9	1.7	2.5	1.8	1.99	20.25
UKCS 255Y											
Y1	2254.50	Carb.cem.sst., oil bearing			17	93.4	3.1	0.5	3.0	11.42	20.24
Y2	2262.65	Carb.cem.sst.		40		96.1	2.7	1.2	0.0	−26.32	19.01
Y3	2273.50	Carb.cem.sst.		43		97.3	2.6	0.0	0.0	−24.82	20.37
Y4	2284.63	Carb.cem.sst.		37		96.9	2.5	0.3	0.1	−17.61	19.96
Y6	2293.50	Porous sst., water bearing	4			58.4	26.1	1.7	13.8	9.62	21.20
Y7	2251.00	Carb.cem.sst., oil bearing			34	94.1	1.9	0.4	3.3	12.37	20.05
Y8	2251.25	Carb.cem.sst., oil bearing				94.0	2.3	0.9	2.9	12.64	19.94
Y9	2252.65	Carb.cem.sst., oil bearing			28	93.7	2.5	0.3	3.4	11.51	20.12
Y10	2254.30	Carb.cem.sst., oil bearing			26	95.6	1.6	0.1	2.6	12.18	20.07
Y11	2254.75	Carb.cem.sst., oil stained			35	96.3	1.2	1.1	1.3	−4.90	18.89
Y12	2254.95	Carb.cem.sst., oil stained			32	96.1	1.4	1.2	1.1	−4.18	20.26
Y13	2255.65	Carb.cem.sst.		49		95.8	3.4	0.8	0.0	−28.05	19.80
Y14	2262.55	Carb.cem.sst.		42		96.6	2.3	1.1	0.0	−25.21	19.38
Y15	2264.50	Carb.cem.sst., oil bearing			31	96.7	0.9	0.4	1.9	12.42	20.88
Y16	2265.70	Carb.cem.sst., oil bearing			7	96.6	1.0	0.4	1.8	12.97	20.02
Y18	2269.30	Carb.cem.sst.		42		97.5	1.6	0.8	0.1	−23.49	19.71
Y19	2269.60	Porous sst., oil bearing	4			57.4	29.4	0.9	12.4	7.64	19.55
Y20	2279.70	Carb.cem.sst.		46		98.0	1.5	0.4	0.0	−26.41	19.18
Y21	2280.90	Porous sst., oil bearing	13			58.0	24.5	1.4	16.1	10.63	19.01
Y22	2281.10	Carb.cem.sst., oil stained			40	97.0	0.9	1.0	1.1	−2.83	20.29
Y23	2281.40	Porous sst., oil bearing	18			58.8	26.3	1.1	13.8	11.98	20.38
Y24	2281.85	Porous sst., oil bearing	18			58.6	28.1	1.1	12.2	12.14	20.88
Y25	2282.40	Carb.cem.sst., oil stained		26	14	97.2	0.9	0.4	1.5	−5.84	18.60
Y25	2282.40	Carb.cem.sst., oil bearing				97.3	2.1	0.5	0.0		
Y26	2283.50	Carb.cem.sst., oil stained	30	9	97.3	1.1	0.6	1.0	−10.62	20.28	
Y26	2283.50	Carb.cem.sst., oil bearing			96.6	3.1	0.3	0.0			
Y28	2270.50	Porous.sst., oil bearing	5			57.6	29.4	0.8	12.1	12.36	20.88
Y29	2282.50	Porous.sst., oil bearing	6			58.0	27.4	1.1	13.5	11.54	20.57
Y30	2282.20	Porous.sst., oil bearing	6			58.1	28.3	1.0	12.6	10.21	20.66
Y32	2287.30	Porous.sst., oil bearing	1			54.7	28.1	1.3	15.7	6.09	17.95
Y33	2283.15	Carb.cem.sst., oil stained				Fe-cal.				−2.75	20.28
Y34	2283.75	Carb.cem.sst., oil stained				Fe-cal.				−12.26	20.81
Y35	2284.00	Carb.cem.sst.				Calcite		No thin sections		−24.00	20.84
Y36	2284.15	Carb.cem.sst.				Calcite		Mineralogy deduced		−24.09	20.30
Y37	2284.25	Carb.cem.sst.				Calcite		by sample staining		−23.95	20.26
Y38	2284.50	Carb.cem.sst.				Calcite				−21.98	20.19
Y39	2282.75	Carb.cem.sst., oil stained				Fe-cal.				−14.18	19.84
UKCS 255Z											
Z1	1957.50	Porous sst., water bearing	3			57.5	30.4	1.0	11.1	1.32	20.96
Z2	1967.50	Porous sst., water bearing	1			58.2	29.1	1.0	11.7	−0.51	20.52
Z3A	1980.00	Porous sst., water bearing	1			58.4	28.1	1.5	12.0	−8.11	20.10
Z3B	1980.00	Carb.cem.sst.		41		98.5	0.7	0.7	0.0	−11.93	20.56
Z4	1956.60	Porous sst., water bearing	7			57.6	30.7	1.0	10.8	2.36	20.41
Z5	1956.75	Porous sst., water bearing	3			57.5	27.2	0.9	14.5	0.04	20.46
Z6	1956.80	Carb.cem.sst.		49		98.1	0.8	1.0	0.0	−16.20	20.57
Z7	1957.00	Carb.cem.sst.		46		98.5	1.0	0.5	0.0	−27.70	20.66
Z9	1969.25	Porous sst., water bearing	4			57.5	30.0	0.8	11.6	3.42	20.10
Z11	1972.30	Porous sst., water bearing	4			58.1	29.9	1.2	10.9	2.17	21.33
Z12	1972.45	Porous sst., water bearing	16			57.6	29.9	1.1	11.4	1.50	20.98
Z13	1973.30	Carb.cem.sst.		46		97.1	1.9	1.0	0.0	−29.48	20.52
Z14	1973.60	Porous sst., water bearing	1			58.3	29.2	1.1	11.3	−9.92	20.12
Z15	1973.85	Porous sst., water bearing	2			57.4	27.9	1.1	13.4	−0.38	18.73
Z18	1975.15	Porous sst., water bearing	1			57.8	28.4	1.1	12.6	0.77	20.22
Z19	1976.10	Porous sst., water bearing	4			58.2	28.7	1.2	11.8	1.04	20.86
Z20	1976.20	Porous sst., water bearing	5			58.2	29.3	1.1	11.3	0.49	19.82
Z21	1976.44	Porous sst., water bearing	3			58.4	28.2	1.2	12.2	1.52	20.68

*Expressed as a % of thin section point count totals where thin sections are available.

Table 2. *Point count % of carbonate, EDS microprobe and $\delta^{13}C_{PDB}$ and $\delta^{18}O_{SMOW}$ results for 16/21a-20, Balmoral*

Sample	Depth (ft)	Carbonate	Sample Position	%Carbonate*	Mol% Mg	Mol% Ca	Mol% Mn	Mol% Fe	$\delta^{13}C_{PDB}$	$\delta^{10}O_{SMOW}$
A4	7322	Non-ferroan calcite	Edge†	48	2.8	96.8	0.5	0.0	8.82	19.45
A4	7322	Ferroan calcite	Edge†		1.1	94.1	1.9	2.7		
A6	7352.5	Ferroan calcite	Centre†	47	1.6	95.3	2.1	0.9	−11.24	20.58
A8	7357	Ferroan calcite	Edge†	45	1.1	95.4	2.4	1.0	−17.32	19.51
A9a	7360.5	Non-ferroan calcite	Edge†	49	3.0	95.3	1.7	0.0	−24.90	20.99
d		Calcite	Intermediate						−15.03	19.61
e		Calcite	Intermediate						− 6.59	20.69
f		Calcite	Centre						− 7.40	19.51
g		Calcite	Centre			Mineralogy quantified by XRD –			− 6.37	19.24
h		Calcite	Intermediate			no probe results available			−12.50	19.77
i		Calcite	Intermediate						−14.92	19.80
j		Calcite	Edge						−25.15	21.69
k		Calcite	Intermediate						−16.20	19.38
l		Calcite	Edge						−23.43	21.68
A10	7362	Non-ferroan calcite	Edge†	48	2.9	95.2	1.9	0.0	−24.61	20.82

* Expressed as a % of thin section point count totals where thin sections are available.
† Denotes thin section available.

Table 3. *$\delta^{13}C_{PDB}$, $\delta^{18}O_{SMOW}$ and $^{87}Sr/^{86}Sr$ results for selected carbonates from UKCS 255Y, Forth*

Sample	Mineralogy	$\delta^{13}C_{PDB}$	$\delta^{18}O_{SMOW}$	$^{87}Sr/^{86}Sr$	$^{87}Sr/^{86}Sr(t)$
UKCS 255Y					
Y1	Fe–Cal	11.4	20.2	0.70826	0.70825
Y2	Non-Fe–Cal	−26.3	19.0	0.70838	0.70836
Y3	Non-Fe–Cal	−24.8	20.4	0.70832	0.70830
Y4	Non-Fe–Cal	−17.6	20.0	0.7.835	0.70833
Y11	Fe–Cal	− 4.9	18.9	0.70859	0.70857
Y13	Non-Fe–Cal	−28.0	19.8	0.70829	0.70827
Y15	Fe–Cal	12.4	20.9	0.70820	0.70818
Y18	Non-Fe–Cal	−23.5	19.7	0.70842	0.70840
Y22	Fe–Cal	− 2.8	20.3	0.70850	0.70849
Y24	Ank	12.2	20.9	0.70835	0.70832
Y28	Ank	12.4	20.9	0.70934	0.70930
Y32	Ank	6.1	17.9	0.70910	0.70895

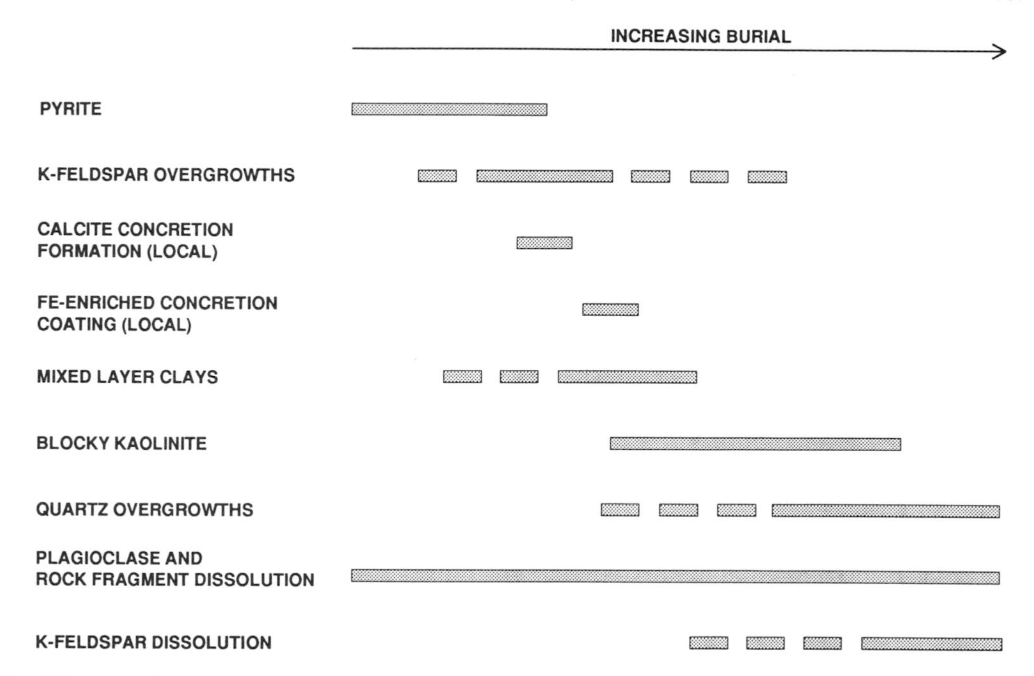

Fig. 8. Paragenetic sequence for Balmoral Field.

dull orange–brown luminescent, Fe-enriched replacive calcite phase (Fig. 6h).

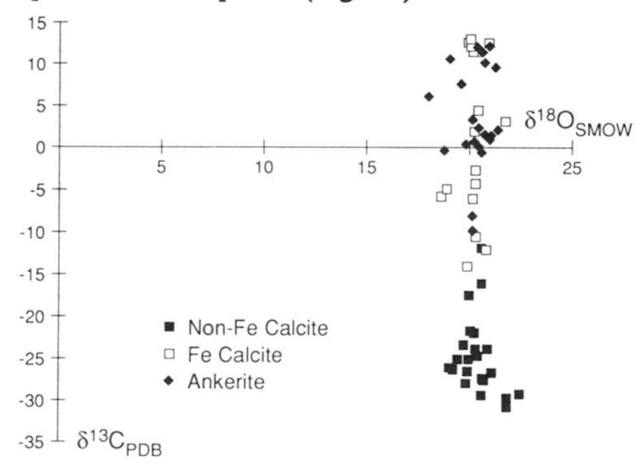

Fig. 9. $\delta^{13}C_{PDB}$ versus $\delta^{18}O_{SMOW}$ for Forth carbonates.

Isotope results

Forth Field carbon and oxygen stable isotopes

All carbon and oxygen isotopic results are reported in Table 1 and plotted in Fig. 9. $\delta^{18}O$ values for non-ferroan calcites, ferroan calcites and ankerites are similar, ranging from 17.9 to 22.4‰$_{SMOW}$ with an average of 20.3‰$_{SMOW}$. The $\delta^{13}C_{PDB}$ values for non-ferroan calcites range from −11.9 to −31.0‰, and all UKCS 184 carbonates gave consistently low $\delta^{13}C$ values, < −25‰$_{PDB}$. $\delta^{13}C_{PDB}$ values of the ferroan calcite cements are between −14.2 and +13 ‰ with 75% of these analyses representing compositions heavier than +2 ‰. Ankerite $\delta^{13}C_{PDB}$ values range from −9.9 to +12.4‰, but 75% of these analyses are >0‰$_{PDB}$.

Samples were taken at intervals of up to 0.5 m vertically throughout one cemented horizon to assess possible spatial or temporal variation in bicarbonate supply, precipitation temperature or pore fluid composition/origin (Fig. 10). Only minor variation in $\delta^{18}O$ is present, but there is a significant centre to edge variation in $\delta^{13}C$ (−2.8 to −24.1‰$_{PDB}$).

Forth Field Rb–Sr isotope data

Both non-ferroan and ferroan calcites have similar $^{87}Sr/^{86}Sr$, ranging from 0.70825 to 0.70857; however, $^{87}Sr/^{86}Sr$ for the ankerites is more varied and in general higher, ranging from 0.70832 to 0.7093. Figure 11 shows the relationship between $\delta^{13}C$ and $^{87}Sr/^{86}Sr$. Despite the wide range of non-ferroan and ferroan calcite $\delta^{13}C$ (−28 to +12‰$_{PDB}$) the $^{87}Sr/^{86}Sr$ ratios are tightly grouped (0.70820–0.70859). In contrast, the three ankerites analysed show a wide variation in

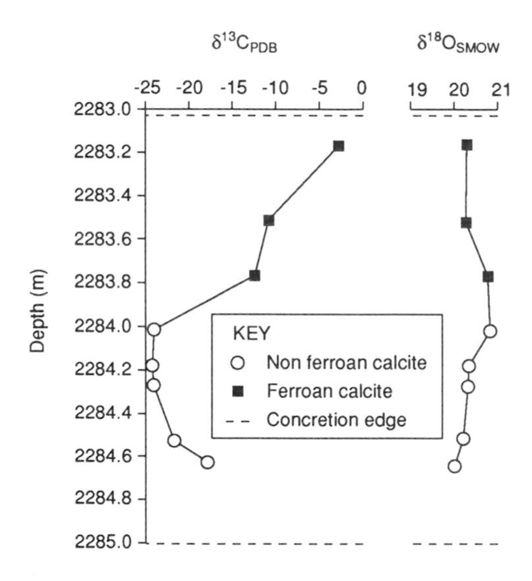

Fig. 10. $\delta^{13}C_{PDB}$ and $\delta^{18}O_{SMOW}$ trends across a concretion from UCKS 255Y, Forth Field.

$^{87}Sr/^{86}Sr$ (0.70832–0.70930) with relatively small differences in $\delta^{13}C$ (+6.1 to +12.4‰$_{PDB}$).

Balmoral Field carbon and oxygen stable isotopes

One concretion was sampled in detail for isotope analysis at regular intervals through its thickness seen in core (samples A9a to l), and other concretions were sampled at their centre or edges.

Calcite $\delta^{18}O$ values recorded for different concretions range from 19.2 to 21.7‰$_{SMOW}$ (Fig. 12),

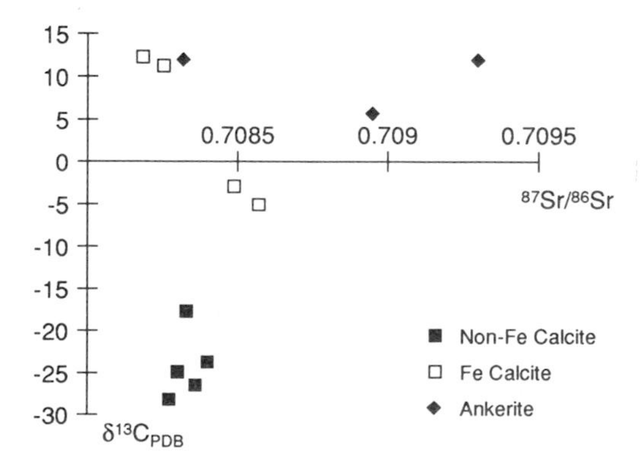

Fig. 11. $\delta^{13}C_{PDB}$ versus $^{87}Sr/^{86}Sr$ for selected Forth carbonates from UKCS 255Y, Forth Field.

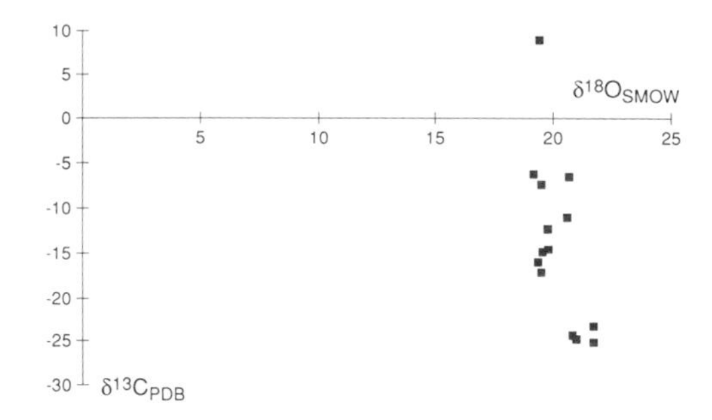

Fig. 12. $\delta^{13}C_{PDB}$ versus $\delta^{18}O_{SMOW}$ for Balmoral carbonates.

and $\delta^{18}O$ values vary by up to 2.5‰$_{SMOW}$ within a single concretion (Table 2). There is a radial centre to edge increase in $\delta^{18}O$ within concretion A9 (Fig. 13).

Sample A4 (which comprised mixed ferroan and non-ferroan calcite) has the highest recorded calcite $\delta^{13}C$ (8.8‰$_{PDB}$), whereas non-ferroan calcites (samples A9a and A10) have

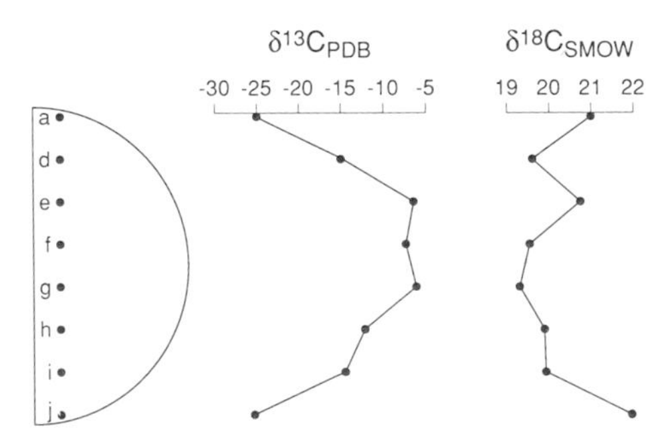

Fig. 13. $\delta^{13}C_{PDB}$ and $\delta^{18}O_{SMOW}$ trends across concretion A9, Balmoral Field.

the lowest recorded $\delta^{13}C$ (–24.9 and –25.6‰$_{PDB}$ respectively). Ferroan calcites (samples A6 and A8) display intermediate $\delta^{13}C$ values of –11.2 and –17.2‰$_{PDB}$ (Fig. 12). Within an individual concretion, the $\delta^{13}C$ of calcite may vary widely from –24.9 to –6.4 ‰$_{PDB}$ (Fig. 13).

Forth Field data interpretation

Timing of concretion growth

Grain packing is loose within all the carbonate-cemented sands, suggesting cementation during shallow burial. Measurement of the total carbonate present with concretions of cemented layers may be used to obtain an estimation of their 'minus cement porosity', and hence an estimate

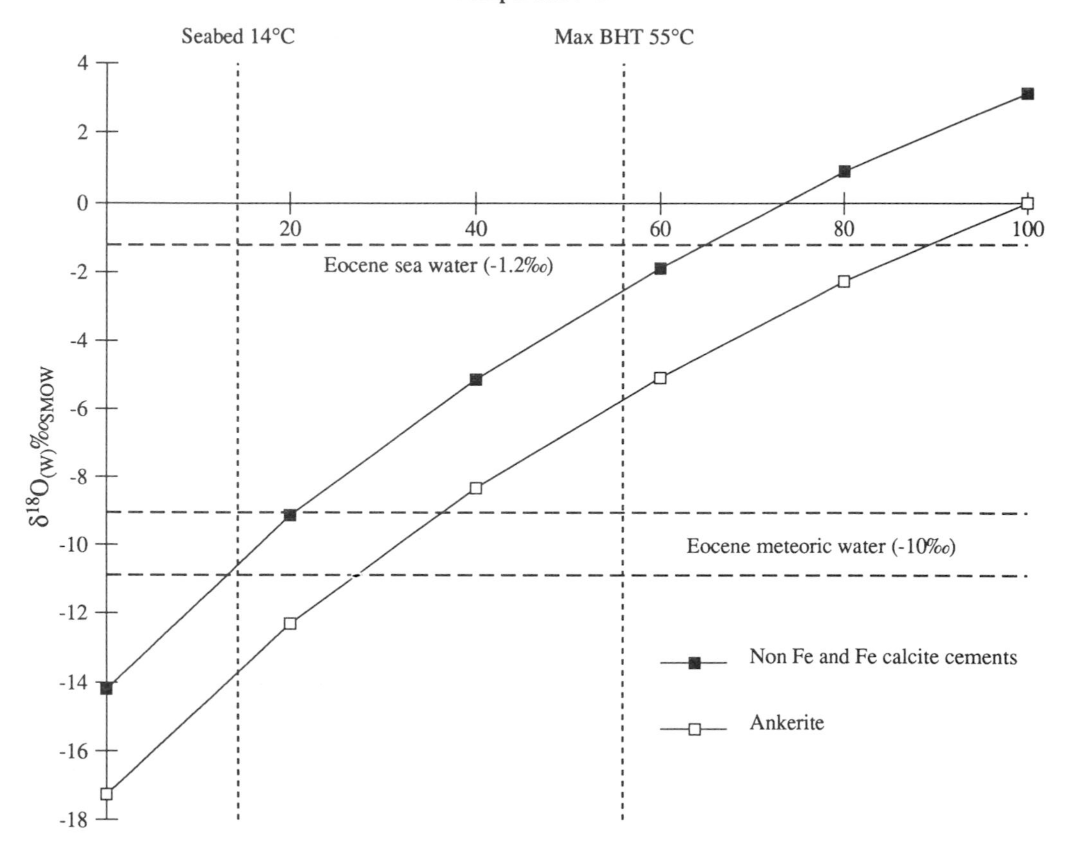

Fig. 14. Temperature v. precipitating water $\delta^{18}O$ for calcites and ankerite (calcites $\delta^{18}O$ = 20.4‰$_{SMOW}$; ankerite $\delta^{18}O$ = 20.2 ‰$_{SMOW}$) in Forth Field.

of their relative depth of cementation, providing there has been no significant replacement of detrital constituents by calcite. Within the studied samples, there has been minor replacement of plagioclase by calcite (1–2%), and no detrital carbonate clasts have been identified. The measured 'carbonate totals' are thus likely to be closely indicative of minus-cement porosities (Watson 1993). Petrographic evidence also suggests that ferroan calcite precipitated at a greater burial depth than the non-ferroan cements. This evidence includes:

(1) Minus-cement porosities within ferroan calcites (37–42%), which are lower than those estimated for non-ferroan calcite cements (44–55%, Table 1). The extremely high non-ferroan calcite minus-cement porosities may have been the result of overpressure, which is likely to have been released within the first 500 m of burial, creating the sandstone dykes characteristic of this Tertiary sequence (Dixon *et al.* 1995). The lower minus-cement porosities for ferroan calcite represent later cementation, probably not exceeding 1000 m burial.

(2) The dominant ferroan calcite phase containing 1.5% $FeCO_3$ is intergrown with up to 4% blocky kaolinite. Kaolinite largely post-dates both ferroan and non-ferroan calcite, but its occurrence as inclusions only within ferroan calcite cements confirms them as a later phase than non-ferroan calcite.

Ankerite post-dates both ferroan and non-ferroan calcite in the diagenetic sequence and is likely to have precipitated at burial depths in excess of 1 km.

Oxygen isotopes

$\delta^{18}O$ isotopic compositions of non-ferroan and ferroan calcites are very similar and the calculated averages for both phases only differ by 0.1‰. This similarity suggests these carbonates precipitated from a common pore fluid and/or at approximately the same temperature.

Fluid inclusion temperatures are not available, therefore other parameters need to be considered in order to estimate carbonate precipitation temperatures and/or possible pore fluid compositions. As these are deep sea sediments, it could be assumed that the original pore fluid was marine. Assuming an initial marine pore fluid with $\delta^{18}O$ of $-1.2‰_{SMOW}$ (Shackleton & Kennett, 1975) precipitation temperatures of *c.* 65°C are calculated for these calcites: using the fractionation 1000 ln$\alpha_{(calcite-water)}$ = 2.78 × 10^6T^2– 2.89; (Friedman & O'Neil 1977). The carbonate $\delta^{18}O$ values in this study are therefore anomalously depleted given their interpreted origin as a shallow burial phenomenon, and the assumed thermal history of the area (i.e. no heating event or raised geothermal gradient during the Tertiary in this region).

Assuming that the cements precipitated in the depth range of 0–1000 m, it is possible to calculate the isotopic composition of the pore fluid using Eocene bottom sea-water temperature. The temperature for 500–1000 m water depth, at a similar latitude (45°N), during the Eocene, is estimated to be 14°C (Shackleton & Kennett 1975). This temperature may appear high for a deep marine environment but during this time there were no polar ice sheets and Britain was experiencing humid, sub-tropical conditions (Anderton *et al.* 1979). Assuming all the cements formed at < 1 km burial depth, using an average geothermal gradient of 2°C per 100 m (based on present-day depths and bottom-hole temperatures), the temperature range for carbonate cementation would be 14–35°C. Using this temperature range the pore fluid isotopic composition was calculated to vary from −10.4 to $-6.1‰_{SMOW}$ (Fig. 14).

The most likely source of isotopically depleted pore fluids is meteoric water. Tertiary meteoric water at a similar latitude on the west coast of Scotland is estimated to have a $\delta^{18}O$ value of *c.* $-12‰_{SMOW}$ (Fallick *et al.* 1985). If carbonates undergo recrystallization, their isotopic compositions can re-equilibrate with ambient conditions, at higher temperatures, but preserve the original early fabric (Dix & Mullins 1987). There is no evidence to suggest that the studied carbonate cements have recrystallized at a higher temperature; however, the uncommon replacement fabrics observed within ferroan calcite phases A and B may represent the remnants of an earlier unstable marine carbonate cement which recrystallized after meteoric invasion. Diagenetic alteration of volcanic ash in the surrounding tuffs (Lawrence *et al.* 1979) may have contributed to the observed ^{18}O depletion; however, it is unlikely that this alternative processes was uniquely responsible.

Ferroan calcite post-dates non-ferroan calcite in the paragenetic sequence observed in the Forth area, yet $\delta^{18}O$ values for these phases are similar. This suggests there was no significant change in temperature or oxygen isotopic composition of the pore fluid between the precipitation of these different calcite cements.

Precipitation temperatures of *c.* 30°C are calculated (see Fig. 14) for ankerite (1000 ln$\alpha_{(ankerite-water)}$ = 2.78 × 10^6T^{-2} + 0.11; Fisher & Land 1986), assuming a meteoric fluid composition of $-10‰_{SMOW}$ and using an average ankerite $\delta^{18}O$ value of $20.2‰_{SMOW}$. Variations in local pore-fluid composition or temperature during precipitation may account for the observed 3‰ range in ankerite results. As for calcite samples, Eocene sea water is regarded an unlikely potential pore fluid as calculated precipitation temperatures for ankerites are in the order of 90°C.

Carbon isotopes

Non-ferroan calcites. Non-ferroan calcite cements are characterized by having consistently low $\delta^{13}C$ isotopic compositions. These carbonate cements contain bitumen inclusions and, therefore it is possible that some of the carbonate may have been sourced via microbial oxidation of hydrocarbons. Carbonates derived from the chemical oxidation or bacterial conversion of hydrocarbons typically have $\delta^{13}C$ values more negative than $-25‰_{PDB}$ (Russell *et al.* 1967; Donovan 1974; Donovan *et al.* 1974; Gould & Smith 1978; Roberts *et al.* 1989). $\delta^{13}C$ values for Viking Graben oils range from −27.5 to $-31.0‰_{PDB}$ (Cornford, 1990), and secondary biogenic carbonate material having low $\delta^{13}C$ could be expected to result from microbial oxidation of Viking Graben crude oil (Gould & Smith 1978). Irwin *et al.* (1977) suggest that bicarbonate activities sufficient to reach carbonate supersaturation are unlikely in the zone of near surface bacterial oxidation because of upward diffusion of bicarbonate into the depositional waters. However, bacterial reactions and carbonate supersaturation can occur at depth in meteoric aquifers especially when oil is present (Connon, 1984). The association of pyrite with non-ferroan calcite suggests that sulphate reduction was also contributing bicarbonate to these cements. Sulphate reducing bacteria cannot utilize petroleum products directly but other organisms associated with oxidation of hydrocarbons can convert oil into suitable materials for metabolism by sulphate reducers (Krouse 1977). Oxidation of methane, derived from deeper in the sediment column, yields extremely isotopically light bicarbonate with $\delta^{13}C$ of -30 to $-60‰_{PDB}$ (Hathaway & Degens 1969; Raiswell 1987), and this cannot be ruled out as a possible minor bicarbonate source.

The majority of non-ferroan calcite carbon isotopic ratios are grouped around $-25‰_{PDB}$, however, less negative values may indicate increased contributions of isotopically heavier carbon released during methanogenic fermentation. Abiotic thermal decarboxylation can produce carbonates with $\delta^{13}C$ of *c.* $-15‰_{PDB}$, however as the cements in this study formed at shallow burial depths and at low temperatures it is unlikely that this process acted as a carbon source. The presence of hydrocarbon inclusions within these cements, indicate that it is possible that CO_2 derived from decarboxylation may have been originally associated with the oil. However, buffering of the CO_2 (Giles *et al.* 1995), by reactions during its long migration from the Kimmeridgian source rock up into the Tertiary, is likely to have almost totally removed this potential bicarbonate source.

Ferroan calcites. Wide ranging $\delta^{13}C$ values are a characteristic of ferroan calcite cements (Mozley & Burns 1993 and references therein). Differential mixing between bicarbonate, derived from oxidation/sulphate reduction and methanogenic fermentation, is the most likely cause of wide-ranging carbon isotope ratios within studied ferroan calcite cements. The predominance of high $\delta^{13}C$ values in ferroan calcites suggests that methanogenic fermentation processes probably acted as the dominant bicarbonate source. Again, the occurrence of bitumen-filled inclusions within these cements which suggests that hydrocarbons were the ultimate organic matter source for these reactions.

$\delta^{13}C$ of bicarbonate within pore waters increases with increasing burial depth, as the sediment column passes through the zones of dominant bacterial oxidation/sulphate reduction and bacterial fermentation (Irwin *et al.* 1977). After meteoric flushing has ceased within an aquifer, it is likely that a similar cement series would precipitate as sources of oxygen and sulphate are depleted. Compared with the non-ferroan calcite cements discussed above, the higher $\delta^{13}C$ values of ferroan calcite cements (Table 1) confirm their later precipitation at a greater depth, post-dating aquifer flushing.

$\delta^{13}C$ Variation during concretion formation. A concretion at 2283 m within well UKCS 255Y (Fig. 10) displays enriched $\delta^{13}C$ isotopic compositions for the ferroan calcite cements towards the top and base of the concretion, reflecting an input of ^{13}C enriched bicarbonate derived from bacterial fermentation. It is possible that this large concretion may have formed by the merging of a number of separate non-ferroan and ferroan calcite concretions which formed at different burial depths.

Ankerites. Ankerites have a wide range in $\delta^{13}C$ similar to the ferroan calcites described above. The possible carbonate producing mechanisms eliminated for ferroan calcites can also be ruled out for the ankerites. A predominance of ^{13}C enriched isotopic compositions suggests that bacterial methanogenesis was the major process responsible for the precipitation of ankerite cements. Mixing of methanogenic derived bicarbonate with bicarbonate derived from bacterial oxidation/sulphate reduction could have produced the lower $\delta^{13}C$ values recorded for ankerite cements.

Strontium isotopes

As $^{87}Sr/^{86}Sr$ ratios are unaffected by kinetic, equilibrium or temperature dependent fractionations during carbonate precipitation (Veizer 1983), Sr isotopes can be used as an important tracer of elemental Sr and related cations in diagenetic systems (Stanley & Faure 1979; Steuber *et al.* 1984). Had carbonate cements in the Forth Field precipitated from marine pore fluids the $^{87}Sr/^{86}Sr$ ratio of the carbonates should be similar to that of Eocene sea water ($^{87}Sr/^{86}Sr$ *c.* 0.7077, Burke *et al.* 1982). Any increase relative to this value must be accounted for by incorporation of Sr from detrital silicate minerals, or older/younger marine carbonate sources (Chaudhuri & Clauer 1993). It is significant that all studied carbonate cements have $^{87}Sr/^{86}Sr$ ratios in excess of that of Eocene sea water indicating that they precipitated from formation waters with modified Sr isotopic composition, significantly different from that of sea water, or of Eocene shell material.

Since there is no linear correlation between $\delta^{13}C$ and $^{87}Sr/^{86}Sr$ isotopic compositions, the sources of carbon and strontium for carbonate cements do not appear to have been interlinked (Fig. 11) (Sullivan *et al.* 1990). A plot of $^{87}Sr/^{86}Sr$ v. 1/Sr concentration (Fig. 15) does not yield a linear relationship, suggesting that these cements did not form by simple mixing of two different component solutions, each having specific $^{87}Sr/^{86}Sr$ ratios and Sr concentrations (Faure 1986). Instead, the scatter of data suggests that these cements may have precipitated from fluids with variable Sr compositions. These variations

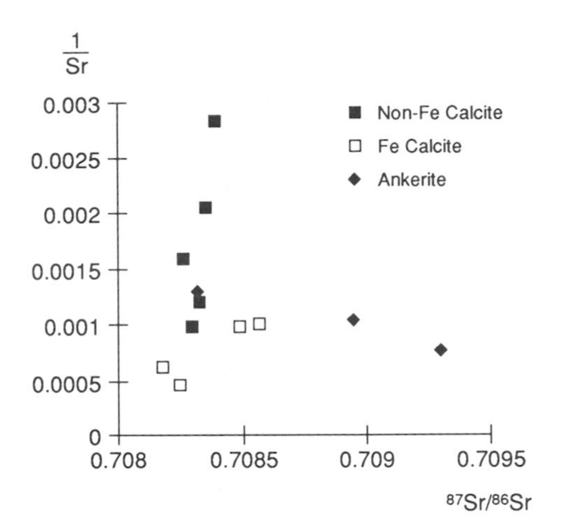

Fig. 15. $^{87}Sr/^{86}Sr$ v. 1/Sr for selected Forth Field carbonates from UKCS 255Y.

were the result of local interaction with radiogenic Sr sourced from detrital minerals (Steuber *et al.* 1984).

Ground water runoff from Dalradian and Caledonian basement should give rise to $^{87}Sr/^{86}Sr$ ratios of 0.72–0.74 (i.e. much higher than is actually recorded in these ferroan and non-ferroan calcite samples) (Hamilton *et al.* 1987). However, there is potential for lowering this ratio through water–rock interaction depending on the lithologies encountered by meteoric water as it travels from the upland regions into the basin. Lawrence *et al.* (1979) reported a $^{87}Sr/^{86}Sr$ ratio of 0.7042–0.7059 for volcanic ash and a value of 0.721 for continental detritus at a DSDP site fairly close to an active spreading ridge. Plagioclase feldspars characteristically have low $^{87}Sr/^{86}Sr$ ratios in comparison with K-feldspars (Chaudhuri & Clauer 1993), and values derived from the analysis of Caledonian plagioclase feldspars range from 0.70656 to 0.70721 (G. Rogers, pers. comm.). Stanley & Faure (1979) studied the sources of Sr in carbonate-cemented sandstones and they concluded that plagioclase feldspars, marine carbonates and volcanic ash all contributed Sr to pore water having low $^{87}Sr/^{86}Sr$ ratios ranging from 0.706 to 0.709. Tuffaceous horizons are abundant in the argillaceous sequence surrounding the Forth sands, and reactive mineral detritus could be leached by low pH meteoric water passing through tuffs and sandstones. Prior to eventual carbonate precipitation, the fluid Sr isotope composition would be decreased (assuming it originally had a Dalradian/Caledonian Sr isotopic signature) by the addition of lower $^{87}Sr/^{86}Sr$ from dissolution and isotope exchange reactions with volcanic air-fall detritus, detrital clays (largely derived from volcanogenics) and plagioclase feldspars. Mixing of meteoric fluids with sea water within shallow buried submarine fan sandstones may also have assisted in lowering the $^{87}Sr/^{86}Sr$ isotope ratio.

K-feldspars appear to have been stable during the early stage of reservoir burial history and are unlikely to have contributed radiogenic ^{87}Sr to pore waters at this time (Fig. 9). However, during ankerite cementation, their decomposition resulted in local increases in ^{87}Sr abundance. Micas are also rich in ^{87}Sr, and mica replacement/alteration could also provide high initial $^{87}Sr/^{86}Sr$ in the replacive ankerite.

Model for carbonate formation in the Forth Field

Oxygen isotope results suggest that a meteoric fluid was present in the reservoir during the formation of ferroan and non-ferroan calcites. Bjørlykke (1988) has argued that early mixing between marine and meteoric waters may be very common in shallow aquifers or coastal sediments, and can occur early in marine sediments if meteoric water intrudes offshore. Early meteoric flushing of nearshore Upper Jurassic turbidite fans, based on carbonate oxygen isotope evidence, has also been suggested for the South Brae oilfield (McLaughlin *et al.* 1991). Long-term flushing by regional meteoric circulation replaces residual waters with geochemically evolved meteoric ground water (Clayton *et al.* 1966). The total flux of meteoric water through sandstones depends on the head as defined by the elevation of the ground-water table, the permeability and the large-scale 'connectedness' of the sandstone bodies (Bjørlykke 1988). Applying the Ghyben and Herzberg equation in its simplest form (Rodda *et al.* 1976), the interface between meteoric water and sea water at 500–1000 m water depth requires the surface of the water table to lie *c.* 12.5–25 m above sea level. During the Palaeocene the Scotland–Shetland Platform was uplifted along its western margin (Anderton *et al.* 1979). Data concerning the extent of uplift and the area of the East Shetland Platform during the Palaeocene and Eocene is not available. However, at this time, the sands were thought to be *c.* 100 km from the East Shetland Platform and were connected to the shelf by a series of delta-plain and delta-slope deposits. These nearshore deltaic sands contain very heavy, severely biodegraded oils (Dixon & Pearce 1995). The association of uplift on the East Shetland Platform during the Palaeocene, connectivity of the shelf with the turbidite reservoirs and the decrease in oil API gravity from the shelf sands to the turbidites all support the findings of the oxygen isotope study. Meteoric water will initially be undersaturated with respect to minerals (e.g. feldspar) and their dissolution will supply excess metal cations which could be incorporated in later cements.

Carbon and oxygen isotope results suggest that the pore-fluid composition within the Forth sands is likely to have remained fairly constant but that the bicarbonate source changed through time. The bicarbonate source in the majority of samples analysed is related to microbial reactions within the sediment. Source-rock maturity calculations confirm reservoir-fill commenced in the Late Palaeocene to Early Eocene (R. Dixon, pers. comm.). Hydrocarbons, migrating into these sands through a series of faults from the mature Kimmeridge Clay Formation of the South Viking Graben, acted as the major source of organic matter required to fuel the reactions (Fig. 16). Stewart *et al.* (1993) mapped the distribution of carbonate cements in the Forth Field and concluded that the largest volumes of cement occur adjacent to the major bounding faults. The biodegraded nature of the oil contained in these reservoirs, and the inclusion of bitumen within non-ferroan and ferroan calcite cements, also suggest that oil was the organic matter-source which triggered biogeochemical reactions. Transformation of crude oil in reservoirs by bacteria requires that a number of conditions are satisfied (Connon 1984): (1) Moving waters – oil biodegradation is generally observed in shallow reservoirs which are flooded by low salinity meteoric waters; (2) sufficient supply of nutrients (nitrate, phosphate) and of dissolved oxygen in moving waters; (3) presence of microbes; (4) oil-water contact – bacteria live in the aqueous phase and do not thrive in oil; bacterial degradation takes place at the oil-water interface; (5) subsurface temperature allowing activity of bacteria – 100°C appears to mark the upper limit of biodegradation when all bacteria become inactive. The above conditions suggest HCO^{3-} generation and carbonate cementation will be concentrated at the oil–water contact (Fig. 17a). Had the oil remained static following emplacement within the reservoir, the oil–water contact would have remained unchanged, and carbonate cementation would be limited to the present-day oil–water contact. However, numerous cement horizons are developed throughout the reservoir units, suggesting that the position of the oil–water contact varied through time. Frequent oil leakage could account for fluctuations in the position of the oil–water contact (Fig. 17b). Evidence for oil leakage includes: (1) residual oil stains within water-bearing sands, suggesting that oil has either moved through or was originally present in these sediments; (2) 'residual' oil columns up to 30 m thickness are developed in some parts of the field; (3) bitumen filled coprolites within the shales surrounding these sands suggest that early hydrocarbon leakage has occurred, and that oil has migrated through poorly consolidated shales and accumulated in these coprolites; (4) The sand units are normally pressured, but overpressure in the shales surrounding the sands is a common phenomenon.

Early oil leakage may have been initiated by soft-sediment deformation, seismic activity or pressure build up and subsequent oil/gas release from a near-surface reservoir. Abundant injected sand structures are present in the shales which surround these sands and within the sand–shale sequences. Seismic activity, which was concurrent with sea-floor spreading in the Norwegian Greenland Sea during the Eocene, may have liquefied the sediment resulting in sand-injection features (Alexander *et al.* 1991; Dixon *et al.* 1995). Sediment liquefaction and 'movement' may have broken the reservoir seal, allowing oil to leak from the reservoir. Post-Eocene oil leakage may have been initiated by tectonic movements related to a NE–SW intraplate Alpine compressional stress regime during the Miocene (Le Pichon *et al.* 1988). Oil and gas accumulation in a shallow reservoir could potentially build up enough pressure to break the overlying shale seal. Oil and gas would then leak upwards into the surrounding shales.

If the carbonate cements were formed at palaeo-oil–water contacts then it is likely that the thicker cemented horizons may be laterally extensive and could possibly act as major barriers to fluid flow within the Forth Field. However, thinner cemented sections in core are obviously tapered and not continuous. These may represent cementation events at 'short-lived' palaeo-oil–water contacts or these cements may be related to localized concentrations of organic matter within the sediment column. Laterally extensive cements could also be accounted for by a high pore-fluid flow rate as concretion elongation increases with increased flow rate (Berner 1980). However, in this study, the concentration of bacterial activity at the oil–water contact is the most likely cause of laterally extensive or elongate cement zones.

A continuous supply of oxygen and nutrients would have been present within the reservoir until the meteoric flushing ceased. Sea water, which may have initially supplied sulphate to

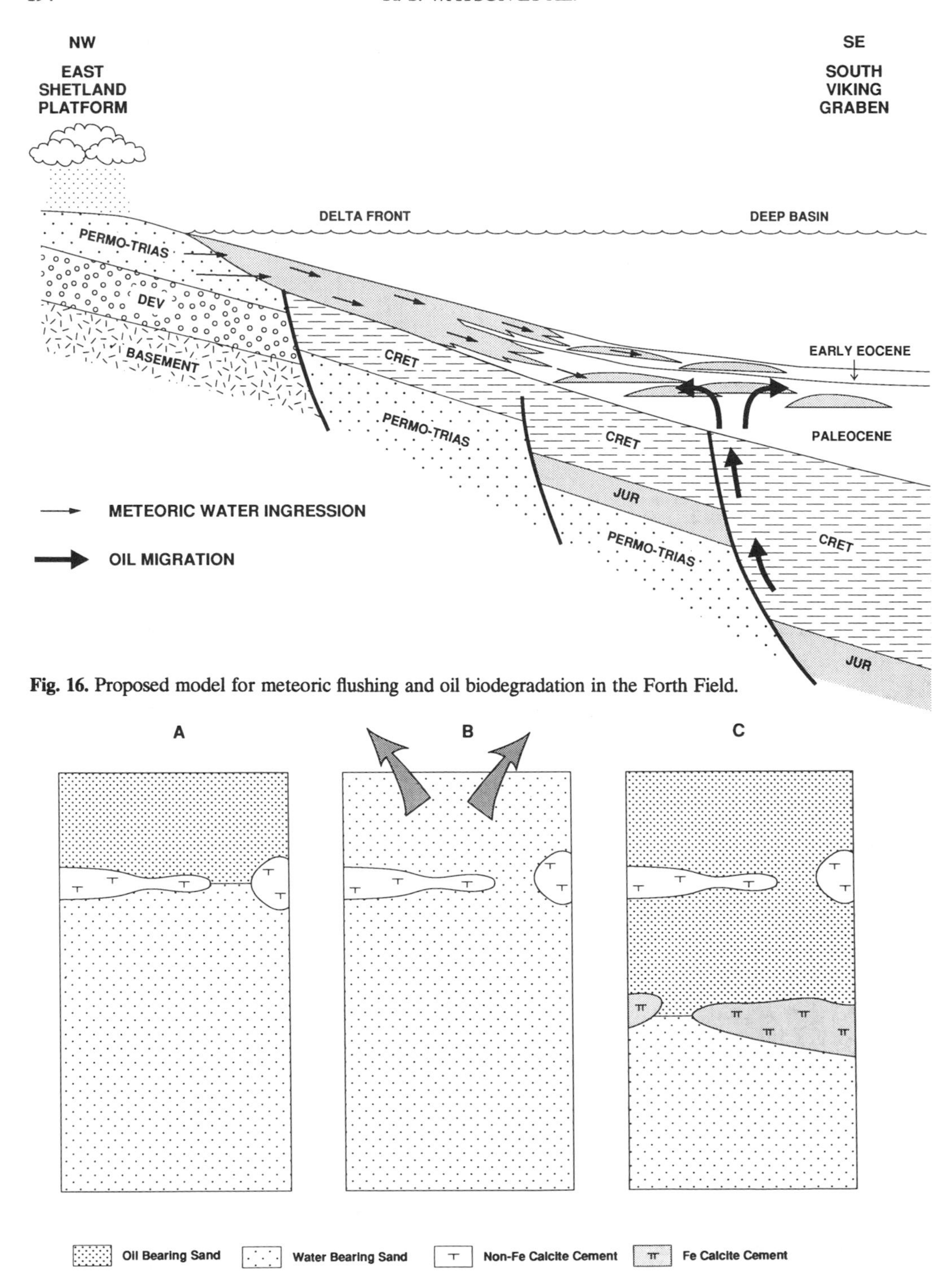

Fig. 16. Proposed model for meteoric flushing and oil biodegradation in the Forth Field.

Fig. 17. Model for the formation of carbonate cements at palaeo-oil–water contacts. **(A)** Precipitation of non-ferroan calcite concretions at palaeo-oil–water contact during early burial. Bacterial oxidation and sulphate reduction were the dominant processes operating. **(B)** Oil leaks from the sands and water resaturation occurs. **(C)** Ferroan calcite cements precipitate at greater burial depths when oxygen and sulphate supplies are exhausted.

pyrite, mixed with and/or was replaced by meteoric water. After meteoric flushing ceased, a closed system began to develop, precipitating carbonate of progressively increasing $\delta^{13}C$. At this point, the activity of oxidizing and sulphate-reducing bacteria would have been restricted and the concentration of dissolved sulphates would have decreased, encouraging bacterial fermentation to commence. This would have resulted in preferential incorporation of Fe into the newly forming calcite cements (Coleman 1985) (Fig. 17c). The edges of the ferroan calcite concretions are more diffuse than those of the non-ferroan calcite cemented sands. This may indicate a gradual waning of calcium or of nutrient supply to the methanogenic bacteria as burial continued.

The alkaline environment induced by bacterial activity (Machel & Mountjoy 1986) and possibly a declining Ca supply was conducive to ankerite precipitation. The Mg, Fe and Mn required for ankerite precipitation may have been derived internally from local dissolution of unstable detrital grains, or externally from the adjacent shale and tuff sequences during compaction or from hydrocarbon-bearing fluids derived from greater burial depths.

Balmoral data interpretation

Timing of concretion growth

In the Balmoral sands, grain packing is loose and minus-cement porosities are high, ranging from 47 to 50%. Quartz surface textures and the occasional presence of over-sized pores suggest some grain-edge or lithoclast replacement may have occurred. The pitted nature of quartz grain surfaces (rather than an embayed surface; Burley & Kantorowicz 1986) suggests there has been only minor grain-edge replacement which is unlikely to affect estimates of original porosity by more than a few per cent. Furthermore, there is no evidence to suggest dissolution and incorporation of detrital carbonate has occurred. The absence of quartz overgrowths and pore-filling kaolinite within the carbonate-cemented sands, as well as the loose packing and limited sediment compaction which preceded carbonate cementation, support concretion formation at shallow burial depths. Taking these factors into account it appears most likely that carbonate concretions within the Balmoral Field formed at < 500 m burial depth.

Oxygen isotopes

Assuming a cementation temperature of 16°C (500–1000 m water depth, 45°N, Palaeocene bottom sea temperature 14°C; Shackleton & Kennett 1975), the Balmoral Field carbonates are predicted to have precipitated from a predominantly meteoric pore fluid ($\delta^{18}O = -9 - -11‰_{SMOW}$). Precipitation from fully marine pore waters can be discounted as calculated temperatures are too high for early, shallow burial cementation. Alternative, less probable explanations for pore fluid ^{18}O depletion are presented in the above Forth Field discussion.

Precipitation temperatures are unlikely to have decreased if concretion formation continued during progressive burial. Therefore, the most likely causes for the observed centre to edge concentric trend towards increasing $\delta^{18}O$ within concretion A9 (Fig. 13) are differences in the relative mixing proportions of sea water and meteoric water present during concretion precipitation, or isotopic modification of pore fluids by water-rock interaction (Longstaffe 1989).

Carbon isotopes

The high positive $\delta^{13}C$ composition within the mixed ferroan and non-ferroan calcite of concretion sample A4 (Table 2) suggests that methanogenic fermentation supplied bicarbonate to this concretion. Low $\delta^{13}C$ values imply sulphate reduction or bacterial oxidation of organic matter or methane acted as the dominant bicarbonate source in samples A6, A8, A9a and A10 (Table 2). The presence of authigenic pyrite within these sediments supports this conclusion. The source of organic matter for these bacterial reactions is likely to have been *in situ* plant material. Oil was not a potential organic matter source because Oligocene oil migration post-dated early concretion formation.

Temporal trends in isotopic composition are frequently preserved in carbonate concretions. The most common trend in mudstone-hosted concretions is one of increasing $\delta^{13}C$ through time (e.g. Hudson 1978; Scotchman 1991). This is often attributed to specific organic processes that can be related to depth of burial from the sediment–water interface (Irwin *et al.* 1977). A systematic decrease in $\delta^{13}C$ during concretion growth in sandstones may be explained by a number of mechanisms:

(1) Dissolution of a marine carbon source or pressure-solution of an underlying limestone could have initially supplied bicarbonate with $\delta^{13}C$ of *c.* $0‰_{PDB}$. This ^{13}C 'enriched' bicarbonate may have progressively mixed with bicarbonate derived from sulphate reduction or oxidation of organic matter ($\delta^{13}C$ of *c.* $-25‰_{PDB}$).

(2) Concretion formation commencing in the zone of methanogenesis and continuing during thermocatalytic decarboxylation (Hennessy & Knauth 1985.)
(3) Concretion growth during progressive meteoric flushing.

Marine carbonate is unlikely to have influenced the isotopic signature of the Balmoral concretions as shell material or calcareous microfossils are absent in these sands and in the enclosing shales. Early carbonate concretions in sandstones frequently enclose shell fragments (e.g. Wilkinson 1991; Walderhaug & Bjørkum 1992), however there is no petrographic evidence to suggest that shell material was ever present in these sands. At the time of concretion precipitation the underlying calcareous Maureen Formation and Chalk Group would have been at approximately 600–800 m burial depth (Fig. 4). Pressure-solution in limestones can begin at 100–150 m but in general pressure-solution becomes effective mainly after burial to *c.* 600–900 m (James & Choquette 1990). This suggests that some of the bicarbonate initially supplied to the growing concretion could have come from the underlying formations. However, $\delta^{13}C$ values of the reworked limestones of the Maureen Formation are not known.

A mixed methanogenic/decarboxylation bicarbonate source is rejected on the grounds that $\delta^{13}C$ values of $-25‰_{PDB}$ are too low to have been produced by decarboxylation ($\delta^{13}C$ typically *c.* $-15‰_{PDB}$; Irwin *et al.* 1977) and these carbonates are likely to have formed at < 500 m burial depth, outside the zone of thermal decarboxylation.

The carbon isotopic values of normal groundwater have a broad range from -30 to $+3‰_{PDB}$ due to varying degrees of interaction with isotopically light organic matter and/or carbonate minerals (Anderson & Arthur 1983). Carothers & Kharaka (1980) reported an average $\delta^{13}C$ composition of $-8‰_{PDB}$ for total dissolved HCO^{3-}(TDC) in shallow ground waters. $\delta^{13}C$ values of -10 to $-3‰_{PDB}$ in sandstones that form part of shallow aquifer systems can be explained by the incorporation of low ^{13}C soil-derived CO_2 into groundwater (Longstaffe 1984). Meteoric water may have contained TDC with an initial $\delta^{13}C$ value of approximately $-5‰_{PDB}$ (this is equivalent to the highest recorded carbonate $\delta^{13}C$ value taken from the centre of a concretion). Alternatively ᶻx1³C enriched bicarbonate could have been supplied by pressure-solution from the underlying Maureen Formation. If meteoric water contained dissolved oxygen it may have oxidised organic matter, contributing HCO^{3-} with $\delta^{13}C$ values of approximately $-25‰_{PDB}$. The initial carbonate cements precipitated would therefore be compatible with $\delta^{13}C$ values of $-5‰_{PDB}$, but as oxidation of organic matter progressed, the $\delta^{13}C$ of cements would have become lower.

With the exception of concretion sample A4, all other samples show similar $\delta^{13}C$ trends dependent upon whether they were taken from the edge or the centre of a concretion. Sample A4, removed from the edge of a concretion, has a $\delta^{13}C$ value of $+8.8‰_{PDB}$. Bicarbonate derived from bacterial fermentation may have continued to be a source for this concretion (A4), after sulphate reduction was complete. The absence ^{13}C enriched compositions within other concretions suggests their growth is likely to have ceased earlier. Sample A4 is also enriched in Fe compared to the other ferroan and non-ferroan calcites studied. Fe is normally incorporated into calcite when there is excess Fe available within pore waters, often after sulphate reduction is complete (Coleman 1985).

Model for concretion formation

Oxygen isotopic results suggest that a meteoric fluid was present in the Balmoral Field reservoir during concretion formation. Meteoric water is likely to have been derived from the East Shetland Platform (*c.* 140 km NW of the field), passing through a series of delta-plain and delta-slope deposits before reaching the Balmoral turbidite sands. This is similar to the hydrological model proposed for Forth Field. Meteoric flushing probably oxidised plant material present within the D Sand of the Balmoral Field Andrew Member, resulting in the precipitation of carbonate cements with progressively lower $\delta^{13}C$ values. When meteoric flushing ceased, sulphate reducing bacteria could have continued producing ^{13}C depleted bicarbonate, and H_2S which reacted with available Fe to form authigenic pyrite when sufficient sulphate was available. After oxygen and sulphate were exhausted, methanogenic fermentation is interpreted to have continued in one of the analysed concretions (A4), forming ferroan calcite towards the concretion margin. The precipitation of an Fe-enriched rim to Balmoral Field concretions suggests there was excess free Fe available within pore waters. Excess sulphate and bicarbonate may not have been available to react with Fe after concretion formation was complete.

Carbonate crystals within the concretions are large and poikilotopic. According to Folk (1974), large poikilotopic crystals form from sparingly supersaturated solutions, however crystal shape is also a function of pressure, temperature and pore-water flow rate. Concretion growth rates and shapes are also dependent upon the rate of pore-fluid flow: growth rates and elongation increase as flow rate increases (Nielsen 1961; Berner 1968, 1980; Johnson 1989). The spherical shape of these concretions suggests that either pore-fluid flow rates were low, or that surface reaction effects overcame the effects of flow (Wilkinson 1992). Surface reaction effects are largely controlled by the Mg/Ca composition of the pore water (Wilkinson & Dampier, 1990). In pore waters with high Mg/Ca (e.g. sea water) surface reaction will control growth rates due to the inhibiting effect of magnesium upon calcite precipitation. Present-day ground water typically has Mg/Ca < 1 (Veizer 1983). Assuming that Mg/Ca of Palaeocene meteoric water was similar to present-day ground water then surface reaction would not be important relative to solute transport, therefore, it may be concluded that pore-fluid flow rates were low. Corrosion of silicate grains enclosed by non-ferroan and ferroan calcite cements suggests that the pore water was not in equilibrium with silicate phases during concretion formation (Saigal & Bjørlykke 1987).

Effects of diagenesis on reservoir quality

High porosities and permeabilities in Tertiary sandstones make them ideal reservoir rocks (Alexander *et al.* 1991), but calcite cements are detrimental to reservoir porosity and permeability. In the Balmoral Field, scattered carbonate concretions reduce the net to gross ratio of the cored reservoir section by only 4% and the tapered concretion outlines suggest that the concretions are non-continuous and therefore, unlikely to affect reservoir continuity. In the Forth Field however, frequent oil leakage during shallow burial is likely to have produced a series of cemented layers. The results of this study suggest that carbonate-cemented sand horizons have the potential to compartmentalize and to reduce the net:gross ratio of the Forth reservoir by up to 30%.

The connection between the location of bacterial activity and the eventual precipitation of laterally extensive carbonate cements across palaeo-oil–water contacts has important consequences for horizontal drilling and reservoir production. Other Tertiary reservoirs may contain cemented layers produced by meteoric flooding and oil biodegradation. Frigg Field oil is biodegraded and the reservoir is affected by 'erratic' calcareous cemented sandstones and concretions (Hertier *et al.* 1981; Conort 1986). The Gryphon Field, situated *c.* 12 km north of Forth, is also similar (Newman *et al.* 1993). These findings tentatively suggest that carbonate-cemented horizons are a characteristic feature of Tertiary reservoirs situated to the east of the East Shetland Platform. Carbonate cemented layers will only develop if oil migration and meteoric flooding are concurrent. Oil migrated into these reservoirs during the Palaeocene and Eocene from the mature Kimmeridge Clay Formation in the South Viking Graben. However, to the south of the East Shetland Platform (in the vicinity of Balmoral) Kimmeridge Clay Formation mudstones in the Witch Ground Graben and Fisher Bank Basin were not mature until the Oligocene (Tonkin & Fraser 1991). Therefore, oil migration post-dated Palaeocene meteoric flooding and carbonate-cemented layers are unlikely to have formed in any Tertiary reservoirs from this region.

Conclusions

The sands of the Forth Field contain pervasive non-ferroan and ferroan calcite cements. Calcite $\delta^{18}O$ and minus-cement porosities suggest meteoric water flushed the reservoir during early burial, carrying with it excess oxygen and microbes, biodegrading the migrated oil. Biodegradation of migrated oil took place at the palaeo-oil–water contact, producing a laterally extensive cementation zone. $\delta^{13}C$ of the calcite cement was dependent upon the amount of oxygen, sulphate and nutrients available to the bacteria. At shallower burial depths (< 500 m), when there was excess oxygen available to bacteria, ^{13}C-depleted non-ferroan calcite cements precipitated. During later burial, when oxygen and sulphate were largely exhausted ^{13}C-enriched ferroan calcite formed. Oil leakage may have occurred a number of times during, and subsequent to, its emplacement, continually changing the position of the oil–water contact, and producing a complex sequence of carbonate cemented horizons which have the potential to act as major barriers to vertical fluid flow. Ankerite is also present in a number of wells and may have precipitated from Eocene meteoric water or from a pore fluid enriched in ^{18}O by water–rock interaction and fluid replacement/mixing.

The location of spherical calcite concretions in Balmoral is controlled by the original distribution of organic matter in the sands; consequently, carbonate concretions have preferentially formed in the D sand of the Andrew Member, which has the highest lignite content. Non-ferroan calcite and ferroan calcite concretions precipitated at < 500 m burial depth, sourced by progressive bacterial oxidation and sulphate reduction of organic matter within a meteoric pore fluid. This meteoric water is thought to have a similar derivation and isotopic composition to the meteoric water which flushed the Forth Field reservoir during the Palaeocene. The scattered distribution, size and tapered or spherical shape of these concretions suggests they are unlikely to significantly affect reservoir production.

The Balmoral and Forth reservoir sands were both deposited as turbidites in a deep marine environment, however, the diagenetic controls relating to carbonate formation are somewhat different. To a large extent, the distribution of carbonate concretions within the Balmoral Field may be controlled by organic matter availability. In the Forth Field, recently-migrated oil provided an almost infinite supply of organic matter; however, element availability and the quantity of nutrients available to bacteria limited the absolute amount of carbonate which precipitated. The episodicity of oil leakage and the relative position of oil–water contacts controlled the distribution and lateral extent of carbonate concretions in the Forth Field.

Laterally extensive carbonate cemented horizons formed in association with oil biodegradation have the potential to compartmentalize a reservoir. The distribution of these cements is likely to be controlled by the relative timing of meteoric flushing and oil migration. Therefore, the likelihood of a reservoir in the vicinity of the East Shetland Platform containing these cemented layers is substantially increased if the oil within the reservoir is biodegraded.

We would like to take this opportunity to thank Corex Services for sponsoring the PhD studentship which led to this research. BP Exploration, Repsol Exploration (UK), Ranger Oil and Sun Oil are also acknowledged for donating core samples and allowing publication of results. Isotopic analyses were carried out at SURRC, which is supported by NERC and the Consortium of Scottish Universities. The comments of the reviewers, Mark Wilkinson and Bruce Sellwood, are also greatly appreciated.

References

ALEXANDER, R. W. S., SCHOFIELD, K. & WILLIAMS, M. C. 1991. Understanding the Forth Field Reservoir UKCS Block 9/23b: myths of deep sea sands exploded. Abstract, EAPG 3rd Conference, Florence, Italy.

ANDERSON, T. F. & ARTHUR, M. A. 1983. Stable isotopes of oxygen and carbon and their application to sedimentologic and palaeoenvironmental problems. *In: Stable Isotopes in Sedimentary Geology,* SEPM Short Course Notes, **10,** 1–151.

ANDERTON, R., BRIDGES, P. H., LEEDER, M. R. & SELLWOOD, B. W. 1979. *A Dynamic Stratigraphy of the British Isles – A Study in Crustal Evolution.* George, Allen & Unwin, London.

BERNER, R. A. 1968. Rate of concretion growth. *Geochimica et Cosmochimica Acta,* **32,** 477–483.

—— 1980. *Early Diagenesis – A Theoretical Approach.* Princeton University Press, Princeton, NJ.

BJØRLYKKE, K. 1988. Sandstone diagenesis in relation to preservation, destruction and creation of porosity. *In:* CHILINGARIAN, G. V. & WOLF, K. H. (eds) *Diagenesis, I.* Developments in Sedimentology, **41,** 531–565.

BURKE, W. H., DENISON, R. E., HETHERINGTON, E. A., KOEPNICK, R. B., NELSON, H. F. & OTTO, J. B. 1982. Variations of seawater $^{87}Sr/^{86}Sr$ throughout Phanerozoic time. *Geology,* **10,** 516–519.

BURLEY, S. D. & KANTOROWICZ, J. D. 1986. Thin section and SEM textural criteria for the recognition of cement-dissolution porosity in sandstones. *Sedimentology,* **33,** 587–604.

CAROTHERS, W. W. & KHARAKA, Y. F. 1980. Stable carbon isotopes of HCO_3- in oil-field waters: implications for the origin of CO_2. *Geochimica et Cosmochimica Acta,* **44,** 323–332.

CHAUDHURI, S. & CLAUER, N. 1993. Strontium isotopic compositions and potassium and rubidium contents of formation waters in sedimentary basins: clues to the origins of solutes. *Geochimica et Cosmochimica Acta,* **57,** 429–437.

CLAYTON, R. N., FRIEDMAN, I., GRAF, D. L., MAYEDA, T. K., MEENTS, W. F. & SHIMP, N. F. 1966. The origin of saline formation waters. *Journal of Geophysical Research,* **71,** 3869–3882.

COLEMAN, M. L. 1985. Geochemistry of diagenetic non-silicate minerals: kinetic considerations. *In:* EGLINGTON, G., CURTIS, C. D., MCKENZIE, D. P. & MURCHISON, D. G. (eds) *Geochemistry of Buried Sediments. Philosophical Transactions of the Royal Society of London,* **A315,** 39–54.

CONNON, J. 1984. Biodegradation in reservoirs. *Advances in Petroleum Geochemistry,* **1,** 299–335.

CONORT, A. 1986. Habitat of Tertiary hydrocarbons, South Viking Graben. *In:* SPENCER, A. M. (ed.) *Habitat of Hydrocarbons on the Norwegian Continental Shelf.* Norwegian Petroleum Society, Graham & Trotman, London, 159–170.

CORNFORD, C. 1990. Source rocks and hydrocarbons of the North Sea. *In:* GLENNIE, K. (ed.) *Introduction to the Petroleum Geology of the North Sea* (3rd Ed). Blackwell Scientific Publications, London, 350.

DEER, W. A., HOWIE, R. A. & ZUSSMAN, J. 1966. *An Introduction to the Rock Forming Minerals* (14th impression, 1982). Longman.

DIX, G. R. & MULLINS, H. T. 1987. Shallow, subsurface growth and burial alteration of Middle

Devonian calcite concretions. *Journal of Sedimentary Petrology*, **57**, 140–152.

DIXON, R.J. & PEARCE, J. 1995. Tertiary sequence stratigraphy and play fairway definition, Bruce–Beryl Embayment, Quadrunt 9, UKCS. *In*: STEEL, R. (ed.) *Sequence Stratigraphy of the North-West European Margin.* Norsk Petroleumsforening, Oslo, in press.

—— SCHOFIELD, K., ANDERTON, R., REYNOLDS, A. D., ALEXANDER, R. W. S., WILLIAMS, M. C. & DAVIES, K. G. 1995. Sandstone diapirism and clastic intrusion in the Tertiary fans of the Beryl Embayment, Quadrant 9, UKCS. *In: This volume*.

DONOVAN, T. J. 1974. Petroleum microseepage at Cement, Oklahoma: evidence and mechanism. *Bulletin of the American Association of Petroleum Geologists,* **58,** 429–446.

—— FRIEDMAN, I. & GLEASON, J. D. 1974. Recognition of petroleum-bearing traps by unusual isotopic compositions of carbonate cemented surface rocks. *Geology,* **2,** 351–354.

FALLICK, A. E., JOCELYN, J., DONNELLY, T., GUY, M. & BEHAN, C. 1985. Origin of agates in volcanic rocks from Scotland. *Nature,* **313,** 672–674.

FAURE, G. 1986. *Principles of Isotope Geology.* (Second Edition), John Wiley & Sons, New York.

FISHER, R. S., & LAND, L. S. 1986. Diagenetic history of Eocene Wilcox Sandstones, South-Central Texas. *Geochimica et Cosmochimica Acta,* **50,** 551–561.

FOLK, R. L. 1974. The natural history of crystalline calcium carbonate: effect of magnesium content and salinity. *Journal of Sedimentary Petrology,* **44,** 40–53.

FRIEDMAN, I. & O'NEIL, J. R. 1977. Compilation of stable isotope fractionation factors of geochemical interest. *In:* M. FLEISHER (ed.) *Data of Geochemistry U.S. Geological Survey,* Professional Paper, 440-KK.

GILES, M. R., BAKKER, P., DE BOER, R. & MARSHALL, J. D. 1995. Diagenesis: a quantitative perspective. *In: Quantitative Diagenesis.* Reading NATO Conference Volume (in Press).

GILES, M. R., STEVENSON, S., MARTIN, S. V., CANNON, S. J. C., HAMILTON, P. J., MARSHALL, J. D. & SAMWAYS, G. M. 1992. The reservoir properties and diagenesis of the Brent Group: a regional perspective. *In:* MORTON, A. C., HASZELDINE, R. S., GILES, M. R. & BROWN, S. (eds) *Geology of the Brent Group,* Geological Society London, Special Publication, **61,** 289–328.

GOULD, K. W. & SMITH, J. W. 1978. Isotopic evidence for microbiologic role in genesis of crude oil from Barrow Island, Western Australia. *Bulletin of the American Association of Petroleum Geologists,* **62,** 455–462.

HAMILTON, P. J., FALLICK, A. E., MACINTYRE, R. M. & ELLIOT, S. 1987. Isotopic tracing of the provenance and diagenesis of Lower Brent Group sands, North Sea. *In:* BROOKS, J. & GLENNIE, K. (eds) *Petroleum Geology of NW Europe.* Graham & Trotman, London, 939–949.

HATHAWAY, J. C. & DEGENS, E. T. 1969. Methane derived marine carbonates of Pleistocene age. *Science,* **165,** 691–692.

HENDRY, J. P. & MARSHALL, J. D. 1991. Disequilibrium trace element partitioning in Jurassic sparry calcite cements: implications for crystal growth mechanisms during diagenesis. *Journal of the Geological Society, London,* **148,** 835–848.

HENNESSY, J. & KNAUTH, L. P. 1985. Isotopic variations in dolomite concretions from the Monterey Formation, California. *Journal of Sedimentary Petrology,* **55,** 120–130.

HERTIER, F. E., LOSSEL, P. & WATHNE, E. 1981. The Frigg gas field. *In:* ILLING, L. V. & HOBSON, G. D. (eds) *Petroleum Geology of the Continental Shelf of North-West Europe.* Institute of Petroleum, London, 380–391.

HUDSON, J.D. 1978. Concretions, isotopes and the diagenetic history of the Oxford Clay (Jurassic) of central England. *Sedimentology,* **25,** 339–370.

IRWIN, H., CURTIS, C. & COLEMAN, M. 1977. Isotopic evidence for source of diagenetic carbonates formed during burial of organic rich sediments. *Nature,* **269,** 209–213.

JAMES, N. P. & CHOQUETTE, P. W. 1990. Limestones – the burial diagenetic environment. *In:* MCILREATH, I. A. & MORROW, D. W. (eds) *Diagenesis.* Geoscience Canada Reprint Series, **4,** 75–112.

JOHNSON, M. R. 1989. Palaeogeographic significance of oriented calcareous concretions in the Triassic Katberg Formation, South Africa. *Journal of Sedimentary Petrology,* **59,** 1008–1010.

KROUSE, H. R. 1977. Sulphur isotope studies and their role in petroleum exploration. *Journal of Geochemical Exploration,* **7,** 189–211.

LAWRENCE, J. R., DREVER, J. L., ANDERSON, T. F. & BRUECKNER, H. K. 1979. Importance of alteration of volcanic material in the sediments of Deep Sea Drilling site 323: chemistry, $^{18}O/^{16}O$ and $^{87}Sr/^{86}Sr$. *Geochimica et Cosmochimica Acta,* **52,** 573–588.

LE PICHON, X., BERGERAT, F. & ROULET, M. J. 1988. Plate kinematics and tectonics leading to the Alpine belt formation; a new analysis. *Geological Society of America, Special Paper,* **218,** 111–131.

LONGSTAFFE, F.J. 1984. The role of meteoric water in diagenesis of shallow sandstones: stable isotope studies of the Milk River aquifer and gas pool. *In:* SURDAM, R. C & MACDONALD, D. A. (eds) *Clastic Diagenesis.* American Association of Petroleum Geologists Memoir, **37,** 81–98.

LONGSTAFFE, F.J. 1989. Stable isotopes as tracer in clastic diagenesis. *In:* HUTCHEON, I. E. (ed.) *Short Course in Burial Diagenesis.* Mineralogical Association of Canada Short Course, **15,** 201–278.

MACHEL, H. G. & MOUNTJOY, E. W. 1986. Chemistry and environments of dolomitization – a reappraisal. *Earth-Science Reviews,* **23,** 175–222.

MCLAUGHLIN, O. M., HASZELDINE, R. S., FALLICK, A. E. & ROGERS, G. 1991. Concretion growth in the South Brae oilfield. *British Sedimentological Research Group Meeting Abstracts.*

MOZLEY, P. S. & BURNS, S. J. 1993. Oxygen and carbon isotopic composition of marine carbonate concretions: an overview. *Journal of Sedimentary Petrology,* **63,** 73–83.

NEWMAN, M. ST. J., REEDER, M. L., WOODRUFF, A. H. W. & HATTON, I. R. 1993. The geology of the Gryphon oil field. *In:* PARKER J. R. (ed.) *Petroleum Geology of NW Europe: Proceedings of the 4th Conference.* Geological Society, London, 123–133.

NIELSEN, A. E. 1961. Diffusion controlled growth of a moving sphere. The kinetics of crystal growth in potassium perchlorate precipitation. *Journal of Physical Chemistry,* **74,** 309–320.

RAISWELL, R. 1987. Non-steady state microbiological diagenesis and the origin of concretions and nodular limestones. *In:* MARSHALL, J. D. (ed.) *Diagenesis of Sedimentary Sequences.* Geological Society, London, Special Publication, **36,** 41–54.

ROBERTS, H. H., SASSEN, R., CARNEY, R. & AHARON, P. 1989. ^{13}C depleted authigenic carbonate buildups from hydrocarbon seeps, Louisiana continental slope. *Transactions of the Gulf Coast Association of Geological Societies,* **39,** 523–530.

RODDA, J. C., DOWNING, R. A. & LAW, F. M. 1976. *Systematic Hydrology.* Newnes-Butterworths, 214.

RUSSELL, K. L., DEFFEYES, K. S. & GOWLER, G. A. 1967. Marine dolomite of unusual isotopic composition. *Science,* **155,** 190–191.

SAIGAL, G.C. & BJØRLYKKE, K. 1987. Carbonate cements in clastic reservoir rocks from offshore Norway – relationships between isotopic composition, textural development and burial depth. *In:* MARSHALL, J. D. (ed.) *Diagenesis of Sedimentary Sequences.* Geological Society, London, Special Publication, **36,** 313–324.

SCOTCHMAN, I. C. 1991. The geochemistry of concretions from the Kimmeridge Clay Formation of southern and eastern England. *Sedimentology,* **38,** 79–106.

SHACKLETON, N. J. & KENNETT, J. P. 1975. Palaeotemperature history of the Cenozoic and the initiation of Antarctic glaciation: oxygen and carbon isotope analysis in DSDP sites 277, 279 and 281. *Initial Reports of the Deep Sea Drilling Project,* **24,** 743–755.

STANLEY, K. O. & FAURE, G. 1979. Isotopic composition and sources of strontium in sandstone cements: the High Plains Sequence of Wyoming and Nebraska. *Journal of Sedimentary Petrology,* **49,** 45–54.

STEUBER, A. M., PUSHKAR, P. & HETHERINGTON, E. A. 1984. A strontium isotopic study of Smackover brines and associated solids, southern Arkansas. *Geochimica et Cosmochimica Acta,* **48,** 1637–1649.

STEWART, R., HASZELDINE, S., FALLICK, A. E., ANDERTON, R. & DIXON, R. 1993. Shallow calcite cementation in a submarine fan; biodegradation of vertically migrating oil. *Abstract, AAPG Annual Convention,* New Orleans, 47.

SULLIVAN, M. D., HASZELDINE, R. S. & FALLICK, A. E. 1990. Linear coupling of carbon and strontium isotopes in Rotliegend sandstone, North Sea: evidence for cross-formational fluid flow. *Geology,* **18,** 1215–1218.

TONKIN, P. C. 1990. The reservoir geology of the Balmoral Field, Block 16/21, UK North Sea. *European Association of Petroleum Geoscientists, Second Conference, Copenhagen, 1990.*

—— & FRASER, A.R. 1991. The Balmoral Field, Block 16/21, UK North Sea *In:* ABBOTTS, I. L. (ed.) *United Kingdom Oil and Gas Fields, 25 Years Commemorative Volume.* Geological Society Memoir, **14,** 237–243.

TREWIN, N. H. 1988. Use of the scanning electron microscope in sedimentology. *In:* TUCKER, M. E. (ed.) *Techniques in Sedimentology.* Blackwell Scientific Publications, London, 229–273.

VEIZER, J. 1983. Chemical diagenesis of carbonates: theory and applications of the trace element technique. *In: Stable Isotopes in Sedimentary Geology, Short course,* **10,** Society of Economic Palaeontologist and Mineralogists, Dallas, 31–100.

WALDERHAUG, O. & BJØRKUM, P. A. 1992. Effect of meteoric water flow on calcite cementation in the Middle Jurassic Oseberge Formation, well 30/3–2, Veslefrikk Field, Norwegian North Sea. *Marine and Petroleum Geology,* **9,** 308–318.

WATSON, R. S. 1993. *The diagenesis of Tertiary sands from the Forth and Balmoral Fields, Northern North Sea.* PhD thesis, University of Aberdeen, UK.

WILKINSON, M. 1991. The concretions of the Bearreraig Sandstone Formation: geometry and geochemistry. *Sedimentology,* **38,** 899–912.

—— 1992. Concretionary cements in Jurassic sandstones, Isle of Eigg, Inner Hebrides. *In:* PARNELL, J. (ed.) *Basins on the Atlantic Seaboard: Petroleum Geology, Sedimentology and Basin Evolution.* Geological Society, London, Special Publication, **62,** 145–154.

WILKINSON, M. & DAMPIER, M. D. 1990. The rate of growth of sandstone hosted calcite concretions. *Geochimica et Cosmochimica Acta,* **54,** 3391–3399.

ZIEGLER, P. A. 1981. Outline of the geological history of the North Sea. *In:* ILLING L. V. & HOBSON G. D. (eds) *Petroleum Geology of the Continental Shelf of NW Europe.* Graham & Trotman, London, 3–42.

Permeability heterogeneity within massive Jurassic submarine fan sandstones from the Miller Field, northern North Sea, UK

D. J. PROSSER,[1] M. E. McKEEVER,[1] A. J. C. HOGG,[23] & A. HURST,[4]

[1] *Z & S Geology Limited, Glover Pavilion, Campus 3, Aberdeen Science and Technology Park, Balgownie Drive, Bridge of Don, Aberdeen AB22 8GW, UK*

[2]. *BP Research, Chertsey Road, Sunbury-on-Thames, Middlesex TW16 L7N, UK*

Present Address: BP Exploration Operating Company Ltd., Blackhill Road, Holton Heath, Poole, Dorset BH16 6LS, UK

[3]. *Department of Geology and Petroleum Geology, Aberdeen University, Meston Building, King's College, Aberdeen AB9 2UE, UK*

Abstract: Analysis of high density (5 cm sample spacing) probe permeameter data has been used to evaluate some of the basic controls on permeability heterogeneity in the Miller Field. Probe permeameter data generally show good correspondence with conventional core-plug permeability data gathered at 25 cm spacings, but record a higher frequency of permeability variation than can be resolved using conventional sampling strategies. Comparison of probe permeameter data gathered from four lithofacies recognizable in core has allowed assessment of their reservoir quality and quantification of their internal permeability heterogeneity. Geostatistical analysis of the permeability data using semivariograms suggests permeability correlation structures and periodicities are related to the frequency of sedimentary phenomena visible within core. In particular, a relationship exists between permeability correlation structure and average bedform or fining upward unit thickness, i.e. the vertical scale at which individual lithofacies packages are stacked. Geostatistical analysis of high sample-density probe permeameter data demonstrates the usefulness of permeability data gathered at more than the conventional 25 cm spacings used for core plugs. High density probe permeameter data can yield important measures of the representative geological 'structures' which could form the basis of geologically realistic two-phase flow upscaling.

Probe permeameter data have been used to investigate permeability heterogeneity of lithofacies types present within the Miller Field. Specifically, it was sought to address the degree of homogeneity within 'massive sandstones' which comprise the best reservoir lithofacies (in the Miller Field), and to evaluate the permeability distribution within more heterolithic lithofacies. The non-destructive nature of probe permeameter measurements permits acquisition of large amounts of data that can be readily compared with geological characteristics observed in core. Statistical analysis, including semivariogram analysis of the permeability data, is presented and the spatial nature of heterogeneity within individual lithofacies is quantified. In particular, relationships between sedimentary bedding and permeability contrast are examined, and the importance of lithofacies classification for definition of reservoir flow units are assessed.

The Miller Field is located at the western margin of the north–south-trending South Viking Graben in UKCS blocks 16/7b and 16/8b (Fig. 1). The geology of the Miller Field has been outlined in McClure & Brown (1991) and by Garland (1993). It has an areal extent of over 45 km^2, occurring to the east of the North, Central and South Brae Fields. The Upper Jurassic Brae Formation forms the oil reservoir at a depth of *c.* 4 km. The Brae Formation is interpreted as a syn-rift submarine fan succession (Fig. 2) sourced from the Fladen Ground Spur lying to the west (Turner *et al.* 1987).

The Miller Field oils were sourced by the Kimmeridge Clay Formation, which interfingers with the Brae Formation, and subsequently onlaps it from the east. To the west, the Miller Field is separated from the South Brae field by a 'structural dome' (Rooksby 1991). In the south, north and northwest, stratigraphic pinchout of the submarine fan sequence into the Kimmeridge Clay provides the trapping mechanism.

From Hartley, A. J. & Prosser, D. J. (eds), 1995, *Characterization of Deep Marine Clastic Systems,* Geological Society Special Publication No. 94, pp. 201–219.

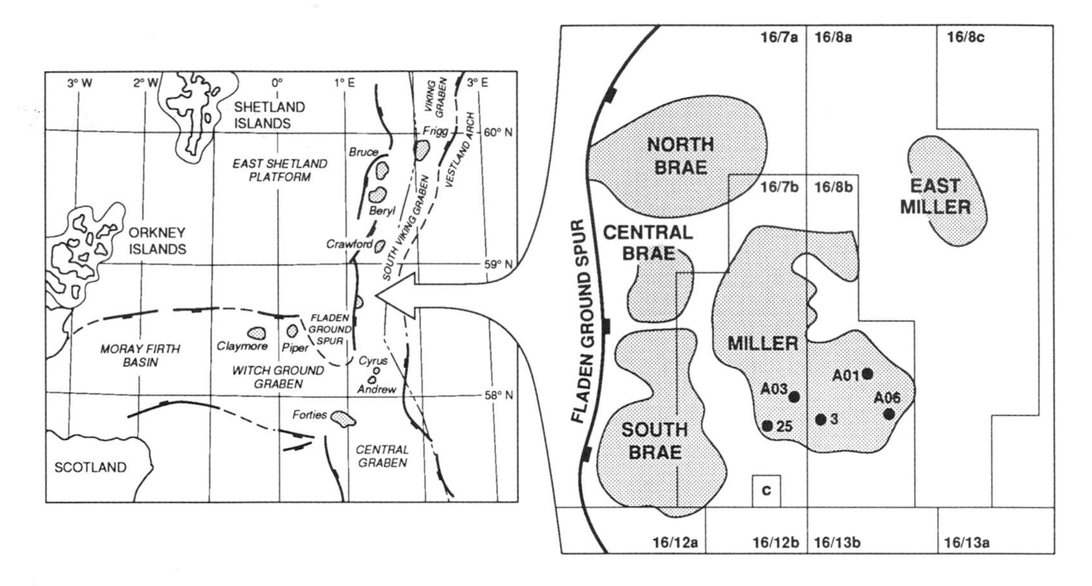

Fig. 1. Miller Field location map (adapted from Rooksby 1991).

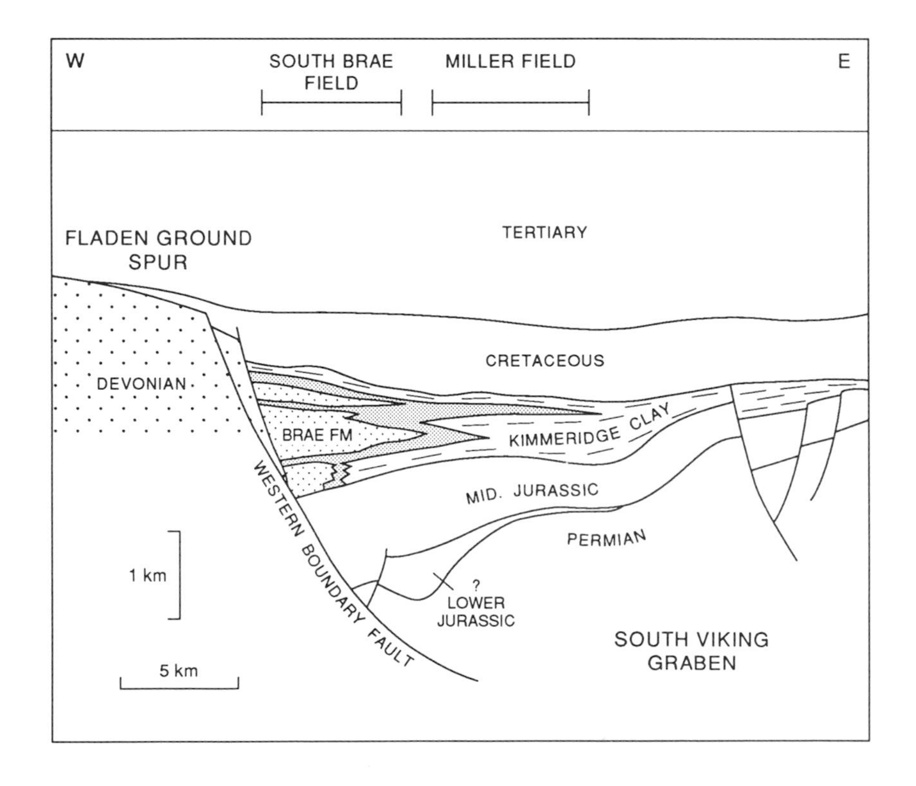

Fig. 2. Cross-section through the Miller and South Brae Fields (adapted from Rooksby 1991).

Table 1. *Lithofacies types within the Miller Field*

Lithofacies	Description	Process	Setting
A	Centimetre to decimetre scale interbedded siltstone, fine to medium argillaceous sandstone and mudstone	Low density turbidity currents and hemipelagic settling	Flows on mid to distal fan, or into regions between fan lobes
B	Medium to coarse sandstones tens of centimetres to several metres thick, with thin (< 10 cm) sparse mudstone interbeds. Structureless and often comprising amalgamated beds up to 30 m thick	High and low density turbidity currents	Channels and flows over proximal to mid-fan lobes
C	Massive mudstone with minor (rare) siltstone or very fine sandstone laminae or interbeds (millimetre to centimetre)	Pelagic/hemipelagic sedimentation	Distal fan and pelagic basin floor
D	*Ortho* and *para*-conglomerates with boulders and pebbles set within a sandstone matrix	High density turbidity currents and debris flows	Channels fills and flows on inner fan

Miller Field geology

The reservoir interval of the Miller Field comprises the middle member of three lithostratigraphic units defined within the Brae Formation by Rooksby (1991). Roberts (1991) and Turner *et al.* (1987) have documented a simple four-fold lithofacies division for the South Brae Field. A similar lithofacies breakdown can be constructed for the Miller Field, and is described in Table 1. These lithofacies comprise: (1) interbedded mudstone and sandstone (Lithofacies A); (2) massive sandstones (Lithofacies B); (3) massive mudstones (Lithofacies C); (4) conglomerates (Lithofacies D). (Note: Lithofacies D is not recognized in the Miller Field, but occurs in South Brae, from which the Miller sands are believed to have been fed).

Lithofacies A is interpreted as mainly representing the deposits of low density turbidity currents as flows on mid to distal regions of the submarine fan, or into regions between fan lobes. Lithofacies B is interpreted as representing the deposits of high and low density turbidity currents within channels, and as flows over proximal to mid-fan lobes. The sediments of Lithofacies C are interpreted as pelagic/hemipelagic deposits in distal fan and basin floor settings. Lithofacies D is interpreted as representing the deposits of high density turbidity currents and debris flows both within channels and as flows over inner-fan regions. This study concentrates upon permeability variation within Lithofacies A and B, which are illustrated in Fig. 3 (modified from Garland 1993). The best quality reservoir sandstones in the Miller Field are the massive sandstones of Lithofacies B, which have a paucity of clearly defined sedimentary structures. Poorer quality, heterolithic units of Lithofacies A typically separate out the more massive units and are potential barriers to vertical fluid flow.

Texture and mineralogy

The Miller sandstones are moderately well to well sorted, medium-grained, quartzose, subarkosic and sublithic arenites. Quartz and polycrystalline quartz form > 90% of the framework mineralogy. Other framework components include K-feldspar (1–2 vol%) with minor plagioclase, rock fragments, mica and traces of glauconite. Rock fragments are commonly mudstone and siltstone intraclasts and metamorphic quartz/mica clasts. In the sandy lithofacies, matrix clay contents are generally < 3% in Lithofacies A and < 1% in Lithofacies B. Small cellular wood fragments are common

Fig. 3. Core photograph illustrating the contrasting lithofacies types studied. Lithofacies B (548–560) is interpreted as mainly representing the deposits of high and low density turbidity currents within channels and as flows over proximal to mid-fan lobes. The remainder of the core illustrated is typical of Lithofacies A (560–572) and is interpreted as having been deposited by low density turbidity currents as flows on mid-distal regions of submarine fans (or into interfan lobe regions), and via hemipelagic deposition. (Core photographs kindly supplied by Chris Garland; modified from Garland 1993.)

in these sandstones and in core these and detrital clay appear as diffuse, wispy laminations < 1 mm in thickness. The quartz grains commonly display pressure solution contacts.

The sandstones are cemented by calcite, quartz with minor illite, pyrite, microcrystalline silica, kaolinite and dolomite. Calcite occurs in early formed concretions of both ferroan and non-ferroan calcite. Quartz cement is found throughout the reservoir as syntaxial overgrowths on detrital quartz grains. Mean quartz cement volumes range from 0 to 10%, however abundances can be as great as 15%. In wells 16/7b-25 and 16/8b-3 there is a general increase in quartz cement volumes from 2 to 13 vol%, with depth in the reservoir.

The principal controls on pore-throat geometry in the Miller sandstones are primary depositional fabric (grain size and sorting), compaction and quartz cementation. In sandstones of Lithofacies B compaction has reduced the theoretical depositional porosities of 40% to *c.* 20%. This compactional porosity has been further reduced *c.* 5% by quartz cement. Locally carbonate concretions and tar mats have reduced intergranular porosities to zero. In uncleaned core slabs, tar mats will greatly reduce interpreted probe permeabilities relative to those measured by conventional Hassler sleeve analyses of cleaned core plugs. No significant tar mats were encountered during this study, and for the most part, probe permeameter data show good correspondence with conventional core plug data.

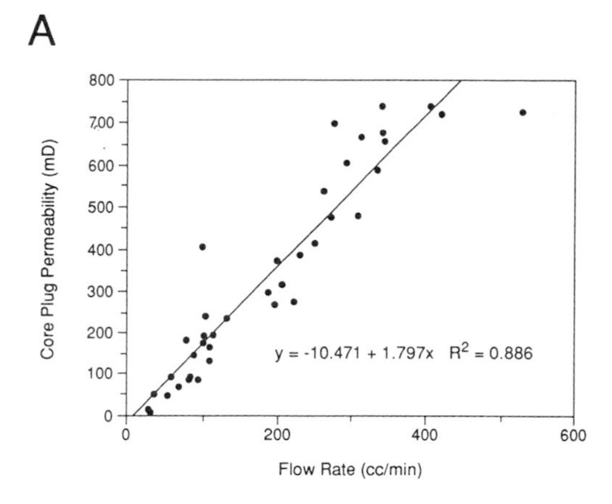

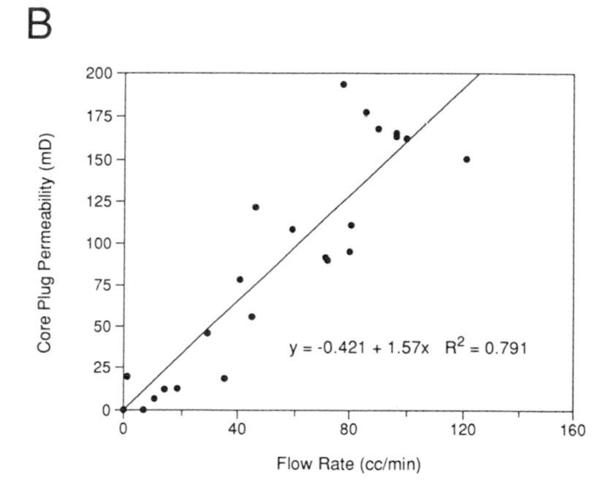

Fig. 4. Simple linear calibration of probe permeameter flow rate with conventional core-plug permeability data. (**A**) Using flow rate of 2000 $cm^3 min^{-1}$; (**B**) using flow rate of 200 $cm^3 min^{-1}$.

Methods

Data gathering

Core intervals with representative sections of the main reservoir units Lithofacies A and Lithofacies B were selected from wells 16/8b-A01, 16/8b-A03, 16/8b-A06, 16/8b-3 and 16/7b-25. Sedimentary descriptions of slabbed cores were made at a 1:10 scale. Gas flow rate measurements were gathered at 5 cm vertical spacing along the middle of core slabs using a portable, steady-state, nitrogen probe permeameter with a 1 cm (external) diameter probe-tip, and maximum nitrogen injection pressure of 5 psi. The 5 cm sample spacing was chosen following visual examination of cores simply because it appeared to facilitate the location of several analyses points between observed bedding features. Even surface texture and partial flushing of the slabbed core surfaces enabled probe permeameter measurements to be taken upon the slabbed cores following a minimum of surface preparation (gentle brushing etc.). Conversion of gas flow rates to permeability was achieved by using calibration curves obtained from comparison of probe permeameter flow rates with conventional Hassler sleeve permeabilities for supposedly homogeneous core-plugs (Fig. 4). [Note: use of different capacity 'flow pathways' (termed laminar flow elements) within the probe permeameter allowed regulation of maximum gas flow rate through the probe tip, and hence enabled the operator to minimize the effects of turbulent flow through samples during analysis.] Calibration curves illustrated in Fig. 4 are for 200cm^3/min^{-1} and 2000cm^3/min^{-1} capacity laminar flow elements. Poor correlation between Hassler

Table 2. *Summary statistics for studied core intervals from the Miller Field*

Well	16/8b–A01	16/8b–A03	16/8b–A06	16/8b–3	16/7b–25
Statistics					
Minimum	9	4	40	0.5	2
Maximum	672	1797	1211	1956	1977
Mean (arithmetic)	223	311	586	355	371
Std deviation	152	325	217	341	348
C_v	0.68	1.05	0.37	0.96	0.94
Mean (geometric)	149	168	539	182	219
Mean (harmonic)	61	64	57	33	62
Median	221	225	591	248	281
N data	128	291	264	505	465
Distribution	Normal	Square root	Normal	Square root	Ln normal
No.	46	111	14	92	88

sleeve and probe permeameter measurements on core slabs is attributable to varying degree of heterogeneity in the reservoir that is reflected in local variations in permeability adjacent to and, sometimes, within the core plugs (Halvorsen & Hurst 1990). During calibration, probe permeameter flow rates were recorded on the core slabs immediately adjacent to core-plug sample points, and compared with conventional Hassler sleeve permeabilities obtained for the cleaned core plugs. Homogeneity of the sandstone at calibration points was evaluated by comparing probe permeameter flow rates gathered for three points on the slabbed core surface immediately adjacent to the core-plug sample point. Because of variations in residual oil saturation in the core slabs, the small-scale heterogeneity of the sandstones, and the different volumes of investigation of the two methods, this technique introduces a somewhat larger degree of 'scatter' into the comparative data than might otherwise occur. However, the correlation coefficients for the regression lines (Fig. 4), give encouraging results that are similar to published data (Robertson & McPhee 1990, Hurst & Rosvoll 1991).

Permeability characterization

Geostatistical analysis

Geostatistical analysis of permeability data using semivariograms has allowed quantification of spatial variation in permeability. Appendix 1 provides a description of the construction and interpretation of semivariograms. Experimental semivariograms constructed for the total core intervals studied are illustrated in Figs. 5–9, which are all scaled to a convenient y-axis (γ^*h) maximum which equals 1.4 times the overall sample variance of the normally distributed dataset.

Not all of the permeability data gathered during this study is normally distributed. For example, the distribution of permeability data departs from normal (i.e. permeability shows square root or log normal distribution), with increasing presence of fine-grained Lithofacies A (Figs 5–9 and Table 2). However, for this study all permeability data have been transformed to approximately normal distributions using natural log or square root transformations, and validated by use of a Lilliefors test (Conover 1971) to test for normality at a 95% confidence level.

Permeability variation within core intervals

Permeability data for each studied core section are summarized in Table 2. Probe-permeability data for well sections studied range from 0.5–1977 mD (Table 2).

The semivariograms are complex, and their interpretation is rarely straightforward. All experimental semivariograms for total studied core intervals show 'moderate' positive-projected y-axis intercepts (Nugget Effect), typically 18–30% of the variogram sill value. High nugget values are interpreted to indicate that sampling a given site more than once would yield different results either because of low measurement precision, or as a result of the presence of permeability variation at a finer scale than the sampling interval (Journel & Huijbregts 1978; Hohn 1988). The moderate nugget effects observed suggest sampling a given site more

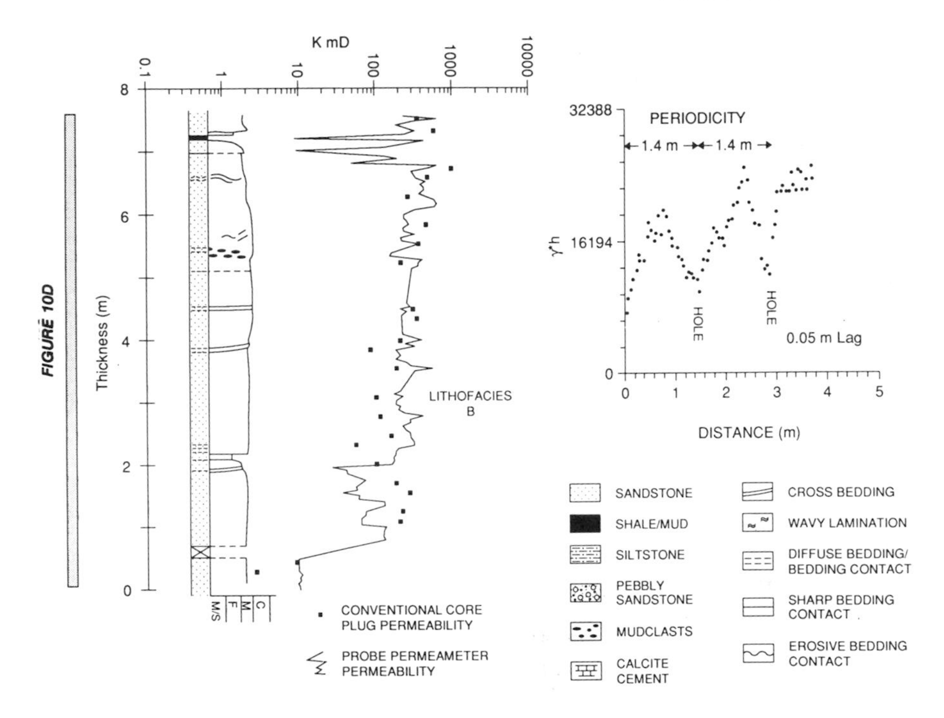

Fig. 5. Core-plug, probe permeameter data, lithological log and experimental semivariogram constructed for probe permeameter data through the total studied core interval from well 16/8b-A01.

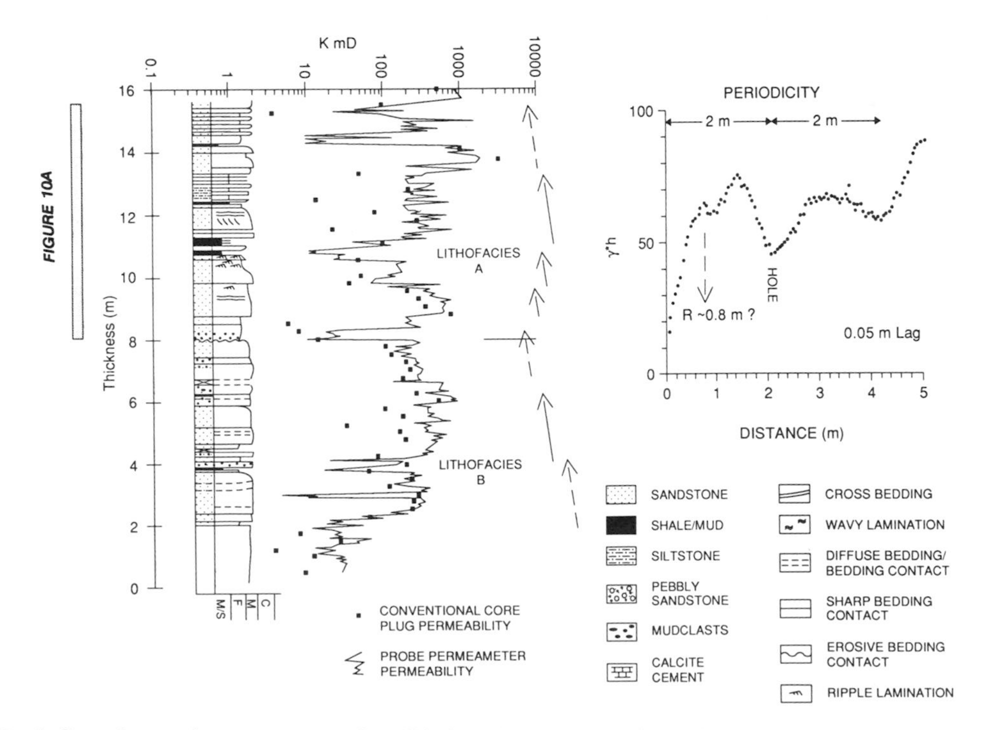

Fig. 6. Core-plug, probe permeameter data, lithological log and experimental semivariogram constructed for probe permeameter data through the total studied core interval from well 16/8b-A03. Note: individual lithofacies analysed are indicated, and referenced to semivariograms in Fig. 10.

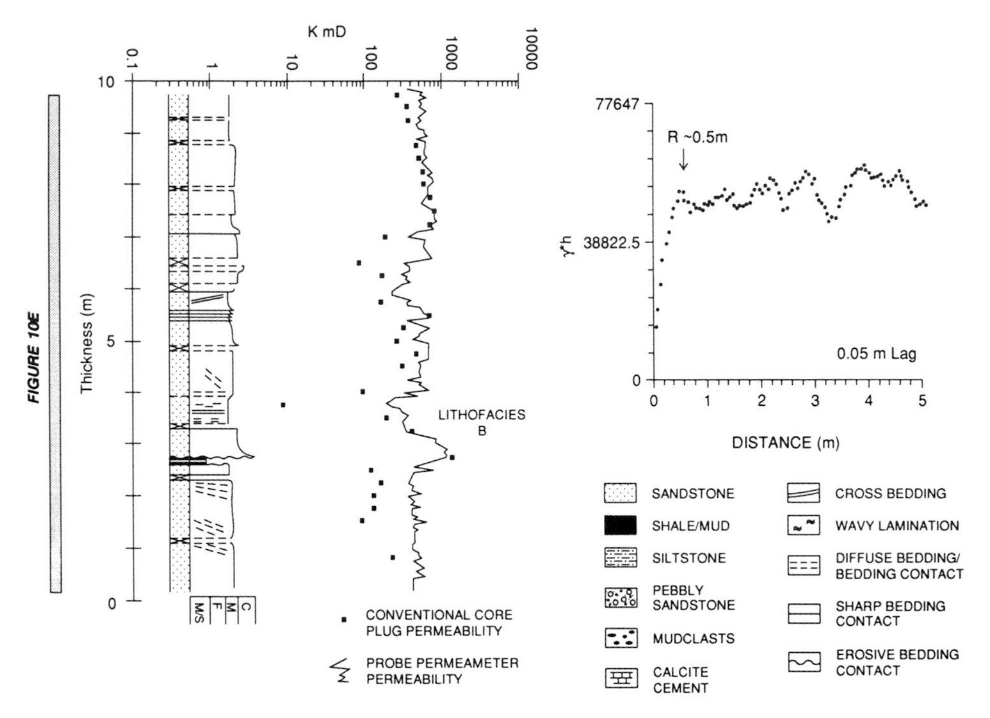

Fig. 7. Core-plug, probe permeameter data, lithological log and experimental semivariogram constructed for probe permeameter data through the total studied core interval from well 16/8b-A06.

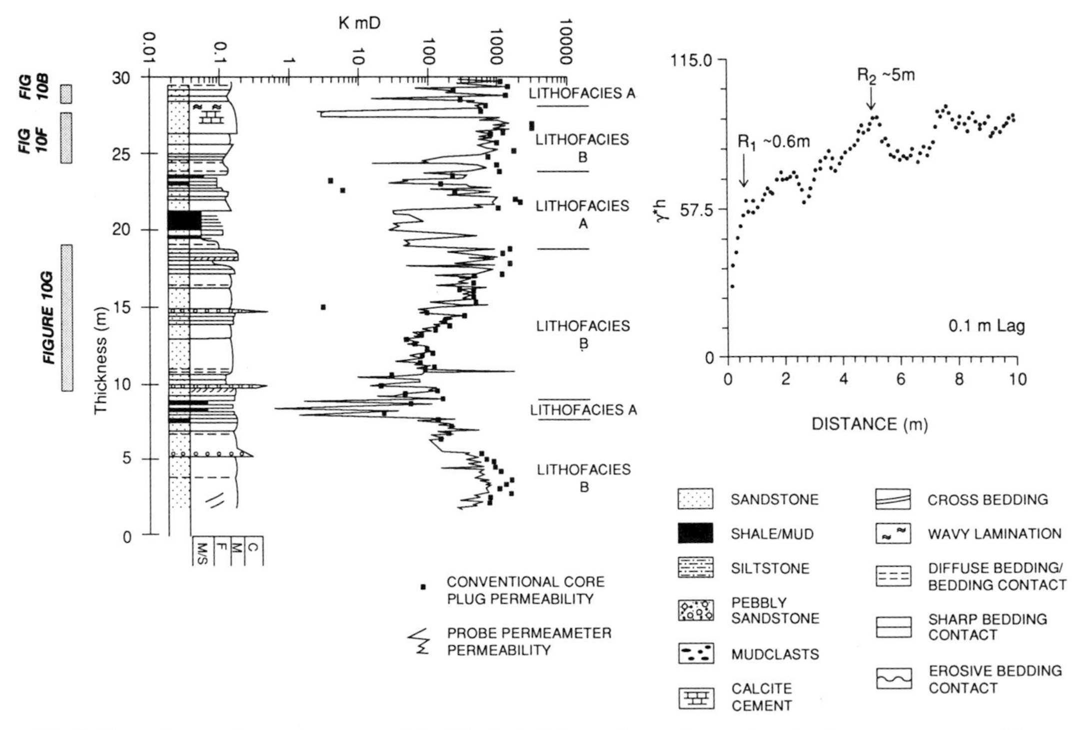

Fig. 8. Core-plug, probe permeameter data, lithological log and experimental semivariogram constructed for probe permeameter data through the total studied core interval from well 16/8b-3. Note: individual lithofacies analysed are indicated, and referenced to semivariograms in Fig. 10.

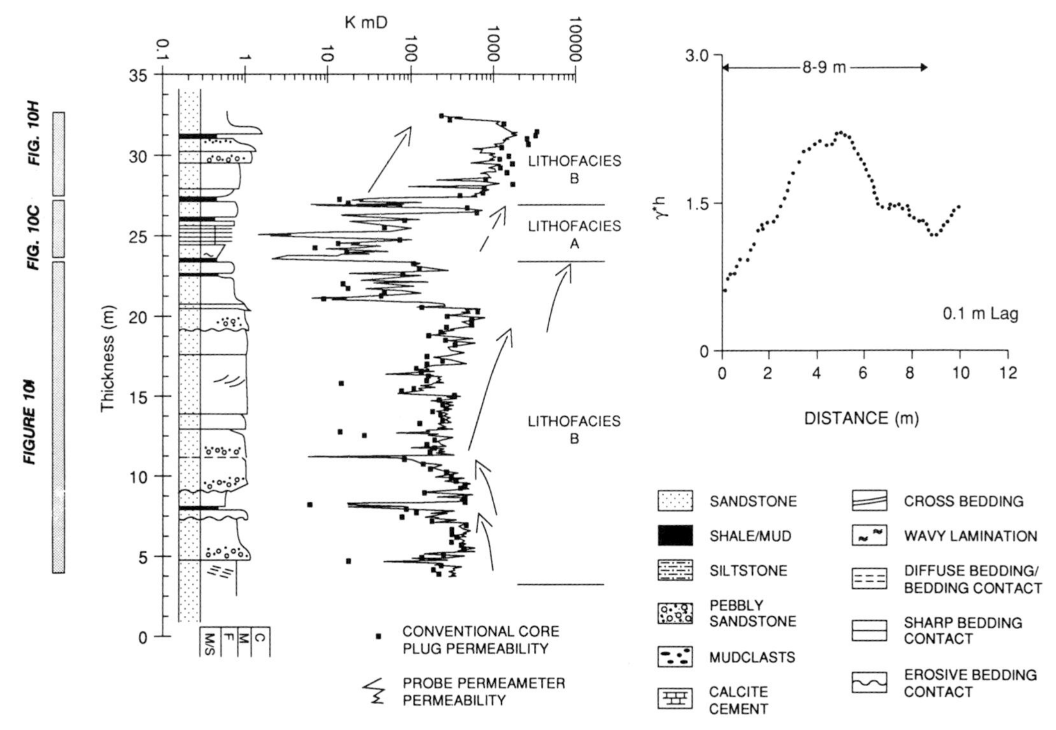

Fig. 9. Core-plug, probe permeameter data, lithological log and experimental semivariogram constructed for probe permeameter data through the total studied core interval from well 16/7b-25. Note: individual lithofacies analysed are indicated, and referenced to semivariograms in Fig. 10.

than once would be likely to yield similar results. The moderate nugget effects could be explained by the presence of centimetre-scale variability (i.e. lamination) within cores, and suggest that finer sample spacings are required to capture this level of heterogeneity.

Experimental semivariograms calculated for some studied intervals, e.g. 16/6b-A01, 16/8b-A03 and 16/7b-25 (Figs 5, 6, and 9), appear to show marked decreases in the calculated semivariogram function at regular multiples of a positive lag spacing. These periodicities (hole effects) result in the semivariograms displaying a marked sinusoidal form about the sill. The controlling mechanism with respect to development of observed periodic spatial variability is interpreted to approximate the scales at which certain sedimentary phenomena are observed in core, occurring at bed thickness, fining-upward unit or lithofacies scale. Periodicities within permeability data also reflect length scales over which a type of permeability 'correlation' exists. Periodicity may thus be a useful guide in selection of input variables for numerical modelling of fluid flow in reservoirs (Corbett & Jensen 1991), and can be utilized as a scaling factor during interpretation of how geological 'structures' are likely to impact upon fluid flow. Corbett & Jensen (1993*a*, *b*) and Corbett *et al.* (1992) have described an 'up-scaling' process (two-phase pseudoization or the derivation of 'Geopseudo's') which incorporates small-scale observations of the type made by the probe permeameter into estimates of reservoir properties (i.e. relative permeability) at scales which are geologically realistic.

The studied core interval from well 16/8b-A01 (Fig. 5) occurs wholly within Lithofacies B. Evaluation of sedimentary data for this core interval indicates that the average spacing (1.1 m) of clay/organic-rich laminae within the analysed core section is similar to the strong periodicity (*c.* 1.4 m) observed within its experimental semivariogram for permeability data. These clay-rich lithologies are typically associated with permeability contrasts in the order of $10–10^2$ mD, and occur at poorly defined bedding surfaces within the massive sandstones which dominate this core interval. They often define the top of broadly fining-upward sediment packages.

Table 3. *Summary results of semivariogram analyses of permeability data from studied core intervals*

Well	Lithofacies	Structure		Interpretation
16/8b–A01				
Total	B	Range 1	–	
(0–7.6 m)		Range 2	–	
		Periodicity	1.4 m	Fining-upward unit thickness
16/8b–A03				
Total				
(0–16 m)	A/B	Range 1	0.8 m	Relict bedding defined by clay/organic laminae
		Range 2	–	
		Periodicity	2 m	Fining-upward unit thickness
(8–16 m)	A	Range 1	0.8 m	Relict bedding defined by clay/organic laminae
		Range 2	–	
		Periodicity	2 m	Fining-upward unit thickness
16/8b–A06				
Total				
(0–19 m)	B	Range 1	0.5 m	Relict bedding defined by clay/organic laminae
		Range 2	–	
		Periodicity	–	
16/8b–3				
Total				
(0–30 m)	A/B	Range 1	0.6 m	Relict bedding defined by clay/organic laminae
		Range 2	5 m	Scale of interdigitation of facies packages
		Periodicity	–	
(24–28 m)	B	Range 1	0.6 m	Relict bedding defined by clay/organic laminae
		Range 2	–	
		Periodicity	1.4 m	Fining-upward packages
(9–19 m)	B	Range 1	0.25 m	Relict bedding?
		Range 2	–	
		Periodicity	–	
(28–30 m)	A	Range 1	–	
		Range 2	–	
		Periodicity	0.5 m	Average bed thickness?
16/7b–25				
Total				
(0–33 m)	A/B	Range 1	–	
		Range 2	–	
		Periodicity	8.0 m	Thickest facies package in studied core section
(23–27 m)	A	Range 1	0.2 m	Average bed thickness
		Range 2	–	
		Periodicity	0.6 m	Average spacing silts/muds defining top of fining upward packages
(27–33 m)	B	Range 1	*c.* 0.5 m	Average bed thickness?
		Range 2	–	
		Periodicity	*c.* 1.2 m	Fining-upward unit thickness
(3–23 m)	B	Range 1	1.2 m	Smaller than apparent bed boundary spacing as defined by clay-rich laminae – relict small-scale bedding?
		Range 2	–	
		Periodicity	–	

The experimental semivariogram for the total studied core interval from well 16/8b-A03 suggests a correlation length of *c.* 0.8 m, with a strong periodicity at 2 m (Fig. 6). This periodicity approximates the average spacing of low permeability horizons, which in this core section are typically associated with permeability contrasts of 10–10^3 mD (arrowed on Fig. 6), and define the tops of fining-upward sediment packages. Within Lithofacies A, the low permeability horizons correspond to thin (decimetre scale) shales and siltstones occurring within very heterolithic (centimetre to decimetre scale interbedded) sandstone–siltstone–shale sediment packages of a metre or so thickness. The heterolithic sediment packages within Lithofacies A, separate thicker fine–medium-grained sandstone beds and amalgamated beds up to a metre or so in thickness. Within Lithofacies B, the low permeability horizons often define individual bed boundaries. Such low permeability boundaries take the form of clay/organic matter concentrations occurring either as matrix material within the sandstones, discrete laminae (as suspension load) at the top of individual beds or fining-upward suspension load sets, or as rip-up clasts/mudstone intraclasts at the base of fining-upward beds (and bed-sets).

The studied core interval from well 16/8b-A06 (Fig. 7) is similar to that of well 16/8b-A01, in that it only samples Lithofacies B. The experimental semivariogram for this interval displays a correlation length of 0.5 m which appears to reflect permeability heterogeneity (10^2 mD) of the individual beds.

A similar 'short' correlation length (0.6 m) was observed for the experimental semivariogram for the studied core interval from well 16/8b-3 (Fig. 8). However, the experimental variogram for this core interval displays a two-fold 'nested' permeability correlation structure, with correlation lengths at both 0.6 and 5.0 m (Fig. 8). As is the case for well 16/8b-A06, the shorter permeability correlation structure appears to reflect the presence of permeability heterogeneities (typically 10–10^2 mD) which occur at spacings approximating 'bed' scale, but the larger correlation structure appears to reflect permeability heterogeneities (10^2–10^3 mD) which occur as a result of the interdigitation of Lithofacies A and and B.

The experimental semivariogram for the studied core interval from well 16/7b-25 (Fig. 9) displays a marked periodicity at *c.* 8 m. This larger scale periodicity is difficult to interpret but again may be related to a larger scale spacing of horizons (mudstones and siltstones, or beds containing increased concentrations of clay/organic matter) associated with permeability contrasts of several orders of magnitude 10^2–10^3 mD.

Note: in wells 16/7b-25 and 16/8b-3, a gradual decrease in permeability down analysed core sections is superimposed upon those permeability trends which can be attributed to variations in depositional fabric. This may correspond to a general increase in quartz cementation (as syntaxial grain overgrowths) with depth within these wells, this feature being a major control on permeability within the Miller Field sandstones. The results of analysis of experimental semivariograms are summarized in Table 3.

Permeability variation within individual lithofacies

Table 4 summarizes probe permeability data for individual lithofacies.

Lithofacies A: interbedded sandstone and shale. Permeability data for probe permeameter traverses through Lithofacies A range from 2 to 1797 mD, with arithmetic average

Table 4. *Summary statistics for individual lithofacies types present within studied core intervals from the Miller Field*

Well depth	Min	Max	Arith. mean	Std dev.	C_v	Geom. mean	Harm. mean	Med.	*N*	Distribution	Lithofacies
16/8b–A01											
0–7.6 m	9	672	223	152	0.68	149	60.5	221	128	Normal	B
16/8b–A03											
8–16 m	9	1797	343	367	1.07	189	74	207	146	Square root	A
2–8 m	4	906	267	172	0.65	192	81	256	111	Square root	B*
16/8b–A06											
0–10 m	40	1211	586	235	0.37	539	57	577	264	Normal	B
16/8b–3											
28–30 m	15	1273	491	38	0.78	321	148	367	33	Ln normal	A
24–28 m	2	1621	636	435	0.68	290	20	626	77	Normal	B
9–19 m	8	1955	222	263	1.18	138	90	110	180	Ln Normal	B
3–7 m	62	780	424	214	0.51	357	282	424	91	Normal	B*
16/7b–25											
23–27 m	2	726	109	195	1.78	34	12	29	56	Ln normal	A
27–33 m	97	1977	965	390	0.40	871	731	938	79	Normal	B
3–23 m	6	809	273	149	0.55	218	128	262	329	Normal	B

* Denotes data traverse not used for geostatistical analysis.

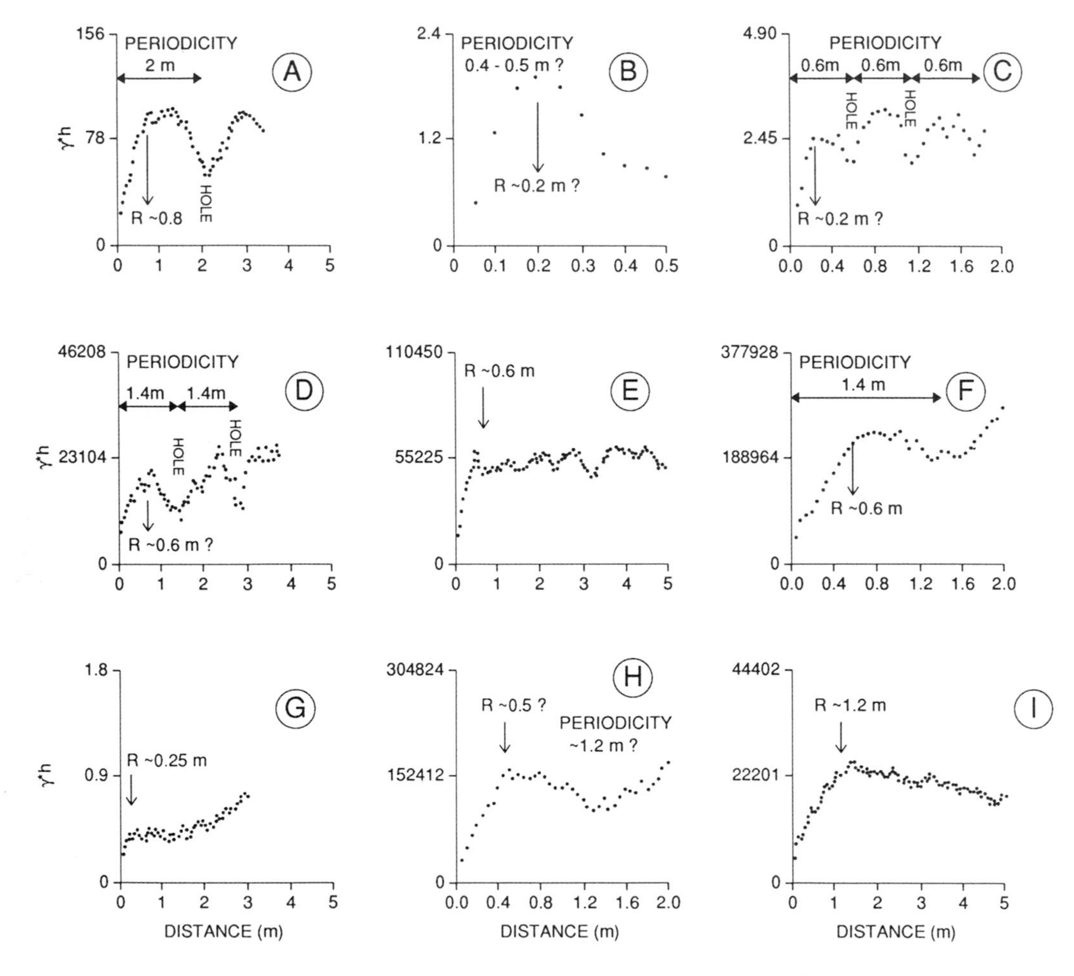

Fig. 10. Experimental semivariograms for probe permeameter traverses through individual lithofacies types present within studied wells. Semivariograms are are all scaled to a *y*-axis (y^*h) maximum which equals twice the overall sample variance of the normally distributed data-set. **(A)** Lithofacies A, well 16/8b-A03, 8–16 m; **(B)** Lithofacies A, well 16/8b-3, 28–30 m; **(C)** Lithofacies A, well 16/7b-25, 23.5–27 m; **(D)** Lithofacies B, well 16/8b-A01, total studied interval; **(E)** Lithofacies B, well 16/8b-A06, total studied interval; **(F)** Lithofacies B, well 16/8b-3, 24–28 m; **(G)** Lithofacies B, well 16/8b-3, 9–19 m; **(H)** Lithofacies B, well 16/7b-25, 27.5–32.5 m; **(I)** Lithofacies B, well 16/7b-25, 4–23.5 m.

permeabilities ranging from 109 to 491 mD. The thin bedded argillaceous sandstones of Lithofacies A have porosities of 8–10%. Experimental semivariograms for permeability traverses through Lithofacies A (Fig. 10a–c) all show strong periodicity (marked hole effects). The semivariogram shown in Fig. 10a, (16/8b-A03, 8–16 m) suggests a correlation length of *c.* 0.8 m, with a strong periodicity at 2 m. This 0.8 m correlation length is approximately equivalent to the average thickness of the sandstone beds within this heterolithic succession. The 2 m periodicity appears to approximate to the larger scale sedimentary 'cycles' within the heterolithic sequence. These cycles are defined by an overall fining-upward grain size trend, and general upward decrease in bedform scale (see Fig. 11 for examples). Correlation lengths of 0.2–0.25 m, (Fig. 10b & c) are interpreted to approximate average bed thickness within heterolithic successions. The semivariograms in Fig. 10b & c indicate a small scale periodicity (0.5–0.6 m), corresponding to the average spacing of low permeability fine-grained sandstone, siltstone or shale interbeds (or packages of beds) which define the upper portions of broad fining-upward 'cycles'.

The coefficient of variation (C_v) or normalized standard deviation (Appendix 2) has been used as a convenient statistical estimate of

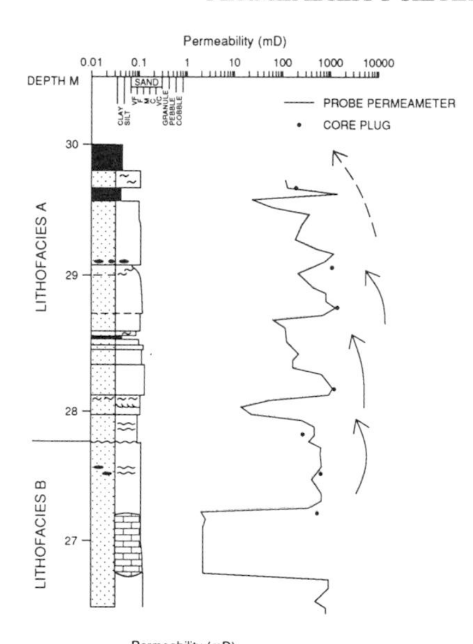

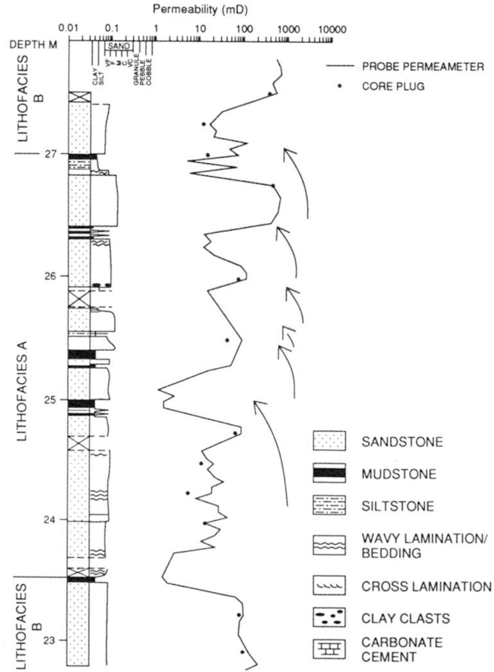

Fig. 11. Bedform and fining-upward 'cycle' scales within Lithofacies A. Arrows define broadly fining-upward grain size trends and corresponding decreasing upward permeability trends. Average bed thickness is *c.* 0.2 -0.25 m. Average scale of fining-upward 'cycles' is 0.5–0.6 m which approximates the periodicity observed in semivariograms for permeability traverses through these sections. (**A**) Well 16/8b-3, 28–30 m; (**B**) Well 16/7b-25, 23.5–27 m

Table 5. Optimum sample number (N_0) and spacing (D_0) calculated for permeability traverses through Lithofacies A and B.

	Well	Depth (m)	N_0	D_0 (m)
Lithofacies A	16/8b–A03	8–16	115	0.06
	16/8b–3	28–30	61	0.03
	16/7b–25	23–27	317	0.01
Lithofacies B	16/8b–A01	0–7.6	46	0.14
	16/8b–A01	2–8	42	0.13
	16/8b–A06	0–10	14	0.90
	16/8b–3	24–28	46	0.08
	16/8b–3	9–19	139	0.08
	16/8b–3	3–7	26	0.14
	16/7b–25	27–33	16	0.18
	16/7b–25	3–23	30	0.99

variability or 'heterogeneity' within permeability data (Goggin *et al.* 1988, 1992; Corbett & Jensen 1992*a,b*, 1993*a,b*). Corbett & Jensen (1992*b*) suggest a simple classification scheme where permeability data are termed homogeneous (C_v < 0.5), heterogeneous (C_v 0.5–1.0) and very heterogeneous (C_v > 1.0). C_v determined for permeability traverses through Lithofacies A range from 0.78 to 1.78 (i.e. heterogeneous to very heterogeneous), which reflect the heterolithic nature of the lithofacies (Table 4, Fig. 12).

Lithofacies B: massive sandstone. Permeability data for traverses through Lithofacies B range from 4 to 1977 mD, and arithmetic average permeabilities for these sections range from 222 to 965 mD. These massive, typically medium-grained sandstones have porosities of 13–17%. Generally, permeability is lowest within fine-grained argillaceous sandstones (typical of

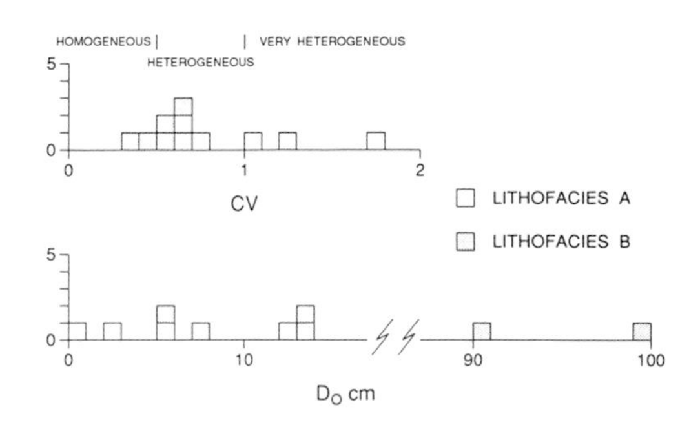

Fig. 12. Histograms of C_v and D_0 calculated for permeability traverses througth individual lithofacies types.

Lithofacies A), and highest within cleaner medium- to coarse-grained sandstones (more typical of Lithofacies B).

Experimental semivariograms for permeability traverses through Lithofacies B are more variable (Fig. 10 d–i). Some (Fig. 10d, f & h) clearly display a strong periodicity between 1.2 and 1.4 m. These periodicities are similar to the average spacings (1.1–1.25 m) of low permeability horizons that often occur at poorly defined bedding surfaces at the top of grossly fining-upward units. The low permeabilities are caused by decreases in grain size, and increase in detrital clay and organic matter content at the top of each fining-upward unit. Interpretation of a correlation length or range within these semivariograms is highly subjective due to their strong periodicity; a range of 0.5–0.6 m is suggested. These correlation lengths often appear smaller than typical bed scale and are very difficult to relate to any sedimentary phenomena observed in the cores. However, this may in part arise because bedding is very difficult to discern in some 'massive' sandstones within Lithofacies B. The sandstones probably represent the amalgamation of several thinner, rapidly deposited beds. Analysis of the spatial separation of subtle grain size contrasts and possible bed boundaries within some sections [e.g. 16/8b-A01 (total studied interval) and 16/8b-3 (24–28 m)] suggests average spacings in the order of 0.5 m (Figs. 5 & 8). This is similar to the ranges interpreted from semivariograms.

C_v determined for permeability traverses through the massive sandstones of Lithofacies B (Table 3), range from 0.37 to 0.68 (i.e. homogeneous to heterogeneous). One permeability traverse through Lithofacies B did, however, yield a C_v of 1.18 (well 16/18b-3; 9–19 m), indicating this particular data set to be very heterogeneous with respect to permeability.

Experimental semivariograms for well 16/8b-A06 (total studied interval, Fig. 10e), and well 16/8b-3 (9.5–19 m, Fig. 10g) have ranges of 0.6 and 0.25 m, respectively. The range for the section from well 16/8b-3 (Fig. 10g) appears to approximate the average spacings (*c.* 0.8 m) of low permeability horizons that occur at bed boundaries within this analysed core interval (Fig. 7). The 0.25 m correlation length for the section from well 16/8b-3 cannot be related to any obvious sedimentary features within core. Furthermore, it is substantially smaller than the scale of bedding features identified within the analysed core interval. The 0.25 m range for this core may reflect permeability variations related to the distribution of primary or diagenetic phenomena (e.g. detrital clay, organic matter or cement content), observed within the massive, amalgamated sandstones.

Figure 10i (well 16/7b-25; 4.0–23.5 m) displays an apparent range of *c.* 1.2 m, but also shows a decrease in the calculated variogram function with increased lag distances, which may be indicative of a larger scale periodicity. The 1.2 m apparent correlation length is very similar to the average spacing (*c.* 1.1 m) of observed bed boundaries. The bed boundaries define the limits of often quite subtle individual fining-upward units.

Evaluation of sampling density

Core plugs used for conventional Hassler sleeve analysis are typically sampled at 25–30 cm sample spacings, which are used for calibration of wireline log data. The probe permeameter enables low cost, non-destructive, rapid analysis of permeability at a much higher sampling density (Halvorsen & Hurst 1990). Thus, it can provide an important supplement to conventional core analysis data. In particular, it allows identification and quantification of laminar-scale permeability heterogeneities (Halvorsen 1993). Jensen (1990) has shown that a better correlation is obtained between the density log (used for porosity/permeability evaluation) and the probe permeameter data than from core plug data in heterogeneous formations, where geological heterogeneity overwhelms measurement precision.

Once the permeability contrast in a reservoir has been identified and related to lithofacies, it is important to quantify the permeability heterogeneity within each lithofacies. If the facies have distinct permeability characteristics they may be useful for defining flow units (*sensu* Ebanks *et al.* 1993) within the reservoir, which in turn, may be incorporated into reservoir simulation models. Hurst (1993) has illustrated the benefit of high density permeability data sampling during evaluation of flow units within the Etive and Rannoch Formations of the Brent Group (UK North Sea). In this study, Hurst (1993) showed that poor characterization of permeability distribution was one of the prime sources of error in flow unit definition, and development of accurate simulation models.

Hurst & Rosvoll (1991) applied the N_0 test (See Appendix 2) to evaluate the optimum number of samples and sample spacing (D_0) that is required to estimate 'mean' permeability. The concepts of N_0 and D_0 testing have been taken further by Corbett & Jensen (1992*a*) and adapted for profiles rather than gridded data.

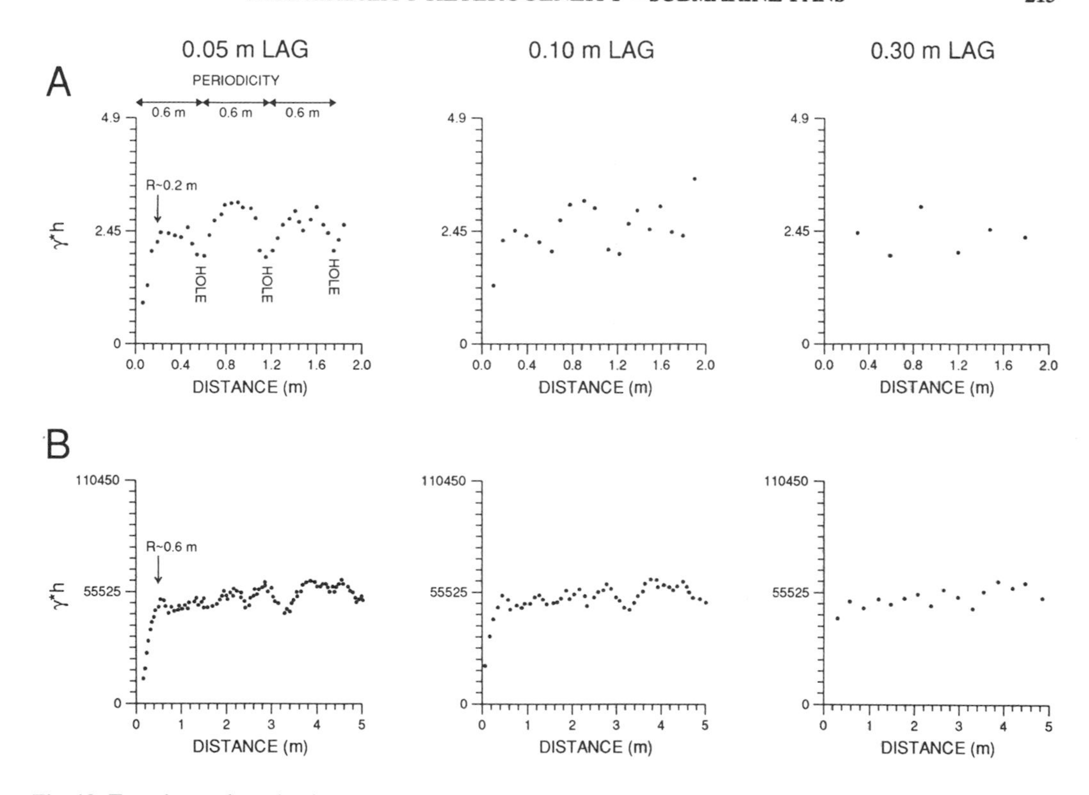

Fig. 13. Experimental semivariograms constructed at 0.05, 0.1 m and 0.3 m sample lag spacings for selected permeability traverses through: **(A)** Lithofacies A, well 16/7b-25, 23.5–27 m; **(B)** Lithofacies B, well 16/8b-A06, total studied interval.

The results of N_0 and D_0 testing of probe permeameter traverses through the heterolithic Lithofacies A (C_v 0.78–1.78) indicate optimum sample spacings of *c.* 0.01–0.06 m (Table 4; Fig. 12). N_0 and D_0 testing of probe permeameter traverses through the massive sandstones of Lithofacies B (C_v 0.37–1.18) indicate optimum sample spacings ranging from 0.06 to 0.99 m (Table 4; Fig. 12). The data (Fig. 12) illustrate that sampling permeability at conventional 25–30 cm spacings, will typically be inadequate for characterization of permeability heterogeneity within the submarine fan sandstones of the Miller Field, regardless of lithofacies. This reinforces the impression given by semivariogram analysis that within individual lithofacies permeability correlates over distances of 0.2–2 m (Fig. 10), and that sampling of permeability at the 25 cm scale, used for conventional core plugs, is unlikely to be adequate for characterization of the reservoir heterogeneity. Figure 13 provides an illustration of the behaviour of experimental semivariograms calculated at 0.05, 0.1 and 0.3 m sampling intervals through example sections of Lithofacies A and B. The experimental semivariograms illustrate that sampling at 0.3 m intervals provides no information as to the spatial nature of permeability heterogeneity within these example traverses. Note: a similar study carried out for shallow marine sandstones (Corbett & Jensen 1992a) provided comparable results. Corbett & Jensen (1992a, b) recommend sampling of permeability at *c.* 1/10 the expected range for modelling of semivariograms. On the basis of this we would recommend a sampling spacing of *c.* 0.05 m to adequately describe the permeability heterogeneity in the high net:gross submarine fan sandstones of the Miller Field.

High density sampling of permeability using the probe permeameter, while providing a wealth of information, is not without its drawbacks. The increased amount of fine-scale permeability data provided by the probe permeameter while filling ‘scale gaps’ in our observations/measurements, is potentially at its most useful where the data can be ‘upscaled’. During upscaling, the data are transformed from closely spaced analyses into a format suitable for simulation studies using ‘grid block’ sizes several orders of magnitude larger. This is not a straightforward process, but might be achieved using two-phase pseudoization or 'Geopseudo's' (Corbett & Jensen 1993b, Corbett *et al.* 1992). This technique utilizes a knowledge of the variety of geological 'structures' present at different scales within a succession and how they are likely to impact fluid flow, in order to produce estimates of relative permeabilities at scales which are geologically realistic.

The probe permeameter data thus provide important measures of the representative geological structures which could form the basis of geologically realistic two-phase flow upscaling (Geopseudo's), i.e. geostatistical analyses of high density probe permeameter data reveal the presence of permeability correlation structures which correspond to the scales at which 'real' sedimentological phenomena (bedding, fining-upward, bed-sets, etc.) are observed in core. These features define the basic building blocks for realistic 'scale up' procedures.

Conclusions

1. Permeability within the submarine fan sandstones of the Miller Field is largely controlled by depositional fabric (i.e. grain size, grain sorting, matrix clay content, bed thickness). The highest permeabilities (up to 1.977 D) occur within clean, upper medium-grained sandstones of Lithofacies B, the lowest permeability (2 mD) occurs within the shaly, heterolithic lithofacies A.
2. Predictably, evaluation of heterogeneity using the coefficients of variation indicate that Lithofacies A (C_v up to 1.78) is more heterogeneous than Lithofacies B. It is significant to note however that massive sandstones in Lithofacies B are also frequently heterogeneous with respect to permeability, (C_v up to 1.18).
3. Semivariogram analysis identifies ranges and periodicities that correspond to observed geological heterogeneity. Ranges are typically similar to the average thickness of bedding features, for example, fining-upward sequences. Periodicity reflects the repetition of units with similar permeability structure and thickness. The analysis of semivariograms has also resulted in identification of bedding or relict bedding within reservoir zones where apparently massive sandstone dominates.
4. Conventional core plug sampling inadequately describes the permeability heterogeneity of all the lithofacies examined. Results of N_0 testing indicate that the optimum sampling density for massive sandstones of Lithofacies B is between 0.06 and 0.99 m to estimate arithmetic mean within ± 20% of that observed. Correlation lengths and periodicities from semivariograms constructed at 0.05 m (ranging from 0.25 to 1.4 m), show that, in most cases, sampling at spacings adequate to estimate sample mean to within ± 20% will be inadequate for geostatistical quantification of the spatial nature of permeability heterogeneity. The spatial heterogeneity shown by bedding features requires a sample spacing determined by, and significantly less than, the average bed thickness. It is probable that the facies studied contain spatial variability at the cm-scale (i.e. caused by lamination) that has not been investigated in this study.
5. Analyses suggest that correlation lengths and periodicities observed within experimental semivariograms for both Lithofacies A and B can be related to the sedimentary structure. These structures arise both as a result of gross variations in sediment grain size, e.g. as a result of a fining-upward profile within an *individual bed* or the presence of thin argillaceous laminae/shale breaks at bed boundaries, and also as a result of variations in the scale or nature of stacking of sedimentary strata within broadly fining-upward *packages* of sediment. That is, in some cases not only do individual beds fine upward, but larger scale fining-upward units may be defined by both an overall fining-upward grain size trend, and upward decrease in bedform thickness. Note: semivariograms cannot distinguish between fining-upward or coarsening-upward grain size profiles.
6. The massive sandstone successions which comprise Lithofacies B are not homogeneous with respect to permeability. They commonly display significant permeability variation (hundreds of milliDarcies), and experimental semivariograms for permeability traverses through Lithofacies B indicate that permeability values will be reasonably correlated over thicknesses of only 0.25–1.4 m. These data lead us to suggest that the massive submarine fan sandstones of the Miller Field should not be assumed to be homogeneous flow units. Accurate evaluation of recovery of hydrocarbons from the sandstones should take account of the observed permeability heterogeneity.
7. The often marked differences in scale of permeability correlation structures within different studied wells suggest that lateral correlation of permeability may also be poor. In

the specific case of the Miller Field, these differences provide some conformation that high permeability 'conduits' are unlikely to be laterally continuous over large distances. This conclusion has been utilized in development planning to minimize the risk of early water breakthrough in production wells, i.e. the observed heterogeneity can be a help during production rather than a hindrance.

Acknowledgements

This paper is published with the kind permission of the Miller Field Partners: BP Exploration Operating Company Ltd, Conoco UK Ltd, Enterprise Oil PLC, Santa Fe Exploration (UK) Ltd. The data utilized in this paper were gathered by Moira McKeever, as part of her MSc Dissertation at Aberdeen University. The authors would especially like to thank Patrick Corbett, Chris Garland and Chris Dodds for their thorough reviews of the original manuscript. Figures were drafted by Barry Fulton.

References

CONOVER, W. J. 1971. *Practical Nonparametric Statistics.* Wiley, New York.

CORBETT, P. W. M. & JENSEN, J. L. 1991. An application of small scale permeability measurements - prediction of flow in Rannoch lithofacies, Lower Brent Group, North Sea. Conference Paper: *Minipermeametry in Reservoir Studies,* Petroleum Science and Technology Institute, Edinburgh, 27th June 1991.

—— 1992*a*. Variation of reservoir statistics according to sample spacing and measurement type for some intervals in the Lower Brent Group. *The Log Analyst,* **33,** 22–41.

—— 1992*b*. Estimating the mean permeability: how many measurements do you need. *First Break,* **10,** 89–94.

—— 1993*a*. Quantification of heterogeneity, a role for the minipermeameter in reservoir characterisation. *In:* NORTH, C. P. & PROSSER, D. J. (eds) *Characterization of Fluvial and Aeolian Reservoirs.* Geological Society, London, Special Publication, **73,** 433–442.

—— 1993*b*. Application of probe permeametry to the prediction of two-phase flow performance in laminated sandstones (lower Brent Group, North Sea). *Marine and Petroleum Geology,* **10,** 335–346.

—— RINGROSE, P. S., JENSEN, J. L. & SORBIE, K. S. 1992. Laminated clastic reservoirs – the interplay of capillary pressure and sedimentary architecture. SPE Technical Paper 24699. *SPE Annual Technical Conference, SPE, 4–7 Oct. Washington DC* 365–376.

EBANKS, W. J. JR, SCHEIHING, M. H. & ATKINSON, C. D. 1993. Flow units for reservoir characterization. *In:* MORTON-THOMPSON, D. & WOODS, A. M. (eds). *Development Geology Reference Manual.* American Association of Petroleum Geologists, Methods in Exploration, **10,** 282–285.

GARLAND, C. R. 1993. Miller Field: Reservoir stratigraphy and its impact on development. *In:* PARKER, J. R. (ed.) *Petroleum Geology of Northwest Europe: Proceedings of the 4th Conference.* Geological Society, London, 401–414.

GOGGIN, D. J., CHANDLER, M. A., KOCUREK, G. & LAKE, L. W. 1988. Patterns of permeability in aeolian deposits: Page Sandstone (Jurassic), North-eastern Arizona. *SPE Formation Evaluation,* **June,** 297–306.

——, ——, —— & —— 1992. Permeability transects in aeolian sands and their use in generating random permeability fields. *SPE Formation Evaluation,* **March,** 7–16.

HALVORSEN, C. 1993. Probe permeametry applied to a highly laminated sandstone reservoir. *Marine and Petroleum Geology,* **10,** 347–351.

—— & HURST, A. 1990. Principles, practice and applications of laboratory mini-permeametry. *In:* WORTHINGTON, P. F. (ed.) *Advances in Core Evaluation: Accuracy and Precision in Reserves Estimation, EUROCAS I.* Gordon and Breach Science, London, 521–549.

HOHN, M. E. 1988. *Geostatistics and Petroleum Geology.* Van Nostrand Reinhold, New York.

HURST, A. 1993. Sedimentary flow units in hydrocarbon reservoirs: some shortcomings and a case for high resolution permeability data. *Special Publication of the International Association of Sedimentologists,* **15,** 191–204.

—— & ROSVOLL, K. 1991. Permeability variations in sandstones and their relationship to sedimentary structures. *In:* LAKE, L. W., CARROLL JNR., H. B. & WESSON, T. C. (eds) *Reservoir Characterisation II:* Academic Press, San Diego, California, 166–196.

JOURNEL, A. G. & HUIJBREGTS, CH.J. 1978. *Mining Geostatistics.* Academic Press, London.

JENSEN, J. L. 1990. *A model for small-scale permeability measurement with applications to reservoir description.* SPE/DOE 7th Symposium on EOR, Proceedings SPE 20265, 891–900.

MCCLURE, N. M. & BROWN, A. A. 1991. Miller Field. A subtle Upper Jurassic submarine fan trap in the Southern Viking Graben, United Kingdom Sector, North Sea. *In: Giant Oil and Gas Fields Of The Decade 1978–1988.* American Association of Petroleum Geologists, Memoir, **54,** 307–322.

MCKEEVER, M. 1992. A minipermeameter study of reservoir intervals of a Northern North Sea Field. MSc thesis, University of Aberdeen, UK.

PROSSER, D. J. & MASKALL, R. 1993. Small scale permeability variation within aeolian sandstone: a case study using core cut sub-parallel to slipface bedding, The Auk Field Central North Sea, U.K. *In:* NORTH, C.P. & PROSSER, D.J. (eds) *Characterisation of Aeolian and Fluvial Reservoirs.* Geological Society, London, Special Publication, **73,** 385–405.

ROBERTS, M. J. 1991. The South Brae Field, Block 16/7a, U.K. North Sea. *In:* ABBOTTS, I. L. (ed.) *United Kingdom Oil and Gas Fields 25 Years Commemorative Volume.* Geological Society, London, Memoir, **14,** 55–62.
ROBERTSON, G. M. & MCPHEE, C. A. 1990. High resolution probe permeability: an aid to reservoir description. *In:* WORTHINGTON, P. F. (ed.) *Advances in Core Evaluation Accuracy and Prediction.* Gordon and Breach Publishers, London, 495–520.
ROOKSBY, S. K. 1991. The Miller Field, Blocks 16/7b, 16/8b, UK North Sea. *In:* ABBOTTS, I. L. (ed.) *United Kingdom Oil and Gas Fields 25 Years Commemorative Volume.* Geological Society, London, Memoir, **14,** 159–164.
TURNER, C. C., COHEN, J. M., CONNEL, E. R. & COOPER, D. M. 1987. A depositional model for the South Brae Oilfield. *In:* BROOKS, J. & GLENNIE, K. W. (eds) *Petroleum Geology of Northwest Europe.* Graham and Trotman, London, 853–864.

Appendix 1

The construction of experimental semivariograms

The semivariogram function $\gamma^*(h)$ at a sample spacing or **lag distance** of h is estimated by:

$$\gamma^*(h) = \Sigma[f(x + h) - f(x)]^2 / 2N(h) \quad \text{(A1)}$$

Where $f(x)$ is the value of the sample at a point x in the core, $f(x + h)$ is the value of the sample at a distance h away from x, $N(h)$ is the number of pairs of samples located at distance h from each other.

The **experimental semivariogram** is constructed by (plotting the quantity $\gamma^*(h)$ (along the y-axis), v. the spatial separation of sample pairs (h) (along the x-axis). Experimental semivariograms constructed for the total core intervals studied (Figs. 5–9), which are scaled to a convenient y-axis (γ^*h) maximum which equals 1.4 times the overall sample variance of the normally distributed dataset. The distance at which spatial correlation of the semivariogram value ceases and the experimental semivariogram levels off to a sill is termed the **correlation length** or **range** (Fig. A1), which is an estimate of the distance over which there will be reasonable correlation between sample permeability values. Positive-projected y-axis intercepts for experimental variograms are termed **Nugget Effects,** which, if high, are interpreted to indicate that sampling a given site more than once

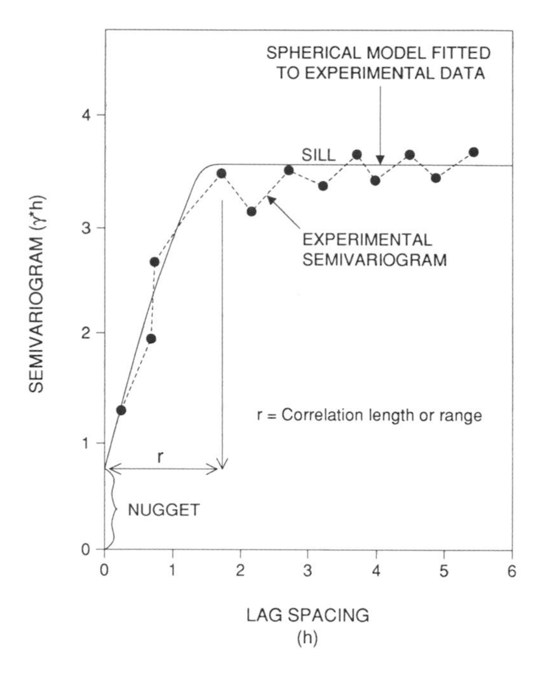

Fig. A1. Elements of the experimental semivariogram. Adapted from Prosser & Maskall (1993).

would yield different results due to low measurement precision, or permeability variation at finer scales than the sampling interval. More detail concerning the calculation and modelling of semivariograms is given by Journel & Huijbregts (1978) and Hohn (1988).

Appendix 2

The N_0 test (Hurst & Rosvoll 1991)

The N_0 test (Hurst & Rosvoll 1991) may be used to evaluate the optimum number of samples and sample spacing (D_0) that is required to estimate 'mean' permeability. The concepts of N_0 and D_0 testing have been further modified by Corbett & Jensen (1992a) and adapted for profiles rather than gridded data. N_0 may be defined as:

$$N_0 = [(t * C_v * 100) / p)]^2 \quad \text{(A2)}$$

Where C_v is the coefficient of variation, t is the critical value of $t_{0.025}$, for $N_s - 1$ degrees of freedom, p is the tolerance level (in this case $\pm$ 20%) or variation allowed around average permeability as defined by the total probe permeameter dataset.

The coefficient of variation (C_v) is defined simply as:

$$C_v = \frac{\text{Sample standard deviation}}{\text{Sample arithmetic mean}} \tag{A3}$$

In order to give an estimate of arithmetic average in typical flow units to within ± 20% of that defined by the total data set, N_o approximates to:

$$N_0 = (10^*C_V)^2 \tag{A4}$$

D_0, the optimum sample spacing for a traverse is then given by:

$$D_0 = (N_s / N_0) * D_s \tag{A5}$$

Where N_s is the original sample number and D_s is the original sample spacing

The N_0 and D_0 tests, as described by Hurst & Rosvoll (1991), assume normal data distributions. Corbett & Jensen (1992*a, b*) however, suggest that in terms of estimates of mean permeability the techniques are both robust and useful, even with non-normally distributed data.

A classification scheme for shale clasts in deep water sandstones

MELISSA JOHANSSON & DORRIK A. V. STOW

Department of Geology, University of Southampton, Highfield, Southampton, SO17 1BJ, UK

Abstract: Shale clasts are widespread in turbidites and associated facies of all ages. They may yield important information regarding the depositional process of the host sandstone and its specific environmental setting. Previous work has tended to generalize the variety of shale clasts that exist and no comprehensive classification has yet been developed. The scheme proposed here is a descriptive classification based on the nature and arrangement of shale clasts within mainly structureless sandstones. Twelve types of clasts are recognized on the basis of the clast's position in the bed and/or characteristic features, and these have been grouped according to whether they have been derived from either basal erosion or disturbance (Group A), flow modification (Group B) or post-depositional disturbance (Group C). These clasts have also been organized into characteristic assemblages which are thought to be indicative of particular depositional environment. The classification scheme when used in association with facies analysis can provide an additional tool for core investigations, contributing to the understanding of permeability barriers and relevant architecture.

Shale clasts are well known from deep-water resedimented successions and have often been described as occurring within thick sandstone turbidites since they were first recognized as such over 40 years ago (e.g. Kuenen & Migliorini 1950; Mutti & Ricci-Lucchi 1972; Walker 1978). During a systematic study of deep-water massive (structureless) sandstones, we have paid particular attention to the presence and nature of shale clasts in a number of field examples around the world. It has become increasingly apparent that previous terminologies within the literature such as 'rip-up clast' (Mutti & Nilsen 1981), and 'floating outsized clasts' (Postma *et al.* 1988) are often used incorrectly and have tended to generalize the wide variety that exists. In fact, shale clasts are highly diverse in their configuration, mode of emplacement and evolution, and therefore warrant a comprehensive classification scheme.

For the purpose of this study, shale clasts are taken as fine-grained (silt–clay grade), coherent sedimentary particles or clasts that occur within a sandstone bed. Compositionally, the clasts may be claystone–mudstone, chalk–micrite or volcaniclastic in nature. The sandstone host beds considered were all deposited in deep-water settings by resedimentation processes. We recognize that similar clasts may also occur in alluvial and pyroclastic deposits and in certain other settings.

The scheme proposed here is a descriptive classification founded on the nature and within bed arrangement of shale clasts. The 12 different clast types can be placed into three categories based on their inferred mode of origin. These include those derived from basal erosion (Group A), erosion and transport (Group B), and post-depositional processes (Group C). Clast types are described by their characteristic features, including average and maximum size, shape, sphericity, position in bed, orientation, lateral extent and density (see below). Their likely method of emplacement is discussed with reference to relevant literature.

It is suggested that correct recognition and classification of shale clasts is important for: (1) better understanding the depositional process of the parent bed; (2) identifying more precisely the depositional setting and proximity to source area; and (3) recognising their potential for influencing wireline log signatures and reservoir permeability.

Classification scheme

The full classification scheme is summarized in Table 1 and described in detail in the following section. The descriptive characteristics used have been developed using the parameters set out below, and are based on the full range of examples studied (Tables 1 and 2).

(1) **Clast size** Where possible an average clast size has been determined for datasets of over 100 individual clasts, according to the Wentworth (1922) grain-size scheme.

From Hartley, A. J. & Prosser, D. J. (eds), 1995, *Characterization of Deep Marine Clastic Systems*, Geological Society Special Publication No. 94, pp. 221–241.

Table 1. *A summary of the characteristic features of the different types of shale clasts*

Type	Average size	Sorting	Shape	Sphericity	Orientation	Position in bed	Lateral extents (m)	Density	Process
A1 Shale clast breccia	Cobble	Very poorly sorted	Very angular	Low sphericity	Random	Whole or base of bed	1–10	High density	Slump debris flow
A2 Scour-lag clasts	Large pebble	Poorly sorted to fairly well sorted	Sub-rounded	Medium sphericity	Transverse to flow	Base of bed	10–20	High density	Traction at base of flow
A3 Rafted clasts	Boulder	Poorly to fairly well sorted	Very angular	Very low sphericity: tabular	Random to sub-parallel	Base of bed	1–10	Low density	Rip-up process
A4 Amalgamation clasts	Small cobble	Fairly poorly sorted	Sub-angular	Low sphericity	Sub-parallel to bedding	Base of bed	10–20	Moderate density	Amalgamation surface
A5 Flame clasts	Small pebbles	Fairly poorly sorted	Sub-angular	Low sphericity	Sub-parallel to bedding	Base of bed	1–10	Low density	Loading and injection
B1 Shale clast conglomerate	Small cobble	Fairly poorly sorted	Sub-rounded	High sphericity	Random	Whole or base of bed	> 20	High density	Cohesive sandy debris flow
B2 Isolated clasts	Cobble	Poorly sorted	Sub-angular to sub-rounded	Low to high sphericity	Random	Anywhere in bed	None	Very low density	Buoyancy
B3 Clustered clasts	Small cobble	Fairly well sorted	Sub-rounded to rounded	High sphericity	Sub-parallel to bedding	Upper mid-bed	> 20	Moderate density	'Floating' on inertia layer
B4 Dispersed clasts	Granule	Well sorted	Sub-rounded	High sphericity	Random/sub-parallel to bedding	Dispersed throughout bed	> 20	Low density	High concentration turbidity current
B5 Lamination clasts	Granule	Well sorted	Rounded	High sphericity	Aligned parallel to lamination/cross-lamination	Base of bed	10–20	Moderate density	Traction at base of flow
C1 Rip-down clasts	Large pebble	Poorly sorted	Very angular	Low sphericity	Sub-paralled to bedding	Top of bed	1–10	Moderate density	Rip-down process
C2 Injection clasts	Granule–cobble	Poorly sorted	Angular	Low sphericity	Random or sub-parallel to the margins of injection	Concentrated near sandstone–shale contact	Localized	High to low density	Ripping process due to sand injection and post-depositional disturbance

(2) **Sorting** This refers to the variation in clast size distribution within any one bed and is described using estimated standard deviations of clast size population (in phi units) as given by Folk & Ward (1957).
(3) **Shape** Shape refers to the degree of roundness or angularity of clasts estimated according to the scheme described by Pettijohn *et al.* (1973).
(4) **Sphericity** Sphericity describes how closely clasts approximate to a sphere, and is divided into low or high sphericity categories (Pettijohn *et al.* 1973). Further qualifying terms such as tabular, elongate, obloid, platy are also given where appropriate.
(5) **Orientation** The orientation of the clasts describes their position (sub-parallel, parallel or random) with relation to bedding features.
(6) **Position in bed** This describes the location of the clasts with reference to the upper and lower boundaries of the bed (e.g. base, middle and top).
(7) **Lateral extent** The lateral extent of the clast zone (where known) has been described as: very extensive, > 20 m; moderately extensive, 10–20 m; little lateral extent, 1–10 m. Isolated clasts or small isolated groups of clasts are described as localized or as having no lateral extent.
(8) **Density** The density or clast concentration within a particular shale clast zone is an estimate of the proportion of core or outcrop surface which comprises shale clasts. The clast area calculated is simply the mean thickness of the clast zone multiplied by a zone length of 100 times the mean long-axis dimension of the clasts. The divisions are as follows: high density, > 60%; moderate density, 30–60%; low density, < 30%.
(9) **Process** The emplacement process for different clast types (as stated briefly in Table 1), is based both on previous work and on interpretation of our own data.

Clast description

Group A

Type A1 – shale clast breccias (Figs 1 & 2). These clasts can vary from pebble to boulder grade (64 to >256 mm) but are generally cobble sized (100 mm). The clasts form very poorly sorted, subangular, matrix-supported, chaotic breccias. These breccias form laterally impersistent zones, orientated subparallel to bedding. The zones occur as shale clast 'nests' between sandstone beds, some of which typically exhibit slump features. In some cases the angular shale clasts can be seen to have been derived from shale beds that have been broken and disrupted by slumping from channel margins.

Type A2 – scour-lag clasts (Figs 1 & 3). These clasts are found in irregular scoured depressions at the basal surface of a bed and are densely packed. The clasts are typically ovoid subrounded, and range in size from small pebbles to large cobbles (> 4 to 256 mm), averaging large pebble size (64 mm). They are poorly to well sorted with random to subparallel orientation and either localized or moderate lateral extent. The zone thickness can vary depending on the depth of the underlying erosive scour and degree of lag deposit development within the host facies. The host sandstone bed may be normally graded with the coarser grain-size fraction forming the shale clast matrix.

Type A3 – rafted clasts (Figs 1 & 4). These clasts can range in size from large pebbles to large boulders (64 to > 256 mm), but are typically boulder sized (> 256 mm). They are found at the base of the host bed and tend to have a low density, occurring singly or in small numbers, with little or no apparent sorting. Clasts are matrix-supported with long-axes tending towards flow-parallel and bed-parallel orientations. They show little evidence of reworking and tend to be tabular in shape, although in some cases partial soft-sediment deformation is apparent. Partially ripped-up clasts also occur still attached to their underlying shale bed. The host sediment tends to have a sharp base with some localized scours.

Type A4 – amalgamation clasts (Figs 1 & 5). These clasts are generally poorly sorted, subangular shaped discs, which range in size from small pebbles to cobbles (> 4 to < 256 mm), but normally average large pebble size (64 mm). The clasts form a matrix-supported layer or zone at the base of the bed often fitting together in a closely knit jigsaw manner to form a competent bed. Sandstones above and below the shale clast zone are typically massive or slightly graded. The clasts can be seen to have been derived from a former shale bed and delineate an amalgamation surface between two sandstone units.

Type A5 – flamed clasts (Figs 1 & 6). These clasts are variable in size ranging from granules to

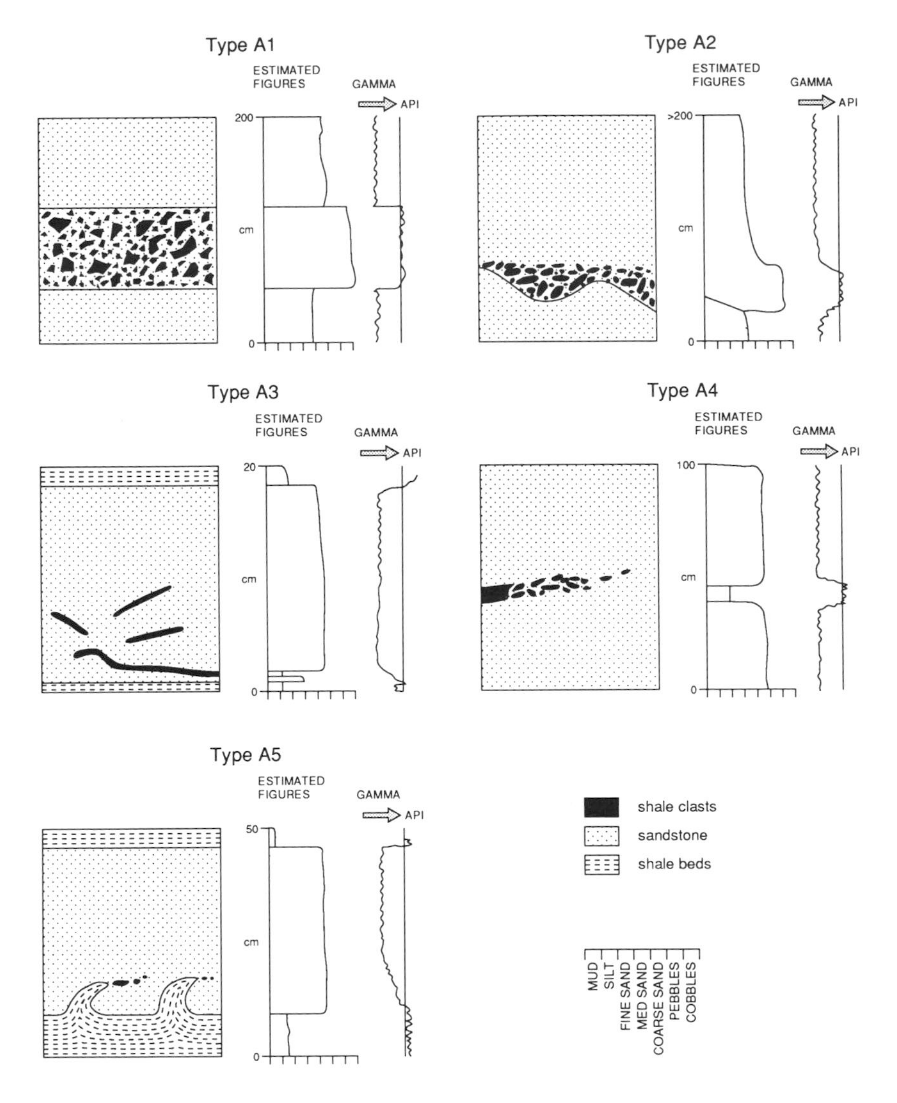

Fig. 1. Group A: clasts formed by erosion disturbance at the base of flow.

very large pebbles (up to 64 mm). They are subangular in shape, often deformed and occur close to their point of detachment from the underlying bed. They are not sorted and have a low density distribution within a matrix-supported sandstone host rock. The host sandstone shows marked loading at its base, which has caused the underlying muddy layer to become over-pressured and squeeze upwards disintegrating in the overlying sandstone.

Group B

Type B1 – shale clast conglomerates (Figs 7 & 8). These conglomerate clasts vary from small pebbles to boulders (4 to > 256 mm) in size, with cobble-sized clasts the most common. They are subrounded, oval-shaped, generally poorly sorted and supported within either a fine sand or muddy matrix. Clasts occur throughout the host bed and clast zones are very laterally extensive. Type B1 clasts differ from Type A1 clasts in being rounded rather than angular in shape. The host bed is generally less chaotic in terms of its internal bedding and lacks slump features. However, some more deformed and angular clasts may also occur with Type B1 clasts, and a complete gradation between shale clast conglomerates and breccias is likely to exist.

Table 2. *Different types of shale clasts found in deep water resedimented sandstone successions*

Host sandstones		
Formation	Locality	Age
Aberystwyth Grits	Dyfyd, West Wales	Upper Llandovery
Albidona	S Italy	Miocene
Andrew Sandstone	Central North Sea (Quad. 30), UK	Palaeocene
Balleny Group	Chalky Island, S New Zealand	Oligocene
Bray Head	Co. Wicklow, Eire	Ordovician
Cabon Conglomerate	Elan Valley, Central Wales	Rhuddianian
Campodarbe (Upper)	S Central Pyreenes, Spain	Middle Eocene
Cantua Sandstone	San Joaquin Valley, California, USA	Early Eocene
Capistrano (Lower)	Dana Point, California, USA	Upper Miocene
Carmelo	Point Lobos, California, USA	Palaeocene
Chatsworth	Los Angeles County, California, USA	Late Cretaceous
Cloridorme	Caspé Peninsular, Quebec, Canada	Ordovician
Conway Castle Grits	Deganwy Quarry, Conway, Wales	Upper Ordovician
Conway Trough	South Island, New Zealand	Neogene
Denbigh Grits	Garn Prys, N Wales	Wenlock, Mid Silurian
Frigg	Norwegian North Sea, Odin Field	Lower Eocene
Gault	Faulknis Nappe, Tschingl, Austenburg	Aptian–Albian
Goldenville	Nova Scotia, Canada	Cambro–Ordovician
Grès d'Annot	Annot, SE France	Lower Oligocene
Grindslow	Ladybowers Res. Pennines, UK	Namurian
Gryphon Sands	Crawford Ridge, northern North Sea	Late Palaeocene–Early Eocene
Habrim Formation	Pohang Basin, SE Korea	Miocene
Hareelv	E Greenland	Upper Jurassic
Heimdal	Sleipner East Field, Norwegian North Sea	Palaeocene
Jackforth (Upper)	DeGray Spillway, Arkansas, USA	Pennsylvannian
Ksiaz	Central Sudetes, SW Poland	Devonian–Carboniferous
Las Vacas	W Argentina	Ordovician (Late Llanvirn)
Loiano Sandstone	Bologna, Apennines, Italy	Middle Eocene
Maesan Fan Delta	Pohang Basin, SE Korea	Miocene
Marnosa Arenacea	Romagna, Apennines, Italy	Early Messinian
Matilija	Santa Ynez Mts, S California, USA	Eocene
Modelo	Los Angeles County, California, USA	Upper Miocene
Mississippi Fan	Louisiana, USA	Upper Pleistocene
Monowai	W Southland, New Zealand	Middle Miocene
Montery Fan	S Central California, USA	Modern
M. Sacro	Campania, S Italy	Upper Miocene
Numidian Flysch	Contrada di Romano, N Sicily	Miocene
Numidian Flysch	Ponte Finale, NE Sicily	Miocene
Otakura	S Otago, New Zealand	Jurassic
Quebrada de las Lajas	E Precodillera Province, Argentina	Carboniferous
Ranzono Sandstone	Emilia, Apennines, Italy	Late Eocene–Middle Oligocene
Reitano Flysch	Cap d'Orlando, NE Sicily	Early Miocene
Rhuddnant/Aberystwyth Grit	Central Wales	Upper Llandovery, Silurian
Rocchetta	NW Italy	Oligocene–Miocene
Sandstone Turbidite (unnamed)	Nizhniye Sergi, W Urals, Russia	Middle Devonian
Shelters Cove	San Francisco Peninsula, California, USA	Upper Cretaceous–Palaeogene
Solitary Channel, Tabernas Basin	Tabernas, SE Spain	Miocene
Sognefjord	Viking Graben, North Sea	Upper Jurassic
S. Mauro, Cliento Unit	Cilento/Campania, Southern Italy	Langhian, Miocene

Clast types											
A1	A2	A3	A4	A5	B1	B2	B3	B4	B5	C1	C2
✓	✓			✓			✓			✓	
✓							✓				
✓						✓	✓	✓			✓
	✓						✓				
								✓			
✓			✓			✓	✓				
					✓		✓				
						✓		✓			
	✓					✓	✓				
✓	✓						✓				
	✓		✓	✓		✓	✓				
					✓						
	✓						✓				
	✓										
						✓	✓				
								✓			
					✓						
					✓		✓	✓			
			✓		✓	✓	✓				
	✓					✓					
											✓
					✓	✓					
			✓				✓				✓
	✓										
✓					✓	✓					
						✓					
	✓						✓		✓		
					✓	✓	✓				
✓						✓	✓				
					✓		✓	✓			
						✓	✓				
		✓	✓								
					✓						
✓	✓			✓		✓	✓	✓			
	✓				✓	✓	✓				
						✓	✓				
						✓					
✓						✓	✓				
						✓	✓				
						✓	✓				
					✓		✓				
					✓			✓			
						✓					
							✓				
	✓										
					✓						
✓		✓	✓	✓		✓	✓	✓			
	✓						✓				
						✓					

Fig. 2. Type A1 – Mudstone Breccia from the Palaeocene, Carmelo Formation, Point Lobos, California, pen for scale (courtesy of B. Cronin).

Fig. 3. Type A2 – Scour-Fill Clasts from the Numidian Flysch, Ponte Finale, Sicily, hammer for scale.

Fig. 4. Type A3 – Rafted-up Clasts from the Oligocene, Grès d'Annot Formation, St Antonin, France, lens cap for scale.

Type B2 – isolated clasts (Figs 7 & 9). These clasts are variable in size, ranging up to and > 5 m in diameter. They are typically subrounded to subangular and ovoid to tabular in shape, with a tendency to become increasingly more tabular with increasing clast diameter. Clasts occur singly or in low density clusters and are of very limited lateral extent. They may occur anywhere within a sandstone bed, but are most commonly found in a lower to mid-bed position, typically along coarse-grained stringers within the host sandstone. Long-axis orientation is parallel to bedding. The host bed generally comprises structureless, ungraded sandstone with a sharp base.

Type B3 – Clustered Clasts (Figs 7 & 10). These clasts are of variable shape and size (small pebbles to large cobbles; 4 to > 256 mm) and are predominantly rounded. They occur in a thin zone along a common plane, orientated parallelto bedding and have a moderate lateral extent. This zone tends to be positioned at a grain-size boundary, normally within an upper mid-bed region. Individual clasts are more randomly orientated within this zone, although with a tendency towards sub-parallel long-axis alignment. The sandstone host beds tend to be structureless and ungraded below the shale clast zone and positively graded above. The basal contact of the sandstone is sharp and can be erosive.

Type B4 – dispersed clasts (Figs 7 & 11). This type of clast is characterized by its small size (2–4 mm), rounded to subrounded nature and equant to ovoid shape. However, dispersed clasts up to 20 cm in diameter have also been observed. Clasts are typically size sorted (i.e. positively graded) throughout the bed, with bed-parallel orientation. Although in most cases the clasts form the coarse-tail of a normally graded bed, dispersed clasts have also been observed in reverse-graded and ungraded sandstones.

Type B5 – lamination clasts (Figs 7 & 12). These clasts are, on average, granule sized (2–4 mm), well sorted, well rounded and of a high sphericity. Clasts tend to be associated with the coarse fraction of the host sandstone and may be aligned along any primary sedimentary structure (e.g. cross- or parallel-lamination).

Fig. 5. Type A4 – Amalgamated Cluster Clasts from the Oligocene, Grès d'Annot Formation, Annot, France, hat for scale (bottom centre).

Fig. 6. Type A5 – Flamed-Up Clasts from the Miocene, Muddy System, Tabernas, Spain, person for scale.

Group C

Type C1 – rip-down clasts (Figs 13 & 14). This clast type is characteristically small in size, generally no larger than small cobbles (100 mm). Clasts are extremely angular, with very low sphericity and are often sheared resulting in a contorted or wavy appearance. They are generally poorly sorted and commonly show a weak preferred orientation parallel to bedding. The only visible organization is a crude reverse-grading observed within some beds, with the larger clasts in close association with the upper boundary whereas smaller clasts are positioned lower within the bed. The density of these clasts tends to decrease downwards, although the actual number of clasts may increase. In some cases, clasts occur still partly attached to the upper boundary surface and extend into the underlying sandstone bed.

Type C2 – injection clasts (Figs 13 & 15). These clasts range in size from small pebbles to medium cobbles (4–200 mm). They are extremely angular and often strongly deformed. They can occur anywhere within the host sandstone, singly, dispersed or in distinct zones. They can be distinguished from A4, B2 or B3 clast types by their angularity and sheared (deformed) nature and by the discordant boundary contacts of the host sandstone.

Inferred processes and examples

The characteristic features and distinguishing qualities of the different types of shale clast are summarized in Table 1, with the most distinctive of these parameters being size, shape, sphericity and within-bed position. It is thought that the principal control on these parameters is the origin of the clasts and their subsequent modification by flow processes. Clasts, can therefore be subdivided into three categories depending on whether they have formed from base of flow erosion, flow modification or through post-depositional erosion.

Group A: base of flow erosion or disturbance

Group A comprises shale clast types derived from erosion or disruption of the underlying or laterally adjacent beds during the flow that deposited the host sandstone.

Type A1 – shale clast breccia. These clasts form from the total disruption and dislocation of a shale bed or beds due to slumping. When a composite unit of semi-consolidated interbedded shale and sandstone is involved in slump folding, there is a tendency for the more competent mud beds to fracture whereas sandstones are more likely to undergo liquefaction. Therefore, the mud material forms fragmented clasts within a disaggregated sand matrix. Where such slumping and bed disruption occurs independently of turbidity current flow within the channel (e.g. by channel-margin collapse), then transport distance is unlikely to be great, and a lenticular bed or nest of shale-clast breccia will be deposited close to the zone of slumping. Subsequent down-channel flows may partially winnow and erode the irregular mound causing some rounding of clasts at the surface. Similar clasts have been described by Rupke (1976) and Johns *et al.* (1981) from the Roncal Unit in the Eocene Turbidites of the southwestern Pyrennees. Channel margin slumping initiated by and coincident with turbidity current flow within the channel will subject any shale clasts formed to a greater degree of abrasion and potential incorporation into the flow. This can then give rise to Type A2 and various Type B clasts.

Type A2 – scour-fill clasts. These clasts are thought to be mainly derived from basal plucking of the channel floor by a highly competent turbidity current. They are then transported in a base-of-flow high-concentration inertia layer or

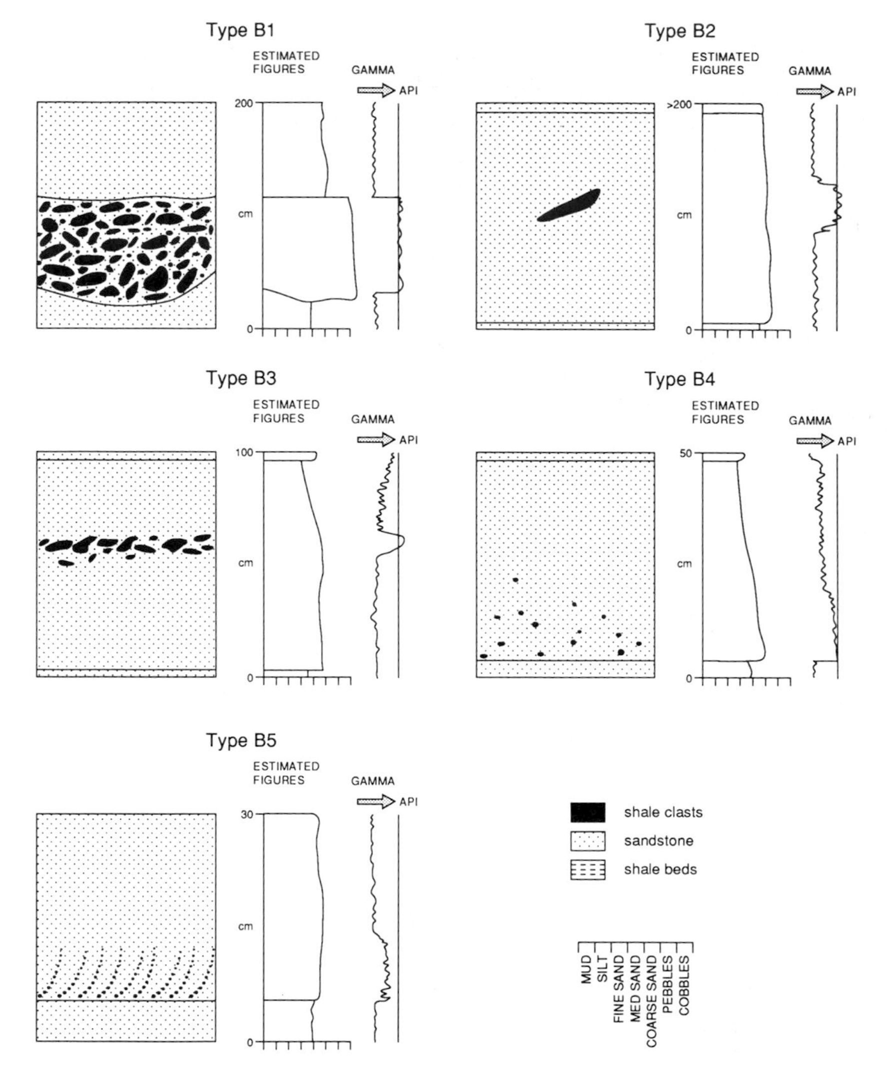

Fig. 7. Group B: clasts modified during within flow transport.

traction carpet, and are not incorporated into the overlying fully turbulent part of the flow. Transport distance is very limited and flow modification slight so they are not classified with our fully flow-modified Type B clasts. Clasts concentrate at the base of thick, channelized sandstone beds and represent the cessation of rolling, saltation or gliding of the shale clast bedload in such high concentration flows. Deposition commonly occurs in deep erosive scours. Type A2 clasts may also form by channel margin slumping and bed disruption coincident with turbidity current flow. Examples of these clasts have been described by Chipping (1972) from Cretaceous and Palaeocene sandstones at Shelters Cove and Devils Slide on the San Francisco Peninsula.

Type A3 – rafted-up clasts. As with the scour-fill clasts, rafted clasts form from erosion at the base of a highly competent sand-laden turbidity current. However, in order to form rafted-up shale 'flaps' that are still partly attached to the underlying bed, the current must erode the shale bed and deposit its load almost simultaneously. Other clasts may become completely detached from the substrate or broken from a larger slab, but are frozen in the depositing flow before being transported any great distance. Their point of origin can often be seen as a tabular erosive scour in the underlying shale. Clasts of this type are well known from many deep-water sandstone formations worldwide [e.g. Modelo Formation, Los Angeles (Sullwold 1960); Solitary Channel, SE

Fig. 8. Type B1 – Mudstone Conglomerate Clasts from the Miocene, Muddy System, Spain, lens cap for scale.

Fig. 10. Type B3 – Cluster Clasts from the Oligocene, Grès d'Annot Formation, Piera Cava, France, lens cap for scale.

Fig. 9. Type B2 – Isolated Clasts from the Oligocene, Grès d'Annot Formation, Piera Cava, France, lens cap for scale.

Fig. 11. Type B4 – Dispersed Clasts from the Numidian Flysch, Ponte Finale, Sicily, Hammer for scale.

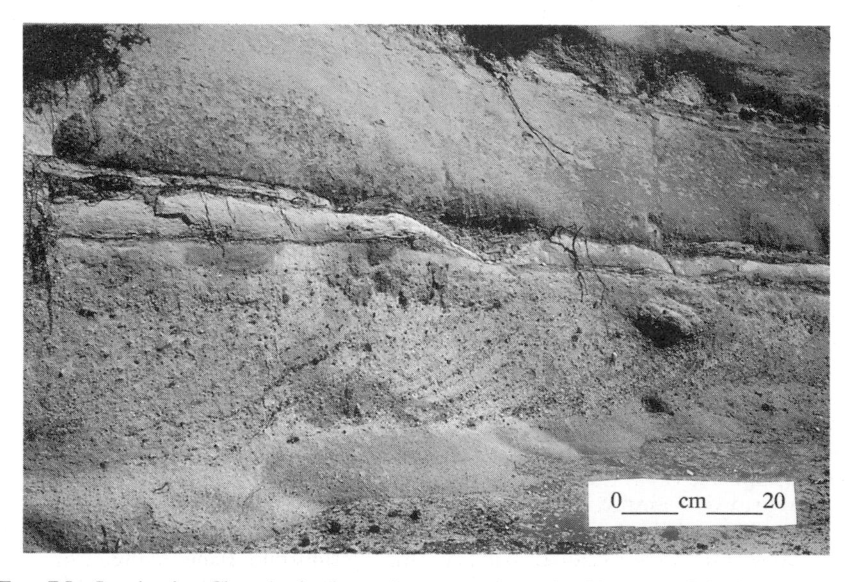

Fig. 12. Type B5 – Lamination Clasts in the form of cross sets from the Oligocene, Grès d'Annot Formation, St Antonin, France, bar for scale.

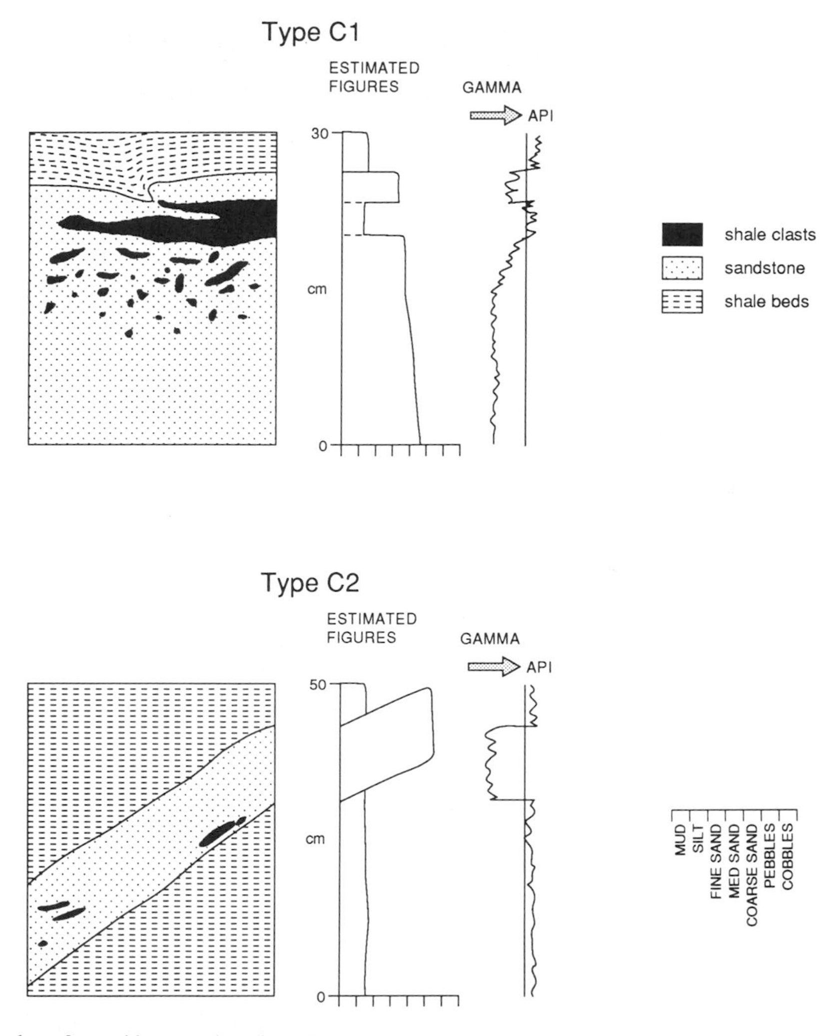

Fig. 13. Group C: clasts formed by post-depositional erosion.

Spain (Kleverlaan 1989)]. They are also known as 'rip-up' clasts.

Type A4 – amalgamation clasts. In general, Type A4 clasts are the product of either: (1) loading by a newly deposited or depositing sandstone unit into and through an intervening semi-consolidated shale bed which overlies a previously deposited sandstone bed; or (2) through the erosive scouring and break-up of a thin shale bed. The chief requirement is that the shale bed is sufficiently competent or brittle to break up into clasts when loaded or eroded. Clasts remain at the base of the flow and are deposited more or less *in situ* along the erosive-loaded contact between the two sandstone units. The distinction between A4 and B3 clasts is that the former can be traced laterally into an undeformed, unbroken shale bed between sandstones, whereas the latter are no longer attached to their parent bed. Examples of amalgamated clasts are seen in the Cretaceous Chatsworth Sands, Southern California in the Simi Hills (Link *et al.* 1981).

Type A5 – flame clasts. These clasts originate from flame structures developed by intense loading of a sandstone into an underlying shale bed. They can be recognized as small isolated clasts associated with the lower bed. Their development is due to the inherent instabilities of sand depositing rapidly on to a soft or semi-consolidated muddy substrate. Experimental work has demonstrated that deformation of an underlying silt and mud layer leads to intrusions into an

Fig. 14. Type C1 – Rip-Down Clasts from the Volgian Humber Group (Upper Jurrasic), South Viking Graben, North Sea.

Fig. 15. Type C2 – Sand-Ripped Mud Clasts from Palaeocene Cores from the Central North Sea (Courtesy of J. Reynolds).

upper sandy bed (McKee & Goldberg 1969). This is mainly due to the sudden deposition of overburden on a highly mobile substrate (Kuenen & Menard 1952; Anketell *et al.* 1970). Further movement of the still mobile depositing sand causes some of the projecting flames to break-up and become incorporated into the lower unit of the depositing sand bed.

Group B: clasts modified by transport within-flow

Flow-modified shale clasts are first incorporated into the flow as a result of processes such as erosion of the channel floor and walls, slumping or sliding of the channel margin or head region, or free fall of rocks and shale blocks from steep slopes. The clasts are thus derived from semi-consolidated sediment at the channel margin, head region or from an underlying substratum. During the flow of a turbidity current, large shear stresses can cause the upward movement of an eroded shale clast (Kano & Takeuchi 1989). During transportation, eroded particles are subjected to dynamic forces within the current. Comparatively large grains, especially when travelling close to the static bed, can be influenced by hydrodynamic lift forces arising from the restriction placed on the motion of the fluid by the proximity of the bed. A particle in relative motion within a sheared fluid moves along the gradient of relative velocity and across the line of flow (Allen 1982 85–88). In most cases these shale clasts become incorporated within the basal denser portion of a sand-rich turbidity current or sandy debris flow. Within this flow the semi-consolidated clasts begin to disintegrate due to clast–clast and clast–matrix interaction. The extent to which a shale clast can tolerate the abrasive qualities of the flow interior is as yet unknown, although it is assumed that the clasts would have a relatively short life span. The 'soft' intraclasts survive either through a process of contemporaneous erosion/influx, minimum transport and rapid deposition, or through rapid transformation in the flow behaviour during transport. It is thought that the increased energy expenditure at the base of the flow during transport of large clasts and the incorporation of the increasing fine fraction could possibly suppress the turbulence within the flow resulting in better clast preservation potential (Postma *et al.* 1988).

Type B1 - shale clast conglomerates. These conglomerates are interpreted as sandy debrites in which the shale clasts are intraformational, most likely derived from slumping or erosion. The clasts are transported by a matrix-supported cohesionless sandy debris flow which may occur in isolation or at the base of a turbidity current. Alternatively (or in addition), the sandy debris flow may be driven by the overriding turbidity current. An example of this clast type has been described by Tanaka *et al.* (1992) from the Chrystalls Beach Complex, Caples Terrane, New Zealand.

Type B2 – isolated clasts and Type B3 – clustered clasts. It is the occurrence of both single isolated shale clasts and small groups or clusters of clasts apparently suspended in a mid-bed position that has proved the most difficult to explain and caused much speculation. Several of the proposed mechanisms described below are valid, and each may apply to different examples observed in the field.

(1) One possible mechanism of formation is that the shale clasts are influxed into the flow and emplaced through the gradual aggradation of sediment (Kneller 1995). The transporting current is a sustained steady or near-steady flow, involving a flow boundary that is dominated by hindered settling. Deposition is thought to continue as long as the downward grain flux to the deposit is balanced by sediment supply from the transport regime. The high downward grain flux generates a dense, non-turbulent zone dominated by hindered settling at the base of the flow, with no sharp interface at the depositing surface and no traction. Deposition occurs incrementally from the base (gradual aggradation of Branney & Kokelaar 1992), and is maintained for as long as the flow continues to provide sediment to the zone. As the sedimentation of the sands need no longer be instantaneous but instead gradational, the outsized clasts that occur mid-bed may always have been close to the depositing surface and are pushed up during incremental deposition. It is thought, therefore, that the clasts would have no visible grain-size boundary above or below them and that their locality is determined by the position within the steady flow of the shale clast influx.

(2) Results from flume experiments suggest that mid-bed shale clasts may settle through a flow and alight on an already developed high density inertia-flow layer (Postma *et al.* 1988). This high density flow layer displays pseudolaminar behaviour due to the suppressed turbulence caused by a high particle concentration (Bagnold 1954; Lowe 1982). The large outsized particles then 'glide' along the top of the underlying pseudolaminar inertia-flow layer. The particles become partly submerged within the inertia-flow layer and are driven by the downflow component of turbulent shear–stress transmitted from the overlying, faster moving turbulent layer. As the inertia-flow layer freezes and a new one forms or as the layer thickens, the gliding clasts may be forced to a progressively higher level within the flow. When the deposit eventually ceases to flow the clasts are prevented from returning to the base due to the underlying sediment and are isolated within the bed. It is thought that the uppermost clast formed in this kind of fluid regime would delineate an interface between a grain-size boundary, with the lower sand being massive with possible tractional evidence, and the upper sand exhibiting positive gradation.

(3) In some cases, shale clasts are ripped-up and become fully incorporated into the basal, denser portion of a sand-rich turbidity current (Mutti & Nilsen 1981). Density differences between the lower density, fluid-saturated clasts and the higher density basal sandy layer of the flow would allow the clasts to float upwards through the basal layer. The buoyant upward movement of clasts within the dense basal portion of the current is accompanied by their progressive disaggregation and erosion. Clasts that are not completely dissagregated accumulate in an apparently aligned zone (i.e. clustered clasts, Type B3). This specific position within the bed is thought to be the boundary between the graded Bouma Tb division and the current laminated Bouma Tc division. This division is thought to correspond to the separation between high and low density flow conditions within the current although field evidence suggests that the clasts lie predominantly between the Bouma Ta and Tb divisions.

(4) Early theories suggested that mid-bed shale clasts were typical of grain-flow deposits (Stauffer 1967), with the mid-bed position attributed to kinetic sieving as a result of dispersive pressures (Bagnold 1954, Jullien & Meakin 1992). However, according to Bagnold (1956), the magnitude of dispersive stress is directly dependent on the grain size of the shearing sediment flow and hence would be relatively low in a sandy grain flow (discussion by Middleton & Southard 1978). This mechanism could, therefore, only account for relatively small shale clasts in coarse sandstones, provided that the grain flow can travel far enough after incorporation of the clast(s) for the process to operate (Lowe 1982).

Type B4 – dispersed clasts. These clasts occur in thick normal and reverse, coarse-tail graded beds, as well as in apparently ungraded units. Normal coarse-tail grading occurs where the size of the coarsest grains decreases upwards within a bed, whilst the size of the host grains remains constant. This grading implies turbulence and effective grain interaction during transport, whereas ungraded beds usually indicate high shear strength or high viscosity, thereby preventing the turbulence and effective grain interaction. The coarse-tail inverse grading probably develops from larger clasts rising upwards due to dispersive pressures or by progressive loss of larger clasts, from the lower, more strongly sheared part of the flow (Allen 1982). This type of clast is present in the Numidian flysh 'externe', Sicily (Braakenburg 1994).

Type B5 – lamination clasts. These shale clasts are small in size and closely associated with the coarser fraction of the sandstone bed. They are considered to have behaved as normal particles within the flow, hydrodynamically equivalent to the coarse sands with which they occur. Those clasts deposited along the foresets of large-scale cross-strata have undergone a period of tractional movement and downcurrent bedform migration immediately prior to deposition.

Group C: clasts formed by post-depositional erosion

Some types of shale clast are derived from post-depositional disturbance of a sandstone unit. In this case a semi-unconsolidated sand is remobilized after burial through dewatering and liquefaction caused by sudden shock, natural slope instability and slumping, or simple overburden pressure. The sand flow or injection that results is thought to be pressurized and cohesive, with an ability to erode or pluck shale clasts from either margin of the flow but with a limited transport potential. The clasts are thus largely unreworked but many exhibit some shearing. The processes generating them can include either the 'ripping-down' of an overlying shale bed (Type C1) or the breaking-up of an interbedded shale bed through sand injection (Type C2). In both cases, angularity and partially attached shale clasts help confirm the origin of those that have become fully detached.

Type C1 – rip-down clasts. These clasts were first described as 'rip-down' clasts by Chough & Chun (1988), who suggested that shearing and fragmentation of an upper shale bed in the late Cretaceous Uhangri Formation was caused by penecontemporanous deformation and liquefaction related to intrastratal flow of the lower bed. In this situation, once a sandstone underlying a shale bed is liquified, it may creep and flow within the strata moving either downslope or down-pressure gradient, and may rip-down parts of the overlying layer (Chough & Chun 1988). The downward movement of clasts within the intrastratal flow may result in the break-up of larger clasts potentially resulting in apparently reverse graded zones. Liquefaction can be enhanced by an impermeable cover, such as an overlying shale bed, raising the pore-water pressures and diminishing the effective stress (Finn *et al.* 1977; Allen 1982).

Type C2 – injection clasts. Burial of thick, massive sandbodies, isolated within mudstones, creates a large potential for injection as sand dykes or sills into the surrounding sediment due to overpressuring of pore waters. Sands are thought to be injected in a fluid state, being mobilized by liquefaction during dewatering (Surlyk 1987; Dixon *et al.* 1995). Clasts are formed through a ripping process resulting from the high fluid pressures of the sand during injection into the surrounding substrate. This clast type has been described from the Upper Jurassic Harleev Formation of East Greenland (Surlyk 1987) and the Tertiary of the North Sea (Dixon *et al.* 1995).

Discussion

Classification scheme

The classification scheme proposed here has concentrated entirely on shale clasts in deep-water successions and has further focused on generally thick-bedded sandstone facies in which shale clasts are particularly common. Micro-shale clasts occur in fine-grained turbidite successions, for example in silt laminae at the base of graded-laminated units (Cremer & Stow 1986, Piper & Stow 1984), and in contourite facies (Mézerais *et al.* 1993). The types and processes of origin are most likely analogous to those we have described for their coarser-grained counterparts. Furthermore, shale clasts are a common component of many deep-water conglomerates, in which their origin and behaviour is exactly the same as for the other pebble components of coarse-grained resedimented deposits (Pickering *et al.* 1989).

The 12 clast types defined here are believed to be a good representation of the diversity of clasts which exists. However, it is by no means always possible to classify any particular shale clast with certainty if its full context is not known. This is especially true of shale clasts recognized in cores recovered from hydrocarbon reservoirs/prospects. The characteristic features given in Table 1 are designed to help with clast identification and hence with interpretation.

Clast Origin

Type A clasts. These are formed by erosive processes including disruption of slumps and slides, rock-fall and bank collapse, the rip-up of underlying beds, the break-up of thin intercalated shales, and the flame-up of soft shales due to loading. Although these processes are understood in general terms, the precise cause and mechanism of base-of-flow erosion and break-up is still not fully understood. It seems likely that a gradation in erosive power of the flow, from greater to lesser, gives rise to scour-lag, raft, amalgamation and flame clasts in that order (i.e. A2, A3, A4, A5). This same order could indicate something about the relative erodibility of the substrate which may, in turn, be influenced by the timescale between flow events.

Type B clasts. Those are clasts that have undergone further modification during transport. Shale clast conglomerates (B1) are best interpreted as sandy debris flows. The relatively common occurrence of these clasts in ancient successions indicates that this process is common. Dispersed clasts (B4) and lamination clasts (B5) are those that have become fully integrated into a flow, following extensive break-up and abrasion, and hence behave as the normal coarse fraction of that flow.

The isolated and clustered clasts (B2 and B3) lie somewhere in between the others in terms of origin. Their specific characteristics and context can give information about the nature of the high energy resedimentation event that was responsible for their emplacement. In most cases their presence supports the concept of a bipartite flow, in which the shale clasts were deposited at the interface of an underlying higher-concentration zone and an overriding lower concentration turbulent flow. Where the basal layer was an inertia flow followed by rapid freezing and deposition, then the underlying sandstone will be massive or show some tractional features and the overlying sandstone will typically be graded. Where there was a basal zone of hindered settling and gradual aggradation from a near-steady flow, then the whole unit may appear massive with no change in grain size or structure across the shale clast zone. Smaller and less dense shale clasts may float upwards through a depositing turbidity current and come to rest between the Bouma Ta and Tb divisions. Each of these slightly different characteristics has been observed in the rock record, so that the inferred flow processes are all likely to occur.

Proximal to distal variation

The types and abundance of shale clasts that occur can be used to infer relative proximality in a deep water setting. More proximal settings with slide scars, canyons and gullies are characterized by relatively large numbers of erosive Type A clasts as well as shale clast conglomerates (B1). Medial settings with channel–levee complexes and interchannel areas are characterized by both Type A and Type B clasts that have been modified by transport. In distal distributary channel–lobe settings, Type B clasts dominate, and most of these do not make it out into the most distal basin plain settings. These are characterized by B4 dispersed clasts, B5 lamination clasts and by microshale clasts in fine-grained turbidites.

The same broad trend is evident within well-established channel systems from their margins towards their axis, although with greater variability. Clasts associated with channel margins tend to be shale clast breccias and scour-lag clasts (A1 and A2), and these pass transitionally into A3 raft clasts, A4 amalgamation clasts and Type B clasts. In coarse, gravel-filled channel thalwegs, few clasts are preserved due to very high energy levels, but remnant shale clast breccias (A1), shale clast conglomerate debrites (B1) and scour-lag clasts (A2) may be present. Immediately adjacent to the channel on the proximal levees, thin sandstones may be characterized by intensely disturbed and balled structures with A5 flame clasts.

Depositional environment

Based on detailed observations of some 12 field examples in which thick structureless sandstone units occur in association with a wide variety of more normal turbidite–hemipelagite facies, we can be even more specific in characterizing specific parts of the deep water environment by their shale clast content and type (Fig. 16). The sedimentary environments in which these deep water sandstones occur with their various shale clast types include: fan-delta (plus coarse-grained slope-apron), muddy slope-apron, fan channel–

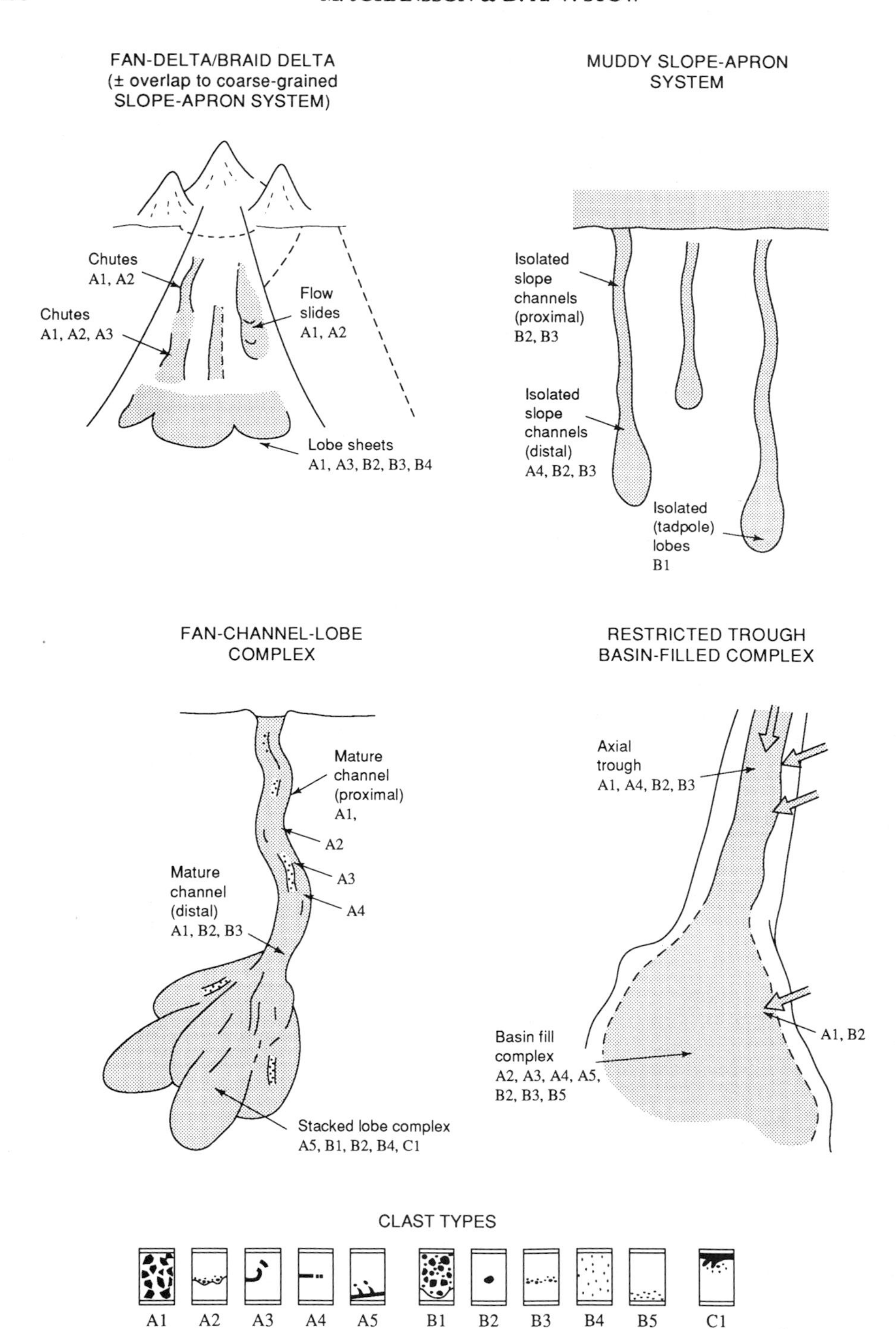

Fig. 16. Model for shale clast distribution within various deep marine environments.

lobe complex and restricted trough–basin-fill complex.

Fan Deltas. These develop where an alluvial fan or braided channel feeds directly on to a subaqueous slope margin. Overlapping fan deltas coalesce to form a coarse-grained slope-apron system. The main processes and features related to this type of environment are chutes, channels, flow slides and lobe sheets. Fan delta chutes and channels are both characterized by A1 mudstone clast breccias and A2 scour-fill clasts. The former clasts are derived from channel-margin collapse and the latter from tractional flow. The channels can be distinguished from the chutes by the presence of an additional clast type, the A3 rafted-up clast. Flow slides are formed by sediment gravity collapse and can be recognized by A1 shale clast breccias and A2 scour-fill clasts. The A2 clasts are thought to be carried as a tractional basal lag. The fan delta lobe sheets are mixed-grade depositional bodies that can be characterized by four types of clast: base-of-bed A4 amalgamated clasts and flow modified B2 isolated clasts, B3 clustered clasts and B4 dispersed clasts. Although there is still some erosion of the underlying shale beds, the majority of clasts are of a depositional nature and are probably derived from previous up-fan erosion.

Muddy slope-apron systems. These are linearly sourced from the platform with mainly fine-grained sediment. The sands form immature channels and lobes which have a tadpole morphology. The proximal isolated slope channels typically contain either B2 isolated or B3 clustered clasts, probably derived from rock-fall or bank collapse. The more distal isolated slope channels have A4 amalgamation clasts as well as B2 and B3 clasts. This is probably due to a slightly lower energy level than the up-channel system. The isolated (tadpole) lobes, however, tend to have B1 debris-flow clasts which have passed through the channel and have deposited due to a reduced carrying capacity within the lobe.

Fan channel–lobe complex. This comprises a high energy, confined, mature channel that feeds into an unconfined, stacked lobe complex. The channel system may contain any of the erosive A-type clasts, with A1–A5 down-channel and across-channel variation in clast types. B1 shale clast conglomerates, B2 isolated clasts and B3 clustered clasts are also common in channels, together being indicative of high energy flows, slumping and tractional processes. The stacked lobe complex is a more variable system with flows switching position on the lobe surface. The environment is predominantly a depositional one, with the main erosional clast forming due to rapid deposition (A5 flame-type clasts). The flow-modified clasts include B1 mudstone clast conglomerates, B2 isolated clasts and B4 dispersed clasts.

Restricted trough–basin-fill complex. This refers to a tectonically confined narrow trough and/or restricted basin which may be variously fed by fan deltas, slope gullies or channel complexes. It can been divided into four subenvironments: the axial trough, basin-fill margin, proximal basin-fill and distal basin-fill complex. The axial trough is characterized by A1 mudstone clast breccias and A4 amalgamated clasts derived from basal erosion, together with B1 mudstone clast conglomerate, B2 isolated clasts and B3 clustered clasts. The basin margin may show a variety of Type A erosive clasts, in particular, A1 chaotic clasts and B2 isolated clasts. The proximal basin-fill complex is close to the mouth of the axial trough and forms a transition between erosion and deposition. The facies have four predominant types of shale clasts, low energy A3 raft-type clasts and A5 flame-type clasts, and flow-modified B3 clustered clasts and B4 dispersed clasts. The more distal basin-fill complex typically displays a greater variety of clast types. The erosive clasts tend to be A2 scour-fill clasts, A4 amalgamation clasts and A5 flame-type clasts. Flow-modified clasts include B1 shale clast conglomerates, B2 isolated clasts, B3 clustered clasts, and B5 lamination clasts. This type of environment is characterized by anastomosing flows, channel switching and cannibalization of former beds, so that the most predominant clasts are those formed by erosion, amalgamation and deposition.

Reservoir implications

Deep water massive sands and their associated facies form significant hydrocarbon reservoirs throughout the world. Thick units of more or less structureless sands recovered in core section are not easy to interpret, largely because of their lack of diagnostic features, so that accurate models for further hydrocarbon exploration or production are difficult to construct.

Shale clasts, however, are one of the few diagnostic features that may be present. It therefore becomes important to be able to recognize the type of shale clast or shale clast association present, and hence to infer the probable mode of origin and likely depositional environment within the spectrum of deep water depositional settings that exist. Proximal–distal, marginal–axial channel, lobe or confined basin systems all have a characteristic shale clast signature. Clearly, this must be used in conjunction with all other available data on facies or geometries before a reliable model can be constructed.

Core sections can be used in conjunction with wireline log response (e.g. gamma ray) to gain some idea of probable lateral extent of any particular shale clast horizon. However, used alone, the gamma-ray curves can be misleading as it is not possible to distinguish between muddy sandstones, interbedded sandstones–shales and zones with abundant shale clasts.

Relatively few of the shale clast types described in this study occur with sufficient density and lateral extent to act as significant permeability barriers within reservoir sections. However, it is important to accurately interpret those that may influence vertical permeability in this way, and to distinguish them from those that will simply have local effect.

This work was carried out, in part, during a larger survey of deep-water massive sands, funded by Enterprise, Fina, Oryx, Shell and Texaco. We also thank colleagues and reviewers for their part in helping improve the final manuscript. Barry Fulton drafted the diagrams.

References

ALLEN, J. R. L. 1982. *Sedimentary Structures: Their character and physical basis. Developments in sedimentology.* Elsevier, Amsterdam.

ANKETELL, J. M., CEGLIA, J. & DZULYNSKI, S. 1970. On the deformational structures in systems with reversed density gradients. *Annals of the Geological Society of Poland,* **40,** 3–30.

BAGNOLD, R. A. 1954. Experiments on a gravity-free dispersion of large solid spheres in a Newtonian fluid under shear. *Proceedings of the Royal Society of London, A,* **225,** 49–63.

—— 1956. The flow of cohesionless grains in fluids. *Philosophical Transactions of the Royal Society of London, A,* **249,** 235–297.

BRAAKENBURG, N. E. 1994. *Anatomy of deep water massive sands using examples from Cyprus, Sicily and California,* PhD thesis, University of Southampton.

BRANNEY, M. J. & KOKELAAR, B. P. 1992. A reappraisal of ignimbrite emplacement: Progressive aggradation and changes from particulate to non-particulate flow during emplacement of high-grade ignimbrite. *Bulletin of Volcanology,* **54,** 504–520.

CHIPPING, D. H. 1972. Sedimentary structure and environment of some thick sandstone beds of turbidite type. *Journal of Sedimentary Petrology,* **42,** 587–595.

CHOUGH, S. K. & CHUN, S. S. 1988. Intrastratal rip-down clasts, Late Cretaceous Uhangri Formation, southwest Korea. *Journal of Sedimentary Petrology,* **58,** 530–533.

CREMER, M. & STOW, D. A. V. 1986. Sedimentary structures of fine-grained sediments from the Mississippi Fan: thin section analysis. *In:* BOUMA, A. H., COLEMAN, J. M., MEYER, A. W., *et al. Initial reports of the Deep Sea Drilling Project,* **XCVI,** Washington.

DIXON, R. J. *et al.* 1995. Sandstone diapirism and clastic intrusion in the Tertiary fans of the Bruce–Beryl Embayment, Quadrant 9, UKCS. *This volume.*

FINN, W. D. L., LEE, K. W. & MARTIN, G. R. 1977. An effective stress model for liquefaction. *Journal of Geotechnical, Engineering Division, ASCE,* **103,** 517–533.

FOLK, R. L. & WARD, W. 1957. Brazos river bar: a study in the significance of grain size parameters. *Journal of Sedimentary Petrology,* **27,** 3–26.

JOHNS, D. R., MUTTI, E., ROSSELL, J. & SÉGURET, M. 1981. Origin of a thick, redeposited carbonate bed in Eocene turbidites of the Hecho Group, south-central Pyrenees, Spain. *Geology,* **9,** 161–164.

JULLIEN, R. & MEAKIN, P. 1992. Three dimensional model of particle-size segregation by shaking. *Physical Review Letters,* **69,** 640–643.

KANO, K. & TAKEUCHI, K. 1989. Origin of mudstone clasts in turbidites of the Miocene Ushikiri Formation, Shimane Peninsula, southwest Japan. *Sedimentary Geology,* **62,** 79–87.

KLEVERLAAN, K. 1989. Three distinctive feeder-lobe systems within one time slice of the Tortonian Tabernas Fan, SE Spain. *Sedimentology,* **36,** 25–45.

KNELLER, B. 1995. Beyond the turbidite paradigm: physical models for deposition of turbidites and their implications for reservoir prediction. *This Volume.*

KUENEN, P. H. H. & MENARD, H. W. 1952. Turbidity currents, graded and non-graded deposits. *Journal of Sedimentary Petrology,* **2,** 83–96.

—— & MIGLIORINI, C. 1950. Turbidity currents as a cause of graded bedding. *Journal of Geology,* **58,** 91–127.

LINK, M. H., SQUIRES, R. L. & COLBURN, I. P. (eds) 1981. *Simi Hills Cretaceous Turbidites Southern California.* Society of Economic Palaeontologists and Mineralogists, Pacific section, Fall Fieldtrip Guidebook.

LOWE, D. R. 1982. Sediment gravity flows II: Depositional models with special reference to the deposits of high density turbidity currents. *Journal of Sedimentary Petrology,* **52,** 279–297.

MÉZERAIS, M. L., FAUGÈRES, J. C., FIGUEIREDO, A. G. & MASSÉ, L. 1993. Contour current accumulation off Vema channel mouth, southern Brazil Badin: pattern of a 'contourite fan". *In:* STOW, D. A. V. & FAUGÈRES, J. C. (eds) *Contourites and Bottom Currents. Sedimentary Geology,* **82,** 173–187.

MCKEE, E. D. & GOLDBURG, M. 1969. Experiments of formation of contorted structures in mud. *Bulletin of the Geological Society of America,* **80,** 231–244.

MIDDLETON, G. V. & SOUTHARD, J. B. 1978. Mechanics of sediment movement, *Society of Economic Paleontologists and Mineralogists, Short Course*, **3**, 242.

MUTTI, E. & NILSEN, T. H. 1981. Significance of intraformational rip-up clasts in deep sea fan deposits. *International Association of Sedimentologists, 2nd European Regional Meeting, Bologna, Italy.*

—— & RICCI-LUCCHI, F. 1972. Le torbiditi dell'Appennino settentrionale: introduzione all'analsi di facies. *Memorie della Societa Geologica Italiana*, **11**, 161–199.

PETTIJOHN, F. J., POTTER, P. E. & SIEVER, R. 1973. *Sand and Sandstone.* Springer-Verlag, Berlin.

PICKERING, K. T., HISCOTT, R. N. & HEINS, F. J. 1989. *Deep Marine Environments.* Unwin Hyman, London.

PIPER, D. J. W. & STOW, D. A. V. 1984. The Laurentian Fan-Sohm Abyssal Plain. *Geo-Marine Letters*, **3**, 141–146.

POSTMA, G., NEMEC, W. & KLEINSPEHN, K. L. 1988. Large floating clasts in turbidites: a mechanism for their emplacement. *Sedimentary Geology*, **58**, 47–61.

RUPKE, N. A. 1976. Large Scale slumping in a flysch basin, southwestern Pyrenees. *Journal of the Geological Society, London*, **132**, 121–130.

STAUFFER, P. H. 1967. Grain flow deposits and their implications, Santa Ynez Mountains, California. *Sedimentary Geology*, **37**, 487–508.

SULLWOLD, H. H. 1960. Tarzana Fan, deep submarine fan of late Miocene age Los Angeles County, California. *American Association of Petroleum Geologists Bulletin*, **44**, 433–457.

SURLYK, F. 1987. Slope and deep gully sandstones, Upper Jurassic, East Greenland. *American Association of Petroleum Geologists Bulletin*, **71**, 464–475.

TANAKA, J., MAEJIMA, W., COOMBS, D. S., LANDIS, C. A., HADA, S., YOSHIKURA, S. & SUZUKI, M. 1992. Depositional mechanism of mudclast conglomerates – an example of the Triassic–Jurrassic in the Caples Terrane, New Zealand, *Earth Science*, **46**, 113–120.

WALKER, R. G. 1978. Deep water sandstone facies and ancient submarine fans: models for exploration for stratigraphic traps. *American Association of Petroleum Geologists Bulletin*, **62**, 932–966.

WENTWORTH, C. K. 1922. A scale of grade and class terms for clastic sediments. *Journal of Geology*, **30**, 377–392.

Index